Integrated Solid Waste Management: a Life Cycle Inventory

Integrated Solid Waste Management: a Life Cycle Inventory
second edition

Forbes R McDougall, Peter R White,
Marina Franke and Peter Hindle

Blackwell
Science

© Procter & Gamble Technical Centres Limited 2001

Blackwell Science Ltd, a Blackwell Publishing Company

Editorial offices:
9600 Garsington Road, Oxford OX4 2DQ, UK
    Tel: +44 (0)1865 776868
Blackwell Publishing Inc., 350 Main Street, Malden, MA 02148-5020, USA
    Tel: +1 781 388 8250
Blackwell Publishing Professional, 2121 State Avenue, Ames, Iowa 50014-8300, USA
    Tel: +1 515 292 0140
Blackwell Publishing Asia Pty, 550 Swanston Street, Carlton, Victoria 3053, Australia
    Tel: +61 (0)3 8359 1011
Blackwell Wissenschafts Verlag, Kurfürstendamm 57, 10707 Berlin, Germany
    Tel: +49 (0)30 32 79 060

First edition published by Blackie Academic and Professional, 1994
Second edition published by Blackwell Science, 2001
Reprinted 2002, 2003, 2005

ISBN-10: 0-632-05889-7
ISBN-13: 978-0-632-05889-1

Library of Congress Cataloging-in-Publication Data
Integrated solid waste management: a life cycle inventory/Forbes R. McDougall ... [et al.]. – 2nd ed.
    p. cm
  New ed. of: Integrated solid waste management/P.R. White, M. Franke, P. Hindle. 1994. Includes bibliographical references and index.
ISBN 0-632-05889-7
  1. Integrated solid waste management. 2. Product life-cycle–Environmental aspects.
I. McDougall, Forbes R. II. White, Peter R. Integrated solid waste management.
TD794.2 .I58 2001
363.72'85–dc21   2001025429

A catalogue record for this title is available from the British Library

Set in Gill and Bookman
by Gray Publishing, Tunbridge Wells, Kent
Printed and bound in Great Britain
by TJ International Ltd, Padstow, Cornwall

The publisher's policy is to use permanent paper from mills that operate a sustainable forestry policy, and which has been manufactured from pulp processed using acid-free and elementary chlorine-free practices. Furthermore, the publisher ensures that the text paper and cover board used have met acceptable environmental accreditation standards.

For further information on Blackwell Publishing, visit our website:
www.blackwellpublishing.com

Contents

Chapter 3 The Development of Integrated Waste Management Systems: Case Studies and Their Analysis   33

Chapter 5 A Life Cycle Inventory of Solid Waste   103

ELEMENTS OF IWM

Preface

Preface to the second edition

'The life cycle of waste can be considered to be a journey from the cradle (when an item becomes valueless and, usually, is placed in the dustbin) to the grave (when value is restored by creating usable material or energy); or the waste is transformed into emissions to water or air, or into inert material placed in a landfill.'

So began the preface to the first edition of this book, which first appeared at the start of 1995. Since then, the whole subject of Integrated Waste Management (IWM), and the use of life cycle tools to assess waste systems have travelled a considerable way on their own journey.

The journey so far . . .

In 1995, many debates were raging on the benefits of recycling versus energy recovery, and on how to implement kerbside collection schemes for recovering recyclable or compostable waste fractions. Today, there is growing acceptance that a combination of integrated options, is needed to handle all materials in municipal solid waste in an effective way. There are also now excellent examples of integrated waste systems on the ground, as detailed in this book. In fact, the debate has progressed further than the Integrated Waste Management advocated in the first edition, and is now focused on sustainable resource and waste management. There is recognition that waste needs to be regarded more as a resource, and that its management needs to be environmentally effective, economically affordable and socially acceptable. If this is achieved then such systems will contribute to the sustainable development of society.

But how can we assess the sustainability of waste management systems? In 1995 decision makers relied on the hierarchy of waste management options, which ranked treatments in order of preference, but which was not based on any scientific or technical evidence. The first edition of this book provided an alternative approach by modelling the whole solid waste system, including any combination of options, to provide both an environmental and economic overall assessment. This was one of the first attempts to apply the tool of Life Cycle Inventory (LCI) to solid waste management to produce a tool for waste managers, policy makers, regulators and other decision makers.

Since 1995, this idea of using LCI tools for solid waste has travelled far too. The model provided in the first edition (IWM-1) has been applied at local, regional and national levels. It has been used by municipalities to assess Integrated Waste Management systems in many countries in Europe and elsewhere. It has been used by waste management companies to assess the tenders they submit to municipalities, and by the municipalities to assess such tenders. It has been used by consultants in reports on waste management strategy for the European Commission (Coopers & Lybrand, 1996).

Since the appearance of the first model, we have also seen the development of a number of other, more sophisticated LCI tools for solid waste management. In the UK, the Environment Agency has launched WISARD – a software package for use by municipalities; in the USA, the Environmental Protection Agency is completing its own model, while in Canada, two industry organisations – CSR (Corporations Supporting Recycling) and EPIC (the Environment and Plastics Industry Council) – have launched a further model specific to Canadian conditions. LCI models for solid waste systems are also available from several consultants.

This mushrooming of interest in the application of LCI to solid waste suggests that this is an idea whose time has come. Talking to users shows that there is growing experience of using the tools for several different functions:

1   A planning tool – to do 'What if . . .?' scenarios of possible future systems.
2   A system optimisation tool – to model existing systems and look for improvements.
3   A communications tool – the tool has been used in public meetings to explore, with all stakeholders, the possible ways in which a community's waste could be handled, and the environmental and cost implications of such options.
4   A source of data – for use in other tools or assessments.

There is also now an International Expert Group on Life Cycle Assessment for Integrated Waste Management, supported by the UK Environment Agency and the US Environmental Protection Agency, where workers in this field can discuss applications and resolve issues. Two of the authors of this book are members of this International Expert Group.

Why write a second edition?

The first edition proposed a vision of IWM, and the use of tools such as LCI to provide a way to assess the environmental and economic performance of waste systems. We now see actual examples of IWM systems on the ground, and published accounts of how LCI models for solid waste have been applied. This seemed a good time, therefore, to stop and take stock of what has been achieved, and to draw out the lessons learned. For that reason, a significant part of this edition focuses on case studies – both of IWM systems, and of where LCI has been used to assess such systems.

The second reason for a new edition was to provide a more user-friendly model (IWM-2) for waste managers. The feedback we received from readers of the first edition was that while the book effectively conveyed the concepts of IWM and the application of LCI to solid waste, only computer experts felt comfortable with the spreadsheet tool provided. To make the tool more widely accessible, this edition provides a new tool in Windows format, with greatly improved input and output features, and the ability to compare different scenarios. A significant part of this edition provides a detailed user's guide, to take the reader through the use of the IWM-2 model, step by step.

Finally, the whole field of LCI has progressed over the past 5 years, with the acceptance of ISO standards (14040 Series on Environmental Management) which stress the need for transparency. The new model, IWM-2, presented here allows for total transparency as to how it calculates results, and as to the sources of data used.

Do we need another computer model?

When the computer model IWM-1 was released in the first edition of the book, it was a relatively novel concept. Today, however, as listed above, there are more sophisticated LCI models for solid waste available, so where does this IWM-2 model fit? It is designed to be an 'entry level' LCI model for solid waste – user-friendly and appropriate to users starting to apply life cycle thinking to waste systems. More expert users may find many of the advanced features of the IWM-2 model helpful, but in time they will probably graduate onto one of the more sophisticated models, with perhaps more geographically relevant data. If IWM-2 helps introduce waste managers to the concept of Integrated Waste Management and the need to take an overall approach, it will have served its purpose.

Why did Procter & Gamble write this book?

As explained in the previous edition, Procter & Gamble (P&G) is concerned with solid waste because some of our products, and most of our packages enter the solid waste stream. Our products are found in 140 countries around the world. Our consumers want us to do everything we can to make sure that our products and packages are sustainable, in environmental, social and economic terms. This involves us in constantly seeking improvements in the design and manufacture of our own products, but in addition we have been working with others in many countries to help develop improved Integrated Waste Management systems that are environmentally effective, economically affordable and socially acceptable. As part of this, P&G has set up a Global Integrated Solid Waste Management Team, made up of its experts around the world – many of whom have contributed to this book. The aim of the team is to promote effective integrated systems for municipal solid waste; this book forms part of that ongoing effort.

Who are the intended readers?

The intended audience is large and diverse:

- *Waste managers* (both in public service and private companies) will find an holistic approach for achieving sustainable solid waste management, together with an improved modelling tool to help assess the environmental and economic aspects of their own, or proposed schemes.
- *Producers of waste* will be able to understand better how their actions can influence the operation of effective waste systems.
- *Designers of products and packages* will benefit by seeing how their design criteria can improve the compatibility of their product or package with Integrated Waste Management systems.

- *Policy makers* will see examples of effective approaches to waste management and the tools needed for their implementation.
- *Regulators* will see the impact of existing and proposed regulations on the development of more sustainable Integrated Waste Management systems.
- *Politicians* (trans-national, national or local) will see how specialists in many areas are combining their expertise to seek better ways of handling society's waste. They will find data and management approaches that they can use and support as they seek to provide direction to the social debate on the emotive issue of solid waste.
- *Waste data specialists* (whether in laboratories, consultancies or environmental managers of waste facilities) will appreciate the importance of their data, and the ways in which its scope, quality and quantity can be improved to facilitate better management of solid waste.
- *Life Cycle Assessment specialists* will see an LCI tool that has already been used in many countries to support decisions on Integrated Waste Management.
- *Environmentalists* (whether or not in environmental organisations) will see how the application of science, financial management and social involvement can be combined in the search for solid waste systems that do not cost the earth.
- *Concerned citizens* will see some of the efforts being made to improve solid waste management around the world, and the tools used to assess this progress. They will also see recognition that science and management do not have all the answers. In the democratic process, there is a role for the concerned citizen to influence developments and to ensure that reasoned decisions are made.

Acknowledgements

We are indebted to many different people, both inside and outside P&G, who have contributed to the content of this book. To start with there are the P&G staff, past and present, who contributed to the writing of the first edition: Derek Gaskell, UK; Mariluz Castilla, Spain; Klaus Draeger, Germany; Dr Roland Lentz, Germany; Philippe Schauner, France; Dr Chris Holmes, Belgium; Willy van Belle, Belgium; Dr Nick de Oude, Belgium; Dr Celeste Kuta, USA; Keith Zook, USA; Karen Eller, USA; Dr Bruce Jones, USA; Dr Eun Namkung, Japan; and Tom Rattray, USA.

For this second edition, we would like to thank the members of P&G's Global Integrated Solid Waste Team, and in particular, AnaMaria Garmendia, Mexico; Joaquin Zepeda, Venezuela; Dr Rana Pant, UK; Arun Viswanath, India; Mine Enustun, Turkey; Klaus Draeger, Germany; Glenn Parker, Canada; Kim Vollbrecht, USA; Briseida Paredes; Mexico; and Suman Majumdar, India.

Externally we would like to acknowledge the considerable contribution to the subject of Integrated Waste Management by Jacques Fonteyne and the European Recovery and Recycling Association (ERRA), especially Elizabeth Wilson who is now with the US EPA, and also that of the Organic Reclamation and Composting Association (ORCA) and the European Energy from Waste Coalition (EEWC), all based in Brussels. Each of these three organisations started off by focusing on one specific recovery method (recycling, composting and energy from waste), but soon realised that there was no single solution, and that a combination of options is required. To match this, from March 2000, the activities of the three organisations have been integrated into ASSURRE – the Association for the Sustainable Use and Recovery of Resources in Europe (see www.assurre.org). We owe special thanks to Andrew Richmond of Richmond Design Programming, whose expertise and patience made the development of the IWM-2 software possible.

Finally, we thank the contributions of the members of the Peer Review panel who reviewed the IWM-2 model and the associated user's guide: Professor Dr J. Jager, University of Darmstadt, Germany; Terry Coleman, Environment Agency, UK; Dr Matthias Fawer, Eidgenoessische Materialpruefungsanstalt (EMPA), Switzerland: Dr Susan Thorneloe, US Environmental Protection Agency; and Dr Keith Weitz, Research Triangle Institute, USA.

FMcD, PRW, MF, PH.

Currency conversion values

Country/currency	Amount for 1 euro
Argentina – peso	1.0436
Austria – schilling	13.7603
Belgium – franc	40.3399
Brazil – real	1.8054
Canada – Canadian dollar	1.5456
China – yuan	8.6395
Denmark – krone	7.4418
Finland – markka	5.9457
France – franc	6.5595
Germany – mark	1.9558
Holland – guilder	2.2037
Hungary – florint	249.2850
Iceland – krone	77.7795
India – rupee	44.7183
Ireland – punt	0.7875
Italy – lira	1936.2700
Luxembourg – franc	40.3399
Mexico – peso	10.1203
Norway – kroner	8.2434
Poland – zloty	4.1462
Russia – rouble	25.8395
Spain – peseta	166.3860
Sweden – krona	8.9783
Switzerland – franc	1.5929
Turkey – lira	424568.0000
UK – pound sterling	0.6510
USA – dollar	1.0436
Venezuela – bolivar	625.6380

Values as of 1 June 1999.
Source: http://www.oanda.com/convert/fxhistory.

CHAPTER I

Introduction

Summary

The concept of waste as a by-product of human activity and the current concerns over waste disposal are discussed. From these the objectives for sustainable waste management are formulated. Current approaches to reaching these objectives rely on both end-of-pipe and strategic legislation, and voluntary initiatives such as Eco-Efficiency and Design Waste Out. The principles of, and difficulties with, present legislation are discussed. An alternative approach, Integrated Waste Management, is introduced as the underlying theme of this book.

The aims of the book

This second edition of the book *Integrated Solid Waste Management: A Life Cycle Inventory* has four key aims.

1. To provide data, in the form of case studies, that support the concept of Integrated Waste Management (IWM) as a sustainable method of managing solid waste.
2. To provide data, again in the form of case studies, that support the use of Life Cycle Inventory (LCI) as a tool for the environmental and economic optimisation of solid waste management systems.
3. To introduce and describe in detail a new LCI computer model for Integrated Waste Management. This model allows the development of Integrated Waste Management systems in practice. It is easy to use, transparent and contains a range of default data to help the modelling process.
4. To present detailed descriptions and data on current waste management practices, such as waste generation, collection, sorting, biological treatment, thermal treatment, landfill and recycling.

What is waste?

Definitions of 'waste' invariably refer to lack of use or value, or 'useless remains' (*Concise Oxford Dictionary*). Waste is a by-product of human activity. Physically, it contains the same materials as are found in useful products; it only differs from useful production by its lack of value. The lack of value in many cases can be related to the mixed and, often, unknown composition of the waste. Separating the materials in waste will generally increase their value if uses are available for these recovered materials. This inverse relationship between degree of mixing and value is an important property of waste (Box 1.1).

1. The relationship between waste and value:

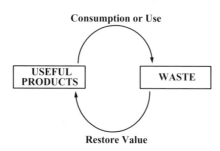

2. The relationship between value and mixing:

$$\text{Value} = f \frac{1}{\text{degree of mixing}}$$

3. Possible classifications of waste. These can be by:
   – physical state
   – original use
   – material type
   – physical properties
   – origin
   – safety level.

BOX 1.1 Waste: some key concepts.

Waste can be classified by a multitude of schemes (Box 1.1): by physical state (solid, liquid, gaseous), and then within solid waste by: original use (packaging waste, food waste, etc.), by material (glass, paper, etc.), by physical properties (combustible, compostable, recyclable), by origin (domestic, commercial, agricultural, industrial, etc.) or by safety level (hazardous, non-hazardous). Household and commercial waste, often referred to together as Municipal Solid Waste (MSW), only accounts for a relatively small part (<10% based on a figure of 522 million tonnes reported by the Organisation for Economic Co-operation and Development (OECD), 1997) of the total solid waste stream. Every year the countries of the OECD produce over 5 billion tonnes of waste (OECD, 1997), which includes, in addition to MSW: agricultural and mining wastes, quarrying wastes, manufacturing and industrial wastes, waste from energy production, waste from water purification, construction and demolition wastes.

There are good reasons for addressing MSW. As the waste that the general public (and therefore voters) have contact with, management of MSW has a high political profile. Additionally, household waste is, by nature, one of the hardest sources of waste to manage effectively. It consists of a diverse range of materials (glass, metal, paper, plastic, organics) totally mixed together. MSW composition is also variable, both seasonally and geographically from country to country, and from urban to rural areas. In contrast, commercial, industrial and other solid wastes tend to be more homogeneous, with larger quantities of each material. Thus if a system can be devised to deal effectively with the materials in household waste, it should be possible to apply such lessons to the management of other sources of solid waste.

The concerns over waste

Historically, health and safety have been the major concerns in waste management. These still apply – wastes must be managed in a way that minimises risk to human health. Today, society demands more than this – as well as being safe, waste management must also be sustainable. Sustainability or Sustainable Development has been defined as 'development which meets the needs of the present without compromising the ability of future generations to meet their own needs' (WCED, 1987). This identifies the synergy between economic development, social equity and the environment (see Figure 1.1). Therefore, sustainable waste management must be:

- economically affordable
- socially acceptable
- environmentally effective.

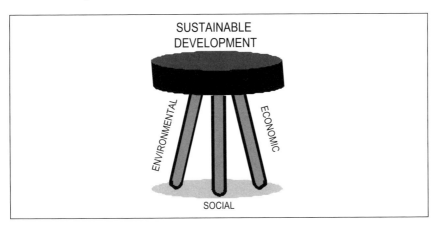

Figure 1.1 The three pillars of Sustainable Development. Equal consideration of each is necessary otherwise the whole system will become unbalanced.

In the past, the economic cost of a waste management system was the major controlling factor in the decision-making process; more recently however environmental considerations have played a more important role in this process. The inclusion of the social aspects of waste management in the decision-making process, although not a new concept in itself, has been limited as research into how to measure social concerns is only just beginning (Nilsson-Djerf, 1999).

Environmental concerns over the management and disposal of waste can be divided into two major areas: conservation of resources and pollution of the environment.

The old concern – the conservation of resources

In 1972 the best-selling book *Limits to Growth* (Meadows et al., 1972) was published. It argued that the usage rates (in 1972) of the earth's finite material and energy resources (non-renewables) could not continue indefinitely. The sequel, *Beyond the Limits* (Meadows et al., 1992), told the same story, but with increased urgency; raw materials are being used at a faster rate than they are being replaced, or alternatives are being found. The result of such reports was the development of the concept of Sustainable Development. This is defined in the Brundtland Report *Our Common Future* (WCED, 1987) as 'development that meets the needs of the present without compromising the ability of future generations to meet their own needs' (see Box 1.2 and Figure 1.2). Sustainability requires that natural resources be efficiently managed, and where possible conserved but not to the detriment of the individual's quality of life.

The original concerns of Meadows et al. about the imminent depletion of natural resources have proved to be incorrect (Beckerman, 1995; UNDP, 1998; UNDESA, 1999) (see Table 1.1). For each raw material the proven reserves in 1989 were greater than the proven reserves in 1970. This is because 'proven reserves' are defined as reserves that could be extracted with today's technology and price structures. Technology and innovation have resulted in most resources being more available, and at a lower extraction cost today than 20 years ago (Meyers and Simon, 1994; Simon, 1996). Consumption has changed in favour of less material-intensive products and services – 'eco-efficiency'. Energy efficiency has improved and technological advances and the recycling of many raw materials have increased the efficiency of material use. These factors have led to material use now growing more slowly than many economies (UNDP, 1998). The *per capita* use of basic materials such as steel, timber and copper has stabilised in most OECD countries (OECD, 1997) – and even declined in some countries for some products. This is not to argue that resources can never be depleted to unacceptable levels, but in most cases, the time period required for this to happen is extremely long, allowing time for the implementation of technological developments.

> 'Our report, Our Common Future, is not a prediction of ever increasing environmental decay, poverty, and hardship in an ever more polluted world among ever decreasing resources. We see instead the possibility for a new era of economic growth, one that must be based on policies that sustain and expand the environmental resource base. And we believe such growth to be absolutely essential to relieve the great poverty that is deepening in much of the developing world. ... sustainable development is not a fixed state of harmony, but rather a process of change in which the exploitation of resources, the direction of investments, the orientation of technological development, and institutional change are made consistent with future as well as present needs.'
>
> *Our Common Future*
> *The World Commission on Environment and Development*
> *Oxford University Press, 1987*

Box 1.2 The World Commission on Environment and Development's view of Sustainable Development as a dynamic process.

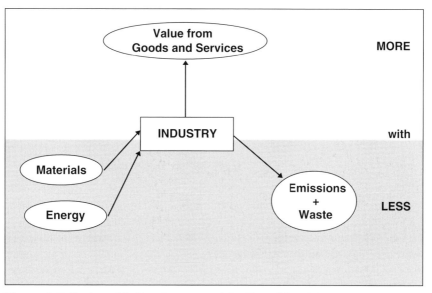

Figure 1.2 Sustainable Development – The Brundtland Report on Sustainable Development (WCED, 1987) introduced the concept of 'More with Less' – the need to produce *more* value from goods and services with *less* raw material and energy consumption, and *less* waste and emission production.

	Reserves 1970	Reserves 1989	Cumulative consumption 1970–89
Aluminium	1170	4918	232
Copper	308	560	176
Lead	91	125	99
Nickel	67	109	14
Zinc	123	295	118
Oil*	550	900	600
Natural gas	250	900	250

Table 1.1 Proven reserves of raw materials (10^6 tonnes unless otherwise stated). In all cases (except lead) reserves in 1989 were greater than reserves in 1970 by more than the amount consumed during this 19-year period. After: Beckerman (1995)
*10^9 barrels of oil equivalent.

Concepts and Case Studies

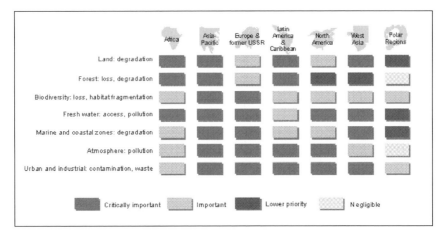

Figure 1.3 Regional concerns: relative importance given to environmental issues by regions. Source: UNEP (1997).

The new concerns – pollution and the deterioration of renewables

The depletion of non-renewables is now not the urgent problem (UNDP, 1998), but two other concerns have become critical with respect to the 'needs of future generations'. These are:

- the generation of pollution and wastes that exceed the ability of the planet's natural sinks to absorb and convert them into harmless compounds and
- the increased deterioration of renewables such as water, soil, forests, fish stocks and biodiversity.

The relative importance of these concerns within and across regions are indicated in Figure 1.3.

It can be seen from Figure 1.3 that urban and industrial contamination and waste are thought to be reaching critical levels in all regions except Africa and the Polar caps. This book aims to address part of the issue of solid waste management.

Sustainable Waste Management

Against the background described above, the production and disposal of large amounts of waste is still seen by many to be a loss of the earth's resources. Putting waste into holes in the ground certainly appears to be inefficient materials management. It needs to be remembered, however, that although the earth is an open system regarding energy, it is essentially a closed system for materials. Whilst materials may be moved around, used, dispersed or concentrated, the total amount of the earth's elements stays constant (with the exception of unstable radioactive elements). Thus although resources of 'raw materials' may be depleted, the total amount of each element present on Earth remains constant. In fact, the concentration of some useful materials is higher in landfills than in their original raw material ores. Such materials could be dug up at a later date.

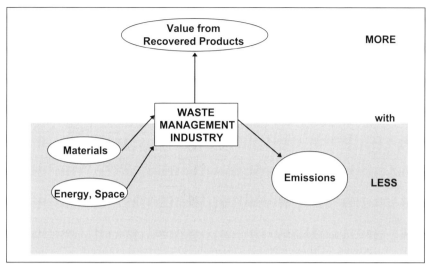

Figure 1.4 Sustainable Waste Management – also calls for 'More with Less' – *more* valuable products recovered from the waste with *less* energy and space consumption and *less* emissions.

Putting waste in holes in the ground, i.e. landfilling, could therefore be considered as long-term storage of materials rather than actual disposal. Is this the most efficient way to manage such materials, however?

Concerns over conservation of resources have led to calls for, firstly, general reductions in the amount of waste generated, i.e. waste minimisation or waste reduction, and secondly, for ways to recover the materials and/or energy in the waste, so that they can be used again. Recovery of resources from waste should slow down the depletion of non-renewable resources, and help to lower the use of renewable resources to the rate of replenishment (see Figure 1.4).

Pollution

Potential or actual pollution is the basis for most current environmental concern over waste management. Historically, the environment has been considered as a sink for all wastes produced by human activities. Materials have been released into the atmosphere or watercourses, or dumped into landfills and allowed to 'dilute and disperse'. At low levels of emissions, natural biological and geochemical processes are able to deal with such flows without resulting changes in environmental conditions. However, as the levels of emissions have increased with rises in human population and activity, natural processes do not have sufficient turnover to prevent changes in environmental conditions (such as the level of atmospheric carbon dioxide). In extreme cases of overloading (such as gross sewage pollution in rivers) natural processes may break down completely, leading to drastic changes in environmental quality.

Just as raw materials are not in infinite supply, the environment is not an infinite sink for emissions. Environmental pollution produced by human activity will come back to haunt society, by causing deterioration in environmental quality. Consequently there is growing awareness that the environment should not be considered as an external sink for wastes from society, but as part of the global system that needs careful and efficient management.

As well as such broad concerns over environmental pollution, specific concerns emerge at the local level whenever new facilities for waste treatment are proposed. Planned incinerators raise concerns over likely emission levels in general, and recently, of dioxin levels in particular. Similarly, landfill sites are known to generate landfill gas. At the global level this has a high Global Warming Potential (GWP), but more immediately at a local level it can seep into properties with explosive consequences. There is also increasing concern of the risk of groundwater pollution from the leachate generated in landfill sites.

Such local environmental concerns (plus, no doubt, concern over the effects on property prices) have given rise to several acronyms describing the attitudes facing waste management planners, which will have to be met with a viable response and a clear long-term strategy (Box 1.3).

Whilst such attitudes are understandable, they ignore our common responsibility for waste management. All human activities generate waste. Each person in the European Union for example, generates an average of 430 kg of Municipal Solid Waste each year (OECD, 1997). This waste has to be dealt with somehow, somewhere. Whilst all the methods for treating and

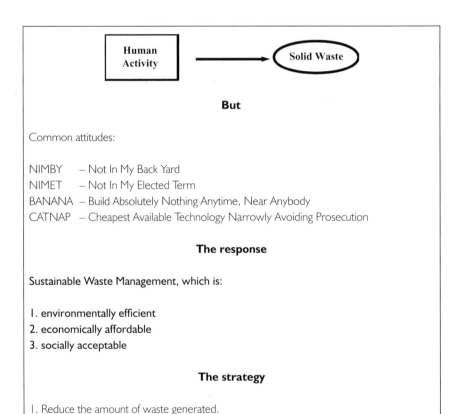

BOX 1.3 The challenge of waste management.

disposing of waste are known to have environmental impacts, the waste still must be dealt with. Waste cannot be moved to 'other people's back yards' indefinitely. What is needed is an overall strategy to manage waste that reduces environmental burdens, at an affordable cost.

Objectives

Clearly, the first objective is to reduce the amount of waste generated. However, even after this has been done, waste will still be produced. The second objective, therefore, is to manage the waste in a sustainable way, by minimising the overall environmental burdens associated with the waste management system (see Figure 1.4). Environmental sustainability addresses both pollution concerns and resource conservation, but can it be assessed, and if so how?

Life Cycle Assessment (LCA) is an environmental management tool that attempts to predict the overall environmental burden of a product, service or function, and it can be applied to waste management systems (White et al., 1993). In this book, the technique of Life Cycle Inventory (LCI) – part of LCA (see Chapters 4 and 5), will be used to look at whole waste management systems, to predict their likely environmental burdens. Then, by selecting appropriate options for dealing with the various fractions of solid waste, the environmental burdens of the whole waste management system can be reduced.

Current approaches – legislation

Waste management in developed countries is governed by legislation, the detail of which could fill this volume several times over. Looking at the basic approach of legislative tools, they fall into two main categories, 'end-of-pipe' regulations and strategic targets.

End-of-pipe regulations

These are technical regulations and relate to the individual processes in waste treatment and disposal. Emission controls for incinerators are a prime example. Such regulations may be set nationally (e.g. the T.A. Luft (1986) emission limit for Germany and US federal Standards of Performance (1995)) or, increasingly, internationally by bodies such as the European Commission (EC). Municipal waste incinerator emissions are the subject of two EC Directives (89/369/EEC for new incinerator plants and 89/429/EEC for existing plants), and there is also an EC Directive (1999/31/EC) on Landfill.

Such regulations are important to ensure safe operation of waste disposal processes. They operate as 'fine tuning' of the system, by promoting best available technology and practices, but as with other end-of-pipe solutions, they are not tools that will lead to major changes in the way the waste management system operates. These will be produced by strategic decisions between different waste management options rather than refining the options themselves.

Strategic targets

Strategic legislation is becoming increasingly used to define the way in which solid waste will be dealt with in the future. This legislation has several common threads: it builds (either explicitly or implicitly) on the 'hierarchy of solid waste management', and within this it sets targets for recovery and recycling of materials. Much current legislation is also directed at specific parts, rather than the whole, of the municipal waste stream. Packaging, in particular, has been targeted. Consequently attention has been diverted away from the overall problem of dealing with all of the waste stream in an environmentally and economically sustainable way. Packaging waste represents between 16–17% (by weight) of the Municipal Solid Waste generated in the European Union (EU) (PricewaterhouseCoopers, 1998). Setting targets for recycling of packaging alone, therefore, only deals at most with this 17% of the Municipal Solid Waste stream (EC Discussion paper, 1999). For example, a 50% reduction in packaging waste would only result in a 9% reduction in total Municipal Solid Waste. Given that Municipal Solid Waste represents 13–15% of the total solid waste production in the EU (OECD, 1997 and EEA, 1998), this means that special measures directed at packaging will have little impact (1.1–1.3% reduction) on the overall solid waste management scenario. At national levels, the German Packaging Ordinance, Dutch Packaging 'Covenant' and UK Environmental Protection Act (1990) all lay down either rules or guide-lines on the waste management options (mainly recycling), which will be used for at least part of Municipal Solid Waste. At the European level, the EC Packaging and Packaging Waste Directive requires member states to set deadlines for meeting value recovery and recycling targets for packaging materials and this has been incorporated into the national legislation of all member states.

The hierarchy of waste management options is headed by source reduction, or waste minimisation. This is the essential prerequisite for any waste management strategy – less waste to deal with. Next in the hierarchy come a series of options: reuse, recycling, composting, waste to energy, incineration without energy recovery and landfill, in some order of preference. At the time of writing, there is increasing recognition within the field of waste management that the hierarchy has several limitations (see Chapter 2). It will be argued in this book that, although useful as a set of default guide-lines, using this hierarchy to determine which options are preferable does not necessarily result in the lowest environmental burdens, nor an economically sustainable system. Different materials in the waste are best dealt with by different processes, so to deal effectively with the whole waste stream, a range of waste management options is desirable. Thus, there are no overall 'best' or 'worst' options; different options are appropriate to different fractions of the waste.

Instead of relying on the waste management hierarchy, this book looks at the whole waste management system, and attempts to assess its overall environmental burdens and economic costs on a life cycle basis.

Recycling targets, similarly, are useful ways to measure progress towards recycling goals, but they may not always measure progress towards environmental objectives. Recycling typically is used to reduce raw material consumption, and in many cases also energy consumption. It must be borne in mind, however, that it is a means to an end, not an end in itself. Rigidly set recycling targets may not produce the greatest environmental benefits. Environmental benefit (e.g. reduction in energy consumption) does not increase linearly with recycling rates (Boustead, 1992; Gabola, 1999). At high levels of recovery, proportionately more energy is needed to collect used materials from diffuse sources, so there is little, if any, environmental gain. In such cases, Life Cycle Assessment could be used to determine the optimal recycling rate to meet defined environmental objectives. A sign that this more flexible approach may be gaining ground is the inclusion in the EC Packaging and Packaging Waste Directive of the provision that 'life cycle assessments should be completed as soon as possible to justify a clear hierarchy between reusable, recyclable and recoverable packaging; ...'. Such LCA studies have now been carried out (RDC/Coopers & Lybrand, 1997). It is worth noting that, as argued by the European Organisation for Packaging and the Environment (EUROPEN, 1999) and others, LCA offers the opportunity to replace rather than reinvent the waste management hierarchy.

At a more fundamental level, how waste is best recovered, treated, or disposed of depends on the nature of the materials in the waste, not on the original use of the discarded object. Whatever legislative instruments (e.g. recycling targets) are used, they should relate to the whole municipal waste stream. Otherwise, as has been seen in Germany, packaging-specific legislation can lead to the development of separate and parallel waste management systems for such packaging waste. Duplication of systems on the basis of original use of the material, or for any other reason, will give rise to increased environmental burdens and economic costs (Staudt and Schroll, 1999).

This book proposes that if real and sustainable environmental improvements are the objective, a single, integrated collection and sorting system, followed by material-specific recovery/ treatment/ disposal represents the most promising approach. It shows that a scheme designed to deal with all fractions of household waste can also deal with other sources such as commercial and some industrial waste. Fragmenting the household waste between different waste management systems rules out an integrated approach, and any economies and efficiencies of scale.

Economic costs of environmental improvements

Environmental improvements to waste treatment and disposal methods should be welcomed, where they are scientifically justifiable. Improvements, however, usually have economic costs associated with them. This is invariably the case with such end-of-pipe solutions as installing new technology to reduce emissions following tighter regulatory control. It can also be the case when introducing strategic solutions aimed at reducing the environmental impact of waste management, such as recycling. Take, for example, the 'Blue Box' schemes that have been introduced for kerbside collection of recyclable materials in many parts of North America, and in some areas of the UK, or the Dual System operating in Germany for packaging materials. Whilst these systems can collect large quantities of high-quality materials, the collection schemes operate in parallel with existing household waste collections. Two vehicles call at each property, where only one did before. These systems, as additions to residual waste collection, clearly involve additional cost. Any other similar 'bolt-on' systems, whether for recovery of materials or collection of compostables will suffer from the same economic problems. This trade-off between economic cost and environmental burden has been seen as a major hurdle to environmental improvements in waste management. However, two concepts that are becoming established can help to overcome this obstacle.

Internalising external environmental costs

Environmental and social costs of waste disposal have historically been seen as external costs. For example, the effects of emissions from burning waste, or the leachate and gas released from landfill sites were not considered as part of the cost of these disposal methods. More recently, however, as emission regulations have become tighter, the costs of emission controls have been internalised in the cost of disposal. Similarly, when legislation (e.g. UK Environmental Protection Act, 1990) requires monitoring of waste disposal sites after closure, and the provision of insurance bonds to remediate environmental problems that may arise in the future, the real (i.e. full and inclusive) cost of each waste management option becomes apparent. Under such conditions, waste management options with lower environmental burdens, which may have appeared more expensive, can become economically viable.

Building environmental objectives into the waste management system

An additional 'bolt-on' system or an end-of-pipe solution will entail additional costs (Box 1.4). If, however, a waste management system is initially designed to achieve environmental objectives

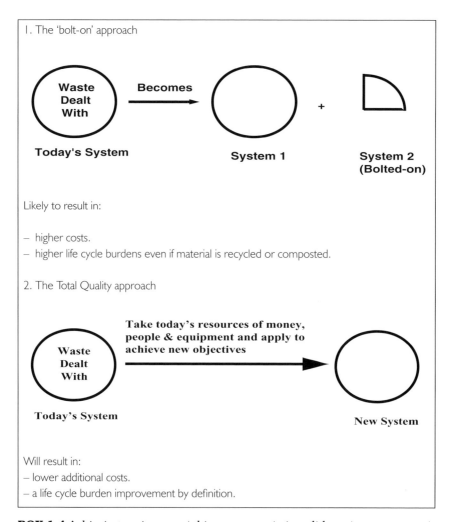

1. The 'bolt-on' approach

Waste Dealt With   Becomes → ○   +   ◱

Today's System   System 1   System 2 (Bolted-on)

Likely to result in:

— higher costs.
— higher life cycle burdens even if material is recycled or composted.

2. The Total Quality approach

Waste Dealt With   Take today's resources of money, people & equipment and apply to achieve new objectives →   ○

Today's System   New System

Will result in:
— lower additional costs.
— a life cycle burden improvement by definition.

BOX 1.4 Achieving environmental improvements in solid waste management.

it may include such features at little or no extra cost compared to existing practices. An integrated system that can deal with all the materials in the solid waste stream represents the Total Quality approach to waste management. Such a system would benefit from economies of scale, since an integrated material-based system allows for efficient collection and management of waste from different sources. The Total Quality objective would be to minimise the environmental burdens of the whole waste management system, whilst keeping the economic costs to an acceptable level. The definition of acceptable will vary according to the group concerned, and with geography, but if the cost is little or no more than present costs, it should be acceptable to most people (Box 1.4).

An integrated approach to solid waste management

An integrated approach to solid waste management can deliver both environmental and economic sustainability.

It is clear that no one single method of waste disposal can deal with all materials in waste in an environmentally sustainable way. Ideally, a range of management options is required. The use of different options such as composting or materials recovery will also depend on the collection and subsequent sorting system employed. Hence any waste management system is built up of many closely related processes, integrated together. Instead of focusing on and comparing individual options, (e.g. 'incineration versus landfill'), an attempt will be made to synthesise waste management systems that can deal with the whole waste stream, and then compare their overall performances in environmental and economic terms.

This approach looks at the overall waste management system, and develops ways of assessing overall environmental burdens and economic costs. As part of this, the various individual techniques for collection, treatment and disposal of waste are discussed. This book gives an overall vision of waste management, with a view to achieving environmental objectives using economically sustainable systems tailored to the specific needs of a region or community.

CHAPTER 2

Integrated Waste Management

Summary

This chapter discusses the needs of society: less solid waste, and then an effective way to manage the inevitable solid waste still produced. Such a waste management system needs to be both environmentally, economically and socially sustainable and is likely to be integrated, market oriented, flexible and operated on a regional scale.

The current Hierarchy of Waste Management is critically discussed, and in its place is suggested a holistic approach that assesses the overall environmental burdens and economic costs of the whole system.

Definition: Integrated Waste Management (IWM) systems combine waste streams, waste collection, treatment and disposal methods, with the objective of achieving environmental benefits, economic optimisation and societal acceptability. This will lead to a practical waste management system for any specific region.

The Key features of IWM are:

1. an overall approach
2. uses a range of collection and treatment methods
3. handles all materials in the waste stream
4. environmentally effective
5. economically affordable
6. socially acceptable.

The basic requirements of waste management

Waste is an inevitable product of society. Solid waste management practices were initially developed to avoid the adverse effects on public health that were being caused by the increasing amounts of solid waste being discarded without appropriate collection or disposal. Managing this waste more effectively is now a need that society has to address. In dealing with the waste, there are two fundamental requirements: less waste, and then an effective system for managing the waste still produced.

15

The generation of less waste

The Brundtland report of the United Nations, *Our Common Future* (WCED, 1987), clearly explained how sustainable development could only be achieved if society in general, and industry in particular, learned to produce 'more from less'; more goods and services from less of the world's resources (including energy), while generating less pollution and waste.

In this era of 'green consumerism' (Elkington and Hailes, 1988; Elkington, 1997), this concept of 'more from less' has been taken up by industry. This has resulted in a range of concentrated products, light-weighted and refillable packaging, reduction of transport packaging and other innovations (Hindle et al., 1993; IGD, 1994; EPU, 1998). Production as well as product changes have been introduced, with many companies using internal recycling of materials as part of solid waste minimisation schemes.

All of these measures help to reduce the amount of solid waste produced, either as industrial, commercial or domestic waste. In essence, they are improvements in efficiency, i.e. 'eco-efficiency', whether in terms of materials or energy consumption. The costs of raw materials and energy, and rising disposal costs for commercial and industrial waste, will ensure that waste reduction continues to be pursued by industry for economic as well as environmental reasons.

There has been interest in promoting further waste reduction by the use of fiscal instruments. Pearce and Turner (1992), for example, suggest ways to reduce the amount of packaging used (and hence appearing as waste) by internalising the costs of waste disposal within packaging manufacture, by means of a packaging levy. It is not clear how effective such taxes would be, however, since they only affect a small section of the waste stream, packaging materials constitute approximately 3% by weight of the UK's total waste stream (INCPEN, 1996) and 3% by weight of the total European waste stream (Pricewaterhouse Coopers, 1998). Furthermore, economic incentives for waste reduction already exist. Experience has often shown that extra taxes add to the total cost but do not reduce the base cost of waste management. Often they become in effect an additional revenue stream rather than being used for specific environmental purposes.

Interestingly, there is one area where there are often no economic incentives for waste reduction – household waste generation. In some communities, notably in the USA (Skumatz et al., 1997; Canterbury, 1998; Horton, 1998), and also in Germany, waste collection charges are on a scale according to the volume or in some cases mass of waste generated, but in most communities, a flat rate collection fee applies. Charging according to waste generation could lead to the generation of less household waste, provided that unauthorised dumping or other alternative disposal routes could be prevented. The issues to be resolved are the level of the charge, and an effective way of managing it. Introducing such charging structures to societies not used to them can be politically unpopular.

'Waste minimisation', 'waste reduction' or 'source reduction' are usually placed at the top of the conventional waste management hierarchy. In reality, however, source reduction is a necessary precursor to effective waste management, rather than part of it. Source reduction will affect the volume, and to some extent, the nature of the waste, but there will still be waste for disposal (see Figure 2.1). What is needed, beyond source reduction, is an effective system to manage this waste.

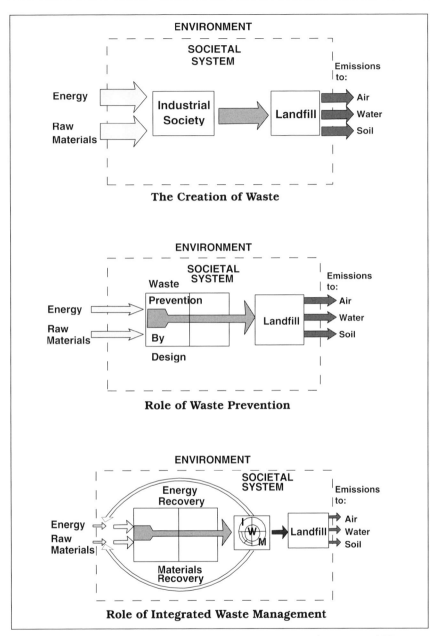

Figure 2.1 The respective roles of Waste Prevention and Integrated Waste Management. In life cycle studies, a 'system' is defined (with boundaries indicated by broken lines). Energy and raw materials from the 'environment' are used in the system. Emissions, including solid waste, leave the system and enter the environment (see Chapters 4 and 5 for a more detailed discussion).

The concept of Sustainable Waste Management

Solid waste management systems need to ensure human health and safety. They must be safe for workers and safeguard public health by preventing the spread of disease. In addition to these prerequisites, a sustainable system for solid waste management must be **environmentally effective, economically affordable** and **socially acceptable.**

1. Environmentally effective: the waste management system must reduce as much as possible the environmental burdens of waste management (emissions to land, air and water, such as CO_2, CH_4, SOx, NOx, BOD, COD and heavy metals).
2. Economically affordable: the waste management system must also operate at a cost acceptable to the community, which includes all private citizens, businesses and government. The costs of operating an effective solid waste system will depend on existing local infrastructure, but ideally should be little or no more than existing waste management costs.
3. Socially acceptable: the waste management system must operate in a manner that is acceptable for the majority of people in the community. This is likely to require an extensive dialogue with many different groups to inform and educate, develop trust and gain support.

Clearly it is difficult to minimise three variables – cost, social acceptability and environmental burden – simultaneously. There will always be a trade-off. The balance that needs to be achieved is to reduce the overall environmental burdens of the waste management system as far as possible, within an acceptable level of cost. Deciding the point of balance between environmental burden and cost will always generate debate. Better decisions will be made if data on environmental burdens and costs are available; such data will often prompt ideas for further improvements.

Characteristics of a Sustainable Waste Management system

A Sustainable Waste Management system is likely to be integrated, market oriented, flexible and socially acceptable. The execution of these principles will vary on a regional basis. A key requirement, is the 'customer–supplier relationship'.

An integrated system

'Integrated Waste Management' is a term that has been frequently applied but rarely defined. Here it is comprehensively defined as a system for waste management that has control over:

1. All types of solid waste materials. The alternative of focusing on specific materials, either because of their ready recyclability (e.g. aluminium) or their public profile (e.g. plastics) is likely to be less effective, in both environmental and economic terms, than taking a multi-material approach.
2. All sources of solid waste. Wastes such as domestic, commercial, industrial, institutional, construction and agricultural. Hazardous waste needs to be dealt with within the system, but in a separate stream. Focusing on the source of a material (on packaging or domestic waste or industrial waste) is likely to be less productive than focusing on the nature of the material, regardless of its source.

An integrated system would include an optimised waste collection system and efficient sorting, followed by one or more of the following options:

1. Materials recycling will require access to reprocessing facilities (see Chapter 14, Materials recycling).
2. Biological treatment of organic materials will ideally produce marketable compost and also reduce volumes for disposal. Anaerobic digestion produces methane that can be burned to release energy (see Chapter 11, Biological treatment).
3. Thermal treatment (such as incineration with energy recovery, burning of Refuse-Derived Fuel (RDF) and burning of Paper and Plastic-Derived Fuel (PPDF)) will reduce volume, render residues inert and should include energy recovery (see Chapter 12, Thermal treatment).
4. Landfill. This can increase amenity via land reclamation but a well-engineered site will at least minimise pollution and loss of amenity (see Chapter 13, Landfilling).

To manage all solid waste arisings in an environmentally effective way requires a range of the above treatment options. Landfill is the only method that can manage all types of waste; since recycling, composting and thermal treatment all leave some residual material that needs to be landfilled. In a landfill the organic fraction of solid waste can be broken down if the appropriate conditions for the growth of aerobic and then anaerobic bacteria occur. These relatively uncontrolled biological processes can take several years to start in a landfill and continue many decades after the landfill has been closed. Methane emissions arise from the breakdown of organic material and groundwater pollution may occur due to leaching of toxic materials from the solid waste. Landfilling operations also require large amounts of space. Use of the other options prior to landfilling can both divert significant parts of the waste stream and reduce the volume and improve the physical and chemical stability of the final residue. This will reduce both the space requirement and the environmental burdens of the landfill.

Market oriented

Any scheme that incorporates materials recycling, biological or thermal treatment technologies must recognise that effective recycling of materials and production of compost and energy depends on markets for these outputs. These markets are likely to be sensitive to price and to consistency in quality and quantity of supply. Managers of such schemes will need to play their part in building markets for their outputs, working with secondary material processors, and helping set material quality standards. They must also recognise that such markets and needs will change over time, so such standards should not be rigid and based in prescriptive legislation, but be set as part of a customer–supplier relationship.

Flexibility

An effective scheme will need the flexibility to design, adapt and operate its systems in ways which best meet current social, economic and environmental conditions. These are likely to change over time and vary by region.

Using a range of waste management options in an integrated system gives the flexibility to channel waste via different treatments as economic or environmental conditions change. For example, paper can either be recycled, composted or incinerated with energy recovery. The option used can be varied according to the economics of paper recycling, compost production or energy generation at the time.

Concepts and Case Studies

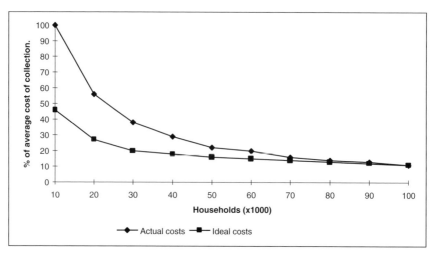

Figure 2.2 The percentage of the average cost of MSW collection in The Netherlands (100% = 250 guilders) compared to the number of households served. Source: Doppenberg (1998).

Scale

The need for consistency in quality and quantity of recycled materials, compost or energy, the need to support a range of management options, and the benefit of economies of scale, all suggest that Integrated Waste Management should be organised on a large-scale, regional basis. This is certainly the case for waste collection, which can cost up to 75% of the total waste management budget for an area (ERRA, 1998). The cost benefit of the economy of scale with regard to collection is shown in Figure 2.2; this benefit is mirrored in the construction and operation of both treatment (EC, 1997; Biala, 1998; Reimann, 1998; Kern et al., 1999) and disposal facilities.

Experience is beginning to suggest that an area containing upwards of 500,000 households is a viable unit (White, 1993) where large-scale additions to infrastructure are necessary. This may not correspond to the scale on which waste management is currently administered. In many cases, therefore, implementation of such schemes will require local authorities to work together. Schemes on a smaller scale can be viable (Schauner, 1996), but essentially have to rely on recycling using source-separated material, bring systems (see Chapter 9) or composting. These less expensive treatment options can still result in a significant diversion of material from final disposal to landfill.

Social acceptability

For Waste Management Systems to operate effectively public participation is necessary. Whether simply putting bins by the kerbside for collection on the correct day, taking paper or glass to streetside recycling containers or sorting out all recyclable material from their household waste, individuals must understand their role in the Waste Management System and co-operate with the local authorities for the system to work. Low participation rates in recycling schemes can be improved by communication strategies. Objections to waste management facility siting can be minimised through considerate facility design and by public consultation and education.

Public support can also be enhanced by stable waste management systems where changes to waste sorting habits or to the waste collection system are rarely required. Where system changes are necessary, effective communication material will inform the public as to the benefits of the new system (environmental, economic and social), thus increasing the chance that they will accept the changes. Public perception of waste management facilities will also have an effect on the acceptance of the overall waste management system. Any facilities that are seen to be dirty, polluting, sources of odours or litter are likely to result in reduced acceptability of the whole waste management system. Clean, well-managed facilities that are known to be operating up-to-date pollution control technology meeting high regulatory standards are more likely to be accepted by the public. The integrity of the waste management system is also essential: the public must be confident that any material they source separate for recycling, is sent for recycling and not landfilled or incinerated. If source-separated material is found to have been sent for incineration or directly to landfill, the credibility of the waste management system is reduced and public support will fall. Good public support is as essential to a waste management system as good planning and good management.

Development of the Integrated Waste Management concept

A systems approach to waste management was proposed by W.R. Lynn in 1962. This approach was described as 'viewing the problem in its entirety as an interconnected system of component operations and functions' and therefore recognised the full complexity of waste management practices. The acceptance that systems analysis and mathematical modelling were necessary to optimise waste management operations and strategy development was the first step on the road to the concept of IWM.

The concept further evolved in 1975 when the newly formed Solid Waste Authority of Palm Beach County, Florida presented their first mission statement. This statement proposed that the Authority would 'Develop and implement programs in accordance with its Comprehensive Plan by *integrating* solid waste transportation, processing, recycling, resource recovery and disposal technologies.'

The penultimate step in the evolution of the concept was made by R.M. Clark, a systems analyst with the US Environmental Protection Agency in 1978, when he observed that '[Waste] Management methods, equipment, and practices should not be uniform across the country since conditions vary, and it is vital that procedures be varied to meet them.' This simple statement of fact has been ignored by supporters of a hierarchical approach to waste management and by certain policy makers who try to develop a generic national or even international approach to waste management.

The final and most significant definition of IWM took place in 1991, when a task force from the Economic Commission for Europe published a Draft Regional Strategy for Integrated Waste Management that defined IWM as a 'process of change in which the concept of waste management is gradually broadened to eventually include the necessary control of gaseous, liquid and solid material flows in the human environment.' This brought all waste arisings under the umbrella of IWM.

The concept of IWM now included all waste types, the option of using a range of treatment technologies depending on the situation and an overall approach being taken with respect to the analysis, optimisation and management of the whole system. These basic points were refined only slightly to give the definition of integrated solid waste management presented in the first edition of this book (White et al., 1995).

The United Nations Environmental Programme (UNEP, 1996) recognised the importance of Integrated Waste Management, which it defined as 'a frame of reference for designing and implementing new waste management systems and for analysing and optimising existing systems'. This UNEP document also described Integrated Waste Management as 'an important element of sound waste management practice' because of the reasons presented in earlier sections.

Implementing Integrated Waste Management

The operations within any waste management system are clearly interconnected. The collection and sorting method employed, for example, will affect the ability to recover materials or produce marketable compost. Similarly, recovery of materials from the waste stream may affect the viability of energy recovery schemes. It is necessary, therefore, to consider the entire waste management system in an holistic way. What is required is an overall system that is both economically and environmentally sustainable (see Table 2.1). Much effort has been directed towards schemes focusing on individual technologies, e.g. recycling, or on materials from one source only (e.g. the German Green Dot system to collect packaging). From the perspective of the whole waste management system, such schemes often involve duplication of efforts, making them both environmentally and economically ineffective.

1. Aim for the following:	Environmental effectiveness:	reduce environmental burdens
	Economic affordability:	drive costs out

2. The system should be:	Integrated:	in waste materials
		in sources of waste
		in collection methods
		in treatment methods
		  anaerobic digestion
		  composting
		  energy recovery
		  landfill
		  recycling

	Market oriented:	materials and energy must have end uses and generate income
	Flexible:	for constant improvement
3. Take care to:	Define clear objectives Design a total system against those objectives Operate on a large enough scale	
4. Never stop looking for improvements in overall environmental performance and methods to lower operating costs. Remember that there is no perfect system.		

Table 2.1 Designing a sustainable solid waste management system

The relevance of looking at the whole system could be challenged, since waste management is often split up into many different compartments. Collection of Municipal Solid Waste is usually the duty of local authorities, though may be contracted out to private waste management companies. Disposal often comes under the jurisdiction of another authority, and perhaps another private company. Different operators may contribute to recycling activities – in the case of material collection banks, these may be the material producers. Similarly, thermal treatment, biological treatment and landfill operations may all be under the control of different operating companies. Each company or authority only has control of the waste handling within its operation, so what is the feasibility of taking an overall systems approach when no-one has control over the whole system?

The holistic approach has three main advantages:

1. It gives the overall picture of the waste management process. Such a view is essential for strategic planning. Handling of each waste stream separately is inefficient.
2. Environmentally, all waste management systems are part of the same system – the global ecosystem. Looking at the overall environmental burden of the system is the only rational approach, otherwise reductions in the environmental burdens of one part of the process may result in greater environmental burdens elsewhere.
3. Economically, each individual unit in the waste management chain should run at a profit, or at least break even. Therefore, within the boundaries controlled by each operator, the financial incomes must at least match the outgoings. By looking at the wider boundaries of the whole system, however, it is possible to determine whether the whole system operates efficiently and whether it could run at break even, or even at a profit. Only then can all the constituent parts be viable, provided that income is divided up appropriately in relation to costs.

The importance of a holistic approach

To achieve fully, Integrated Waste Management will require major system changes from the present situation. The objective of an integrated system is to be both environmentally and economically sustainable. This is a Total Quality Objective (Oakland, 1994, 1995); it can never be reached, since it will always be possible to reduce environmental burdens further, but it will lead to continual improvements to processes and systems.

Application of Total Quality thinking can be of further use in waste management. To reach a Total Quality objective one builds a system to achieve this objective. To deliver environmentally and economically Sustainable Waste Management requires building a system designed for this purpose. This is a key point. When designing or re-designing a waste management system, an approach must be taken that considers the system in its entirety.

Different components of the system are inter-connected so it is necessary to conceptually re-design a whole new system rather than constantly making minor changes to the old system. For example, simply adding recycling to a waste management system just adds the cost of the new recycling system to the cost of the original waste management system. A systems approach would ensure that the cost of recycling was kept to a minimum by operating fewer collections for restwaste (the material remaining in MSW after the removal of recyclable material), as there should be less restwaste remaining once the recyclable material has been removed. Further system changes based on a holistic viewpoint may also allow previous economic inefficiencies to come to light, which can be used to offset any increased costs.

Paying for Integrated Waste Management

All beneficiaries of the Integrated Waste Management system, the public, the recycling industry and local authority, should pay for waste management services. An Integrated Waste Management system minimises risks to public health and results in a clean, healthy environment for all citizens. The recycling industry benefits from a steady supply of recyclable materials. The system must be affordable for all sections of the community but the full cost of the waste management system must be recovered to ensure that the system is sustainable. This is another of the great challenges of waste management.

Waste management planning and the Hierarchy of Waste Management

Past waste management strategies have relied on the 'Hierarchy of Waste Management'. This has varied in its exact form (see Figure 2.3), but usually gives the following order of preference: waste minimisation; re-use; materials recycling; biological treatment; thermal treatment with energy recovery; thermal treatment without energy recovery; landfilling.

A rigid use of a priority list for waste management options has serious limitations.

1. The Hierarchy has little scientific or technical basis. There is no scientific reason, for example, why materials recycling should always be preferred to energy recovery.
2. The Hierarchy is of little use when a combination of options is used, as in an IWM system. In an IWM system, the Hierarchy cannot predict, for example, whether biological treatment combined with thermal treatment of the residues would be preferable to materials recycling plus landfilling of residues. What is needed is an overall assessment of the whole system, which the Hierarchy cannot provide.
3. The Hierarchy does not address costs, therefore it cannot help assess the economic affordability of waste systems.
4. The Hierarchy cannot account for the wide variety of specific local situations where waste management systems must operate effectively, such as small islands, sparsely populated areas, or popular tourist destinations, where large increases in the population occur on a seasonal basis.

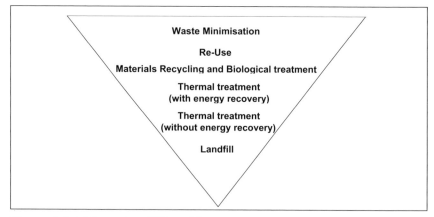

Figure 2.3 A Hierarchy of Waste Management.

The limitations of the Hierarchy of Waste Management are becoming increasingly apparent, especially in relation to IWM systems. The UK Waste Strategy, *Making Waste Work* (DOE, 1995), for example, states that although useful as 'a mental checklist', 'the Hierarchy will not always indicate the most Sustainable Waste Management option for particular waste streams'. The limitations of the waste management hierarchy were also identified in a document by the United Nations Environmental Programme (UNEP, 1996) as 'the hierarchy cannot be followed rigidly, since in particular situations the cost of a prescribed activity may exceed the benefits, when all financial, social and environmental considerations are taken into account'. Similarly, a study comparing different solid waste management options in the European Union concluded: 'the social cost-benefit analysis of MSW management systems in the European Union seems to support the conclusion that the "waste hierarchy" is too simplistic, and that blind adherence to its tenets can lead to welfare losses' (Brisson, 1997).

Rather than a hierarchy of preferred waste management options (as in Figure 2.3) a holistic approach is proposed, which recognises that all options can have a role to play in Integrated Waste Management (see Figure 2.4). The model illustrates the interrelationships of the parts of the system; it does not suggest, for example, that 25% of the collected waste should be treated by each option. The percentage of waste treated by each of the four options will depend on the local situation. Each option should be assessed using the most recent data available, but the overriding objective is to optimise the whole system, rather than its parts, to make it environmentally and economically sustainable and socially acceptable.

Unlike the hierarchy, this approach does not predict what would be the 'best' system. There is no universal best system. There will be geographic differences in both the composition and the quantities of waste generated. Similarly there will be geographic differences in the availability of some waste management options (such as landfill), and in the size of markets for products derived from waste management (such as recovered materials, compost and energy). The economic costs of using different treatment methods will reflect the existing infrastructure (i.e.

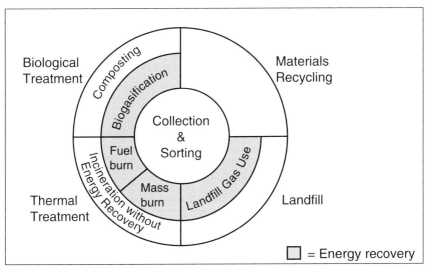

Figure 2.4 The elements of Integrated Waste Management.

whether the plant already exists or needs to be built). This approach allows comparisons to be made between different waste management systems for dealing with the solid waste of the local regions. The best system for any given region will be determined locally.

In essence what is needed is less waste to deal with in the first instance, and then an Integrated Waste Management system to manage the waste that is still produced in an environmentally effective, economically affordable and socially acceptable way. Rather than rely on the waste hierarchy, the environmental management tool of Life Cycle Inventory (see Chapter 4) can be used to help reach this objective.

Integrated Waste Management in countries with developing economies

The management of Municipal Solid Waste (MSW) is an integral but much neglected aspect of environmental management in most low and middle income countries. Despite consuming a significant share of Municipal budgets (often between 10 and 50% of operational expenditures), solid waste management services in the towns and cities of most low and middle income countries are unreliable, provide inadequate coverage, conflict with other urban services and have adverse effects on public health and the environment (Bartone, 1999).

The IWM approach is as valid in countries with developing economies as it is in countries with developed economies, but the actual establishment of integrated systems differs considerably. In countries with developing economies the combination of a lack of existing waste management infrastructure and severely limited resources change the approach that must be taken. Under these conditions, a relatively simple but effective Integrated Waste Management system is desirable.

IWM systems for countries with developing economies

The waste management systems that exist in the majority of countries with developing economies are often characterised by inadequate collection services, little or no treatment and uncontrolled dumping. The establishment of IWM systems will require the following:

1. Data collection on waste composition. This is needed for the planning of collection, transport and treatment of MSW. Good data is the foundation of effective IWM systems.
2. Progress from uncontrolled dumping to the use of simple sanitary landfills.
3. Separation of organic waste from MSW, which can then be composted.
4. Formal involvement of scavengers in the collection of recyclable materials.

When available resources and MSW composition are considered, it is apparent that the waste hierarchy is too rigid to be relevant for waste management in countries with developing economies. The flexibility of IWM offers a more realistic opportunity to improve waste management by accounting for local conditions.

Dumping and landfilling

The dumping of MSW into uncontrolled sites is the most common form of waste disposal in the developing world and is the result of both limited technical and financial resources. This method of final disposal is environmentally and socially unacceptable as it does little to protect the environment or public health. Pollution of surface and groundwater by leachate, migration of combustible gases (methane), odours and breeding of disease carriers are all common results. Dumps provide very poor living conditions for scavengers (see Chapter 13, Landfill section) and pose significant health risks today and in the future.

In most situations disposal of waste to landfill is the lowest cost option, even in countries with developed economies where landfills are often highly engineered and include multiple liners on the bottom of the landfill. Moving away from a system based on dumping to one based on sanitary landfills represents the most financially realistic option for developing economies to improve waste management. Although the resources for highly engineered facilities are unlikely to be available, simple, low cost options are possible. Landfills should be sited away from water courses and highly populated areas and built on soils with a relatively low permeability such as clay. Filling should be on a cell by cell basis with, wherever possible, the application of a cover material (soil or compost) at the end of each working day. Organic waste and recyclable material should be separated before entering the landfill whenever possible and the site should be fenced or positioned to limit access by trespassers. Implementation of these steps is likely to reduce or eliminate many of the problems associated with dumping.

Separation and treatment of organic waste

Separation of organic waste from the MSW stream represents an opportunity to reduce the quantity of waste entering landfills in developing countries by up to 50% (by weight). Correct treatment of this waste will also significantly reduce the pollution and health problems previously described by removing the major source of leachate, combustible gases, odours and food for disease carriers.

Separation of organic waste can be done at the households prior to collection, at the landfill prior to final disposal or a combination of both. Motivation of households to collect organic waste is required to ensure a high level of efficiency. Efficient separation of organic materials by householders will only be sustained if the source separation system is convenient, hygienic and beneficial. A comprehensive and simple educational campaign from the municipality or group of municipalities to the householders, before the householders are required to start collection of organic waste, is advisable. Educational materials should be easy to understand and use pictures to explain what to separate. Education must also continue after the launch of the scheme, in an advisory and supportive manner, through the use of waste advisors and school programmes.

Separation at the landfill will require the organisation of labour/scavengers and must occur before final disposal to avoid separation by scavengers occurring in the landfill itself. Once separated the organic waste can be composted or used to produce biogas. Markets and uses for the compost will depend on local conditions and needs. In Bombay, India, 300 tonnes of Municipal Solid Waste is composted to yield 60–70 tonnes of compost per day. The compost process is a simple 'windrow' method (see Chapter 11): the compost is sold to farmers generating a profit of approximately $10 per tonne (Panjwani, 1998). In other countries, the soil displaced to create the landfill may have more value than the compost. In certain parts of Argentina (e.g. around Buenos Aires) where the soil is of very high quality, the selling of compost may not be economically viable. Instead, the compost may be used to cover the landfill and the displaced soil might be sold, which generates more revenue than selling the compost (Franke, personal communication).

Recycling and scavenging

Scavenging occurs at a number of stages within the MSW management stream. 'High quality' recyclables such as entire glass containers, plastic bottles, metals, etc. are often collected door-to-door by individuals or by the waste collectors themselves (Bernache-Perez, 1999). Recyclable materials are also very effectively sorted at the kerbside by people searching through garbage containers and by people scavenging at the landfill/dump (Nagpal et al., 1999). It is this latter form of scavenging that poses the greatest threat to health, represents the poorest living conditions and as such is most urgently in need of improvement (see Chapter 13, Landfill section). Nagpal et al. (1999) has suggested that plastics recycling in India would be improved by the establishment of deposit centers for post-consumer plastic waste. This would help ensure that higher quantities of plastic waste are deposited by users and salvaged by waste-pickers. Once deposited, plastic waste can be taken to licensed recycling units, which have the added benefits that product quality can be monitored and tax evasion avoided.

Scavengers need to be formally involved in the sorting, collecting and recycling of materials. This has been achieved at low cost in the suburbs of Mexico City by the construction of a simple material recovery facility where some of the collected waste (approximately 1500 tonnes per day) is placed on a series of conveyor belts to be sorted by scavengers. Materials recovered by the sorters are further cleaned and processed for sale. All of the various pieces of the equipment in the facility were specially designed so that they could be built and maintained in Mexico (Diaz, 1994). The construction of a materials recovery facility of this type offers a number of benefits including:

1. improved working conditions for the scavengers who no longer have to sort materials on the landfill itself,
2. an opportunity for scavengers to increase their income by pooling recyclable material to be sold in bulk,
3. an increase in recycling and landfill diversion rates,
4. the opportunity for children to attend school rather than work as scavengers, and
5. modest accommodation provided for the scavengers and their families at an affordable rate, which is paid for out of the money they earn for separating recyclable materials (Diaz, personal communication; Garmendia, personal communication).

In Brazil, with the help of non-profit organisations and industry, the training and organisation of scavengers has allowed them to offer a reliable public service as part of the waste management system (CEMPRE, 1999).

Formal organisation of scavengers would also allow them to earn extra income by assisting with the separation of organics, composting operations and covering of waste with soil/compost. A good example of this is the Madras-based non-governmental organisation EXNORA (see Chapter 3, Integrated Waste Management case studies for more details). EXNORA has assigned streets to scavengers who take care of street sweeping, collecting of MSW, sorting of recyclables and disposing of the restwaste to the nearest municipal transfer sites. The scavengers also collect organics in some streets separately. They bring the organics to a backyard composting site. Here the organics are composted using a simple method of containers with holes that allow enough aeration for the composting process to occur. This simple composting process is possible because the collected organic fraction contains very little meat or bones due to the mainly vegetarian diet eaten in India. Meat and bones do not compost well in a simple backyard composting process where optimum composting conditions do not occur (e.g. temperature, moisture) and therefore no sanitisation of compost can be guaranteed. The compost is used for gardening in the streets and yards in which the organic waste is collected separately. EXNORA calls this project the 'Zero Waste' project because the remaining waste that requires disposal is minimised, as both recyclables and organic material have been recovered (Franke, personal communication).

Incineration

Although the incineration of waste is an essential element of many IWM systems in the developed world it is expensive to implement and is unlikely to be a realistic option for countries with developing economies. However, the previously described strategies that can be implemented in such developing regions will serve to lay the foundations for energy recovery from waste in the future as these economies develop. This is particularly important in the case of separation of organic, highly putrescible waste from the MSW stream. Organic waste significantly increases the moisture content and decreases the calorific value of MSW. Its removal is the first step towards preparing a waste stream for the possibility of incineration.

The benefits of IWM to countries with developing economies

Although limited by technical and financial resources, countries with developing economies still have the potential to significantly improve waste management. Implementation of certain elements of IWM as practised in Europe, North America and other developed regions of the world presents the opportunity to establish waste management systems that are both environmentally, socially and economically desirable. Moving from open dumping to simple sanitary landfills in conjunction with separation and composting of organic waste is likely to result in significant benefits. Pollution of surface and groundwater by leachate, migration of combustible gases (methane), odours and breeding of disease carriers can be minimised. Living conditions of scavengers can be improved and health risks reduced. Their formal involvement in the collecting, sorting and recycling of materials can offer these people the potential to supplement their income and increase recycling rates. A careful analysis of market conditions for recyclable materials and compost could be conducted to prevent imbalances that could affect their final prices.

As in countries with developed economies, what is needed in countries with developing economies is less waste to deal with in the first instance, and then an Integrated Waste Management system to manage the waste that is still produced in an environmentally effective, economically affordable and socially acceptable way. Again, the tool of Life Cycle Inventory (see Chapter 4) can be used to help reach this objective.

Modelling waste management – why model?

Optimising the waste system to reduce environmental burdens or economic costs requires that these burdens and costs can be predicted. Hence the need to model waste management systems. Modelling may at first glance appear a purely academic exercise, but further investigation reveals that it has several very practical uses:

1. The process of building a model focuses attention on missing data. Often the real costs, in either environmental or economic terms, for parts of the waste chain are not widely known. Once identified, missing data can either be sought out, or if not in existence, analyses can be carried out to gather the relevant data.

2. Once completed, the model will define the *status quo* of waste management, both by describing the system, and by calculating the overall economic costs and environmental burdens.

3. Modelling allows 'what if ...?' calculations to be made, which can then be used to define the points of greatest sensitivity in the system. This will show which changes will have the greatest effects in reducing costs or environmental burdens.

4. The model can be used to predict environmental burdens and likely economic costs in the future. Such forecasts will not be 100% accurate, but will give rough estimates valuable for planning future strategy. These are useful especially in such long-term processes as the development of markets for secondary materials. Market development is vital to ensure that higher levels of recycling can be sustained. Modelling the waste system will allow prediction of the likely amounts of reclaimed material available, which will in turn allow investment in the necessary equipment to proceed with confidence.

Previous modelling of waste management

Modelling of waste management is not a new idea. Clark (1978) reviewed the use of modelling techniques then available to optimise collection methods, predict the most efficient collection routes and define the optimal locations for waste management facilities. Such models concentrated on the detailed mechanics of individual processes within the waste system. Other models have attempted to take a broader view and have compared alternative waste disposal strategies from an economic perspective (e.g. Greenberg et al., 1976).

More recently, detailed models have been developed to model the economics of materials recovery for recycling, and some of its environmental burdens (Boustead, 1992), as well as broader models including cost, public acceptance, environmental burdens and ease of operation and maintenance of waste management alternatives (Sushil, 1990). The model constructed in the following chapters attempts to predict both the overall environmental burdens and economic cost, since both are crucial, for an Integrated Waste Management system.

Using Life Cycle Assessment for Integrated Waste Management

Modelling may be divided into two areas – model structure (which will determine how the model will work) and data acquisition (for insertion into the model). Recent developments in each of these two key areas have made this work both possible and timely.

Models

Several models for predicting environmental burdens have been produced within the discipline of Life Cycle Assessment (LCA) (see Chapter 4). This is a relatively new branch of applied science, which has resulted in the development of a new environmental management tool (see Chapter 4, Limitations of a Life Cycle Approach). The modelling technique used here is essentially a Life Cycle Inventory of Municipal Solid Waste.

Data

Even the best models are useless without accurate, relevant and accessible data to enter into them. Whilst data have been available for many of the technical processes in waste management, such as incinerator emissions, it is fair to say that information on the costs and burdens of collection and sorting systems has not been readily available. In many countries, especially those with developing economies, source-separated collection and sorting systems only exist at the planning stage, therefore operational data could not be obtained. Fortunately, through the efforts of a range of bodies, many different pilot schemes, both on small and large scales, have been set up, and data are becoming available. For the first time it is now possible to model Integrated Waste Management schemes based on actual data.

The model developed in this book therefore attempts to combine recent developments in Life Cycle Assessment methodology with the stream of hard data beginning to emerge. This should allow prediction of the overall environmental burdens and economic costs of different executions of Integrated Waste Management.

CHAPTER 3

The Development of Integrated Waste Management Systems: Case Studies and Their Analysis

Summary

This chapter highlights the underlying principles of an integrated approach to waste management. A summary of the findings from European case studies (based on the European Recovery and Recycling Association (ERRA) research project carried out in 1998), one North American case study and a case study from India are presented. Each case study includes a detailed description of the waste management systems in operation. The case studies from North America and India are included to demonstrate that an integrated approach to waste management is not just a European phenomenon or an approach only suitable for developed economies. The principles of IWM each case study has adopted are also identified.

The key points learnt from the case studies were:

1. The quantitative data gathered from each of the systems were not comparable (*differences*).
2. Common driving forces that helped to shape each system could be identified (*similarities*).
3. Prescriptive legislation prevents the development of IWM systems, whereas *enabling legislation* allows the development of IWM systems.
4. An integrated approach to waste management begins at a *local level*; the waste hierarchy has little practical use at a local level.
5. A trend of waste management systems *evolving* to become part of a more comprehensive 'resources management system' was also identified.

Introduction

Each case study presented demonstrates at least one (but often several) of the main principles of an integrated approach to waste management (see Chapter 2). These principles are highlighted in *italics* in each case study where they are considered to be important. Table 3.1 lists the best examples of the principles of IWM demonstrated by each of the case studies.

Case study format

The case studies are presented in two different formats. The European case studies and the case study from North America include a description of waste collection, treatment and disposal, additional relevant information and a schematic diagram. The schematic diagrams represent the material flow of the basic waste management system and show details of the

33

Programme	Treatment technology	Key IWM characteristics (see Chapter 2)	Page no.
Brescia, Italy	Composting, Recycling	Economy of scale, Stability, Long-term perspective	48
Copenhagen, Denmark	Composting, Recycling, Incineration	Economy of scale, Stability, Control of all waste arisings, Continuous technology improvement, Enabling legislation	71
Hampshire, UK	Composting, Recycling	Economy of scale, Flexibility, Public support	51
Helsinki, Finland	Composting, Recycling	Economy of scale, Control of all waste arisings	54
Lahn-Dill-Kreis, Germany	Composting, Recycling	Flexibility	58
Madras, India	Composting, Recycling (informal)	Vision, Public support	79
Malmö, Sweden	Composting, Recycling, Incineration	Economy of scale, Control of all waste arisings, Continuous technology improvement	64
Pamplona, Spain	Recycling	Economy of scale	43
Prato, Italy	Composting, Recycling	Control of all waste arisings	45
Seattle, USA	Composting, Recycling	Control of all waste arisings, Public support	74
Vienna, Austria	Composting, Recycling, Incineration	Stability, Public support	61
Zürich, Switzerland	Composting, Recycling, Incineration	Economy of scale, Stability, Long-term perspective, Continuous technology improvement, Polluter pays	67

Table 3.1 Case study summary

amount of material collected, treated and disposed of by the overall system. This is a relatively simple method of presenting an overview of what are often complicated systems and allows for quick comparisons between systems that would otherwise require lengthy descriptions in the accompanying text. Instead of this, the text describes the existing local infrastructure and highlights the characteristics of the integrated approach to waste management demonstrated by each of the case studies. As the waste management system described in the Indian case study handles only a (relatively) small fraction of the total MSW arisings, a schematic diagram would provide little useful information, so it is presented as text only.

Case studies

The findings of the European case studies and those from North America and India were analysed and the results are presented below. There is no single simple formula that can be applied to every waste management system that will make that system integrated. The wide range of different circumstances specific to each area's waste management system means that a 'one size fits all' approach is not appropriate. The variation between geography, demographics, politics, legislation, existing waste management infrastructure and public opinion within countries makes a single solution impossible. This variation within countries clearly increases significantly between countries (see Table 3.2).

As described above, the unique set of circumstances that define how a specific waste management system operates means that the development of an Integrated Waste Management system cannot follow a simple set of rules. However, the development of an Integrated Waste Management system can follow the principles described in Chapter 2. A range of these principles (depending on circumstance) were identified as being fundamental to the evolution of each of the case studies that are described in the following text.

Difficulty of comparison

The difference in scale between the programmes was significant. The number of inhabitants served by each programme varied from 190,000 in Brescia to 1,700,000 in Hampshire. This difference of an order of magnitude was also seen in the total amount of waste managed by the different programmes, for example between 90,000 tonnes in Prato, and 870,000 tonnes in Copenhagen (see Table 3.2).

The large differences observed in the ratio 'kg of waste managed/inhabitant/year' (390 kg/year in Pamplona up to 1600 kg/year in Copenhagen) were explained when the definitions of Municipal Solid Waste (MSW) from each of the programmes were examined. Definitions of MSW vary across Europe. Some programmes define MSW as just household and assimilated (light commercial) waste, while other programmes include industrial waste (either on a voluntary or obligatory basis), hazardous waste and construction waste (see Table 3.3).

Major differences in existing infrastructure were expected and observed (see Table 3.4). Significantly, all of the programmes operated recycling schemes and all of the programmes with the exception of Pamplona also operated composting facilities. Pamplona is currently considering the addition of either a composting facility or an anaerobic digestion facility to its waste management infrastructure. Energy from Waste (EfW) incinerators linked to district heating

Programme	Number of inhabitants served	Total waste managed (net tonnes)	Kg of waste managed/ inhabitant/year
Brescia, I	190 000	113 000	595
Copenhagen, DK	555 000	**867 000**	**1562**
Hampshire, UK	**1 700 000**	753 000	443
Helsinki, FI	905 800	790 000	872
Lahn-Dill-Kreis, D	260 000	268 000	1031
Malmö, SE	500 000	554 000	1108
Pamplona, E	282 000	110 000	**390**
Prato, I	**168 000**	**90 000**	536
Seattle, USA	533 000	725 107	1358
Vienna, A	1 640 000	890 000	543
Zürich, CH	360 000	239 000	664

Table 3.2 Difference in programme scale

schemes were in place in four of the programmes, treating between 15% and 52% of the waste stream. In Copenhagen, their state of the art facility is viewed favourably by the Danish Society for the Conservation of Nature (WMIC, 1995) due to its stringent emissions limits. Landfills, still necessary for final disposal, received between 5% and 90% of all MSW arisings depending on the programmes' other operational infrastructure. All of the programmes operated landfills with gas collection systems, and most had plans to retrofit old landfills with gas collection systems in acknowledgment of the contribution of methane to global warming.

Several different methods of paying for MSW management were also observed (see Table 3.5). These differences made comparisons of actual service costs between programmes impossible. Payment methods included: bin size and frequency of collection; bin size; frequency of collection and household type; property value; apartment size; per inhabitant charge; number of rooms (in an apartment) and waste bag price.

The difficulty in comparing these payment systems was compounded by the very different accounting and financing methodology used by each of the programmes. Variations in allocation of revenues from recycling and energy sales from Energy from Waste (EfW) plants, subsidies or grants, certain treatment facilities operated by other municipal departments or private companies, and even certain waste management operations carried out by different departments within a single local authority further compounded the problems associated with a comparative analysis.

Common drivers

The following similarities between IWM systems were identified during the development of the case studies. These common features (individually or collectively) contribute significantly to the development of an integrated approach to waste management.

1. Good system management. This is as necessary in waste management as in any other business; decision making in both the long and short term must be data based.

Programme	Household	Assimilated*	Industrial	Hazardous household waste	Construction	Agricultural
Brescia, I	✓	✓	Obligatory	✓		
Copenhagen, DK	✓	✓	Obligatory	✓	✓	
Hampshire, UK	✓	✓				
Helsinki, FI	✓	✓	Voluntary	✓		
Lahn-Dill-Kreis, D	✓	✓	Voluntary	✓		
Malmö, SE	✓	✓	Obligatory	✓		
Pamplona, E	✓	✓	Voluntary			
Prato, I	✓	✓				
Seattle, USA	✓	✓				
Vienna, A	✓	✓	Obligatory	✓		
Zürich, CH	✓	✓	Voluntary	✓		

Table 3.3 Difference in definition of MSW (grey areas represent waste streams included in each municipality definition of MSW)

*Assimilated waste – similar in composition to household waste and includes most commercial wastes

Programme	Recycling	Composting	Energy from Waste incinerators (EfW)	Incineration without energy recovery	Landfill	Other
Brescia, I	11%	8%	Operational since 1998, DH*	–	80%	–
Copenhagen, DK	64%	2%	27% DH	–	4%	3%
Hampshire, UK	9%	6%	–	10%**	76%	–
Helsinki, FI	29%	11%	–	–	60%	–
Lahn-Dill-Kreis, D	40%	11%	–	–	49%	–
Malmö, SE	37%	5%	29% DH	–	28%	–
Pamplona, E	11%	–	–	–	89%	–
Prato, I	10%	5%	–	–	85%	–
Seattle, USA	36%	7%	–	–	57%	–
Vienna, A	27%	11%	31% DH	–	31%	–
Zürich, CH	19%	6%	52% DH	–	23%	–

Table 3.4 Difference in programme infrastructure

*DH = District heating; **Facility closed in 1997

Programme	MSW fee base
Brescia, I	Apartment size
Copenhagen, DK	Bin size and frequency of collection
Hampshire, UK	Property value
Helsinki, FI	Bin size, frequency of collection, household type
Lahn-Dill-Kreis, D	Per inhabitant charge
Malmö, SE	Bin size and frequency of collection
Pamplona, E	Property value
Prato, I	Apartment size
Seattle, USA	Bin size (garden waste collection is extra)
Vienna, A	Bin size and frequency of collection
Zürich, CH	Number of rooms and waste bag price

Table 3.5 Difference in MSW payment systems

2. Vision. It is essential that an individual or a team has a well defined and clear long-term strategy.
3. Stability facilitates the development of a long-term strategy, and is required both within the waste management department and within the political framework of the local authority.
4. Critical mass (economy of scale), is essential for the development of major infrastructure and to ensure necessary quantities of recycled material or compost are available for viable systems to be established.
5. Landfill space. The availability and cost of landfill space plays a major role in the development of waste management systems. Low cost (therefore often abundant) landfill can restrict the development of an integrated approach to waste management, while higher cost (due to scarcity, or taxes) landfill can make other waste management options more economically viable.
6. Control of all solid waste. Although this is important for economies of scale it is also essential that an established waste management system does not lose control of waste streams. This can change an economically viable waste management system into an uneconomical system by forcing facilities to operate at below their initial design capacity. This increases the cost per tonne of the whole system.
7. Legislation, the effects of which can be both positive and negative. Enabling legislation improves flexibility and promotes an integrated approach to waste management, whereas prescriptive legislation has a restrictive effect.
8. Availability of funding, through grants, subsidies, partnerships or co-operative agreements, are again essential for development of major new infrastructure and upgrading of existing infrastructure.
9. Public opinion. Public support is essential for collection systems to function and for infrastructure development to take place. Communication through education campaigns, public consultation meetings and stakeholder dialogues increases awareness and understanding of waste management issues.

These features were seen to be common to the development of the Integrated Waste Management systems studied.

Legislation

Enabling legislation allows for flexibility (and therefore integration) by making clear definitions of responsibility between players and by setting goals, e.g. 'minimum standards', whereas pre-scriptive legislation restricts flexibility (and therefore integration) by defining the means by which specific goals must be reached. For example, the EC Packaging and Packaging Waste Directive, which prescribes that a fixed percentage of this packaging material must be recycled, was seen to influence the case study programmes in different ways. In some instances it has changed how systems operate, in others it has not. This has depended essentially on the stage of devel-opment of the waste management system.

A system with a well developed infrastructure such as Copenhagen can avoid making changes to its Integrated Waste Management system as the Packaging and Packaging Waste Directive targets are already being exceeded. In a city or region that has a developing infra-structure there is the possibility that, due to the Packaging and Packaging Waste Directive, the waste management system focuses unduly on the recovery of packaging and therefore less on a balanced, integrated approach.

IWM begins at a local level

All programme managers acknowledged the existence of a waste hierarchy and several different hierarchies were described, but it was universally agreed that the hierarchy was only a 'menu' of possible treatment options and should not be used as a rigid set of rules to be followed blindly. It was clear that the programme managers accepted that the waste hierarchy has severe limitations at the operational level. Firstly it has no scientific basis, secondly it cannot consider combinations of treatment technologies and thirdly it does not address cost issues. The waste hierarchy cannot identify the Best Practical Environmental Option (BPEO) with respect to planning waste management systems. Therefore a policy advocating use of the waste management hierarchy and BPEO is contradictory because a hierarchical approach and a BPEO approach are mutually exclusive.

Waste management also needs to be socially acceptable. This can be achieved through a process of public consultation and information campaigns, a costly and time consuming process but a necessary one if new waste management systems are to be successful.

System evolution

From the range of waste management systems documented within this study the variation between waste management systems can be seen to be vast. An evolutionary trend was observed, which begins with waste management primarily addressing the issue of public health and safety. Through an organised system of waste management optimisation this initial approach is superseded by an integrated approach to waste management, where economic and environmental concerns are added to the system. Eventually an Integrated Waste Manage-ment system can itself become part of a resource management system, where all resources such as water, power, CO_2 balance and solid waste are managed within a single optimised sys-tem. This will eventually enable the development of a Sustainable Waste Management system, as is demonstrated schematically in Figure 3.1.

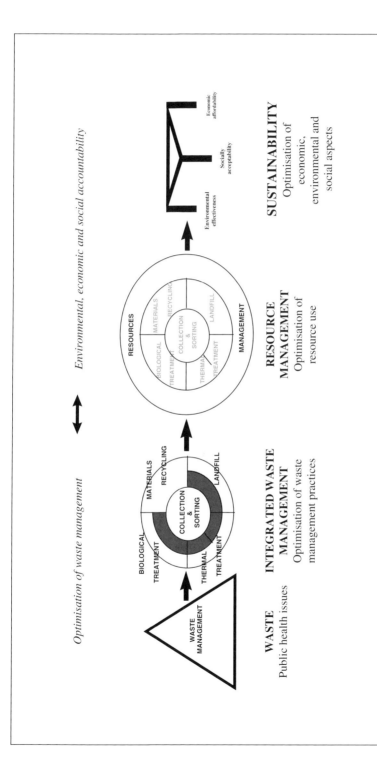

Figure 3.1 The evolution of waste management.

Case study details – schematic diagrams

As described in the preceding text, schematic diagrams are used to represent the material flow of the basic waste management systems. They show details of the amount (tonnages) of material collected, treated and landfilled by the overall system. To present the considerable amount of information necessary to clearly describe each waste management system, abbreviations often have to be used. Full text is used where space permits.

Abbreviations

PA   Paper (where a programme differentiates between paper and cardboard, or between paper and cardboard from packaging, it is indicated in the text)

PL   Plastic (where a programme differentiates between all plastic and plastic packaging, this is indicated in the text)

GL   Glass (where a programme differentiates between all glass and glass bottles, this is also indicated in the text)

ME   Metal

TE   Textiles

LFG Landfill gas

L   Landfill

I   Incineration

R   Recycling

C   Composting

Definitions (see also Chapters 8–14)

Source separation   – also known as home sorting, the most simple form being removal of recyclable materials from the restwaste either for collection or delivery to a recycling point. More extensive schemes require the householder to separate household waste into several different material streams.

Bring system   – material taken from property to collection point by householder.

Kerbside collection   – material collected from property/home.

Separate collection   – collection containers that accept single material types only.

Co-mingled collection – collection containers that accept multiple material types.

Restwaste   – the material remaining after recyclable material (and, if applicable, organic material for biological treatment) has been separated from household waste.

Each schematic diagram also contains a small bar chart in the bottom left-hand corner; this is a summary of the material flow of the waste management system presented in the schematic.

Pamplona, Spain, 1996

Key IWM Characteristics: Economy of scale and social acceptability.

An association of 40 municipalities from the district of Pamplona, the Mancomunidad de la Comarca de Pamplona, was established in 1982 to manage both water and solid waste more effectively than the previously fragmented institutions that were common in Spain at that time.

This *economy of scale* enabled the development of policies and supporting infrastructure, which would be beyond the financial reach of any single municipality (see Figure 3.2) and the separate collection of recyclables began in 1992 *due solely to social and political pressure*. Landfill was (and still is) very inexpensive in Pamplona, as in the rest of Spain, so no real economic incentive for recycling exists.

Summary – Pamplona

Population	282 000
Total tonnes of material managed	110 000
Total operating costs (estimated)	1610 million pesetas
	(9.68 million euros, 10 million US dollars)
Collection	67%
Recycling	20%
Landfill	13%
Collection method	Bring system that requires:
	Source separation of PA, PL, GL, ME,
	TE and Restwaste
	Separate collection of GL, TE and Restwaste
	Co-mingled collection of PA, PL, GL and ME
Total recycled	14%
Total landfilled	86%

Several technologies for the biological treatment of organic waste are currently being assessed by the Mancomunidad (municipality).

Collection

A bring system operates in Pamplona, which requires the residents to source separate recyclable materials from their restwaste and take these two fractions to kerbside collection points. Blue bins are used for co-mingled paper, plastic, metal and glass, while green bins are for restwaste. These bins are emptied six times a week by a contract waste collection company. Containers for glass collection only are also placed at the kerbside collection points.

Treatment

The co-mingled recyclable material is sent to a Materials Recovery Facility (MRF) where it is sorted manually. The contamination rate of 60% is high, but a continued campaign of public information about the system is expected to reduce this figure over time. The glass collected from the containers is sent directly to reprocessors.

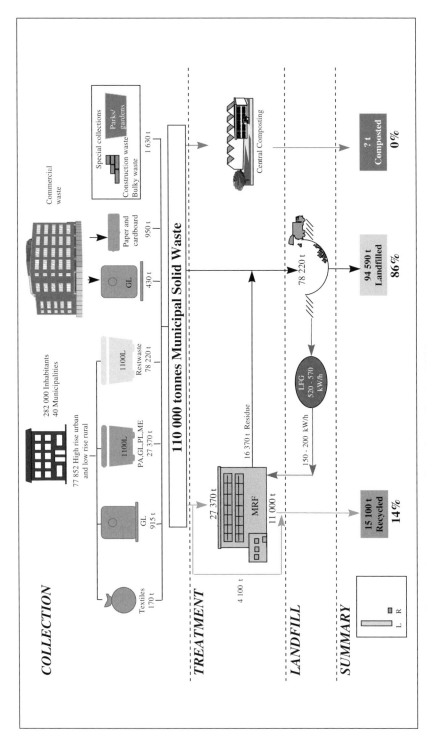

Figure 3.2 Integrated Waste Management in Pamplona, Spain, 1996.

Landfill

The collected solid waste and the residue from the MRF is sent to the Pamplona landfill. This landfill has a solid rock base (of very low permeability) that facilitates leachate collection. All collected leachate it treated on site. Landfill gas is collected and used to produce electricity, which provides power for the MRF. Surplus electricity is sold to the local grid. This revenue from landfill gas will help to keep the cost of landfill down.

Additional information

Solid waste collection and transport costs account for 67% of the total municipal waste management budget. Waste processing accounts for 20%, while landfill accounts for only 13% of the budget due to the very low cost of landfill in the region.

Municipal Solid Waste management in the Pamplona region is still in the process of developing. Original programme changes were politically motivated and part of a larger evolution of the entire municipality. The programme is one of the first of its kind in Spain and *enjoys strong local support*.

Prato, Italy, 1997

Key IWM Characteristics: Control of all waste arisings.

The Azienda Speciale Municipalizzata per l'Igine Urbana (ASMIU) is responsible for the provision of waste management services for the commune of Prato and is *actively seeking to win other waste management contracts from surrounding communes*. ASMIU is also playing a key role in ongoing efforts to *establish a regionally based waste management system* (see Figure 3.3).

Summary – Prato

Population	168,000
Total tonnes of material managed	90,000
Total operating costs	Not available
Collection method	Bring system that requires:
	Source separation of PA, GL, PL, ME, TE and Restwaste
	Two different co-mingled collection systems currently exist for PA, PL and ME
	The original co-mingled collection system is being expanded and the new bins required have been redesigned to reduce contamination levels (see following text)
Total composted	5%
Total recycled	10%
Total landfilled	85%

The municipality of Prato is responsible for the treatment and disposal of a relatively small amount of waste arisings and does not own a landfill. The landfill used by Prato (80 km away near Pisa) has a high gate fee and when this is added to the national and regional landfill taxes, materials recycling and composting become financially viable options.

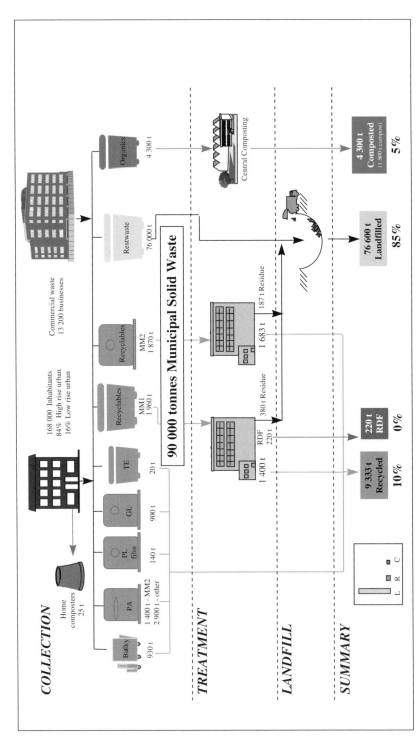

Figure 3.3 Integrated Waste Management in Prato, Italy, 1997.

Collection

A bring system for separate collection was established in Prato in 1992. The initial system comprised collection points with blue (1100 litre) containers for paper, plastic, metals, and glass and black (1200 litre) containers for restwaste. This scheme worked well but high contamination rates and liquid-soaked paper were common. In 1995 the original scheme was expanded but this time the materials requested for collection in the mixed material containers were changed and the openings in the collection containers were designed to be of a similar shape to that of the targeted materials. This resulted in paper and cardboard being collected in yellow (1100 litre) containers with wide but narrow openings and plastic, glass and metal being collected in blue containers with appropriately sized round openings. The total amount of material collected has remained constant while the contamination rate has decreased significantly.

Treatment

ASMIU manages the treatment of all solid waste arisings from Prato, but they do not have control over the management of final disposal at the Pisa landfill. ASMIU operate one waste transfer station where the majority of restwaste is brought to be baled for transport to the landfill.

Two MRFs operate within the Prato area. One is owned by ASMIU and sorts the material collected by the original separate collection scheme. A small fraction of the residue from this plant is sold as Refuse-Derived Fuel (RDF) to a local industrial incinerator (not owned by ASMIU). The second MRF is privately owned and sorts the material from the second (expanded and improved) separate collection scheme.

Organic material is delivered to a privately owned composting facility where high-quality potting compost, which has a stable market, is produced. The composting gate fee is less expensive than landfill disposal. ASMIU are considering buying the composting facility in the future.

Landfill

The residue from the MRFs and all of the restwaste is transported to the Pisa landfill. In addition to the gate fee there is a national tax, which is a fixed rate and a regional non-linear regressive tax rate, which means that the tax decreases as the diversion rate increases. As Prato is able to divert more than 8% of its MSW from the landfill, the municipality receives a 20% discount on the regional tax. This discount will increase as it is able to divert more material from landfill.

Additional information

The double tax and discount based on diversion makes *both composting and materials recycling very attractive options* for the municipality. Economically, it makes good sense to recycle in Prato, at least until the cost of recycling exceeds the landfill tax discount that is currently available.

Brescia, Italy, 1996

Key IWM Characteristics: Stability, public acceptability and economy of scale.

The municipally owned Azienda Servizi Municipalizzati (ASM) has managed waste services in Brescia commune since 1904. This established service has provided a *stable platform* for the development of a *long-term planning perspective*. Paper recycling began in 1974 and since then communication with the public has been ongoing. The current waste management system with its high environmental standards has *wide public support*.

An improvement in *economy of scale* was achieved by the addition of commercial waste to the household waste already managed by ASM in 1994 (see Figure 3.4). This has been further enhanced by the addition of the new Energy from Waste (EfW) incinerator to the existing infrastructure and the subsequent *expansion of ASM's waste management services* to the regions surrounding Brescia. ASM also manage electricity provision, district heating (both of which will benefit from the addition of the EfW incinerator) water supply and sewage treatment. This has resulted in not just an integrated approach to waste management but an integrated approach to the provision of all of these essential services.

Summary – Brescia commune

Population	190,000
Total tonnes of material managed	113,100
Total operating costs (estimated)	35.5 billion Lira
	(18.3 million euros, 19 million US dollars)
Collection	52%
Recycling	14%
Landfill	34%
Collection method	Bring system that requires:
	Separate collection of PA, PL, Organic material and Restwaste
	Co-mingled collection of GL and ME
	Co-mingled collection of TE and Wood
Total composted	9%
Total recycled	11%
Total landfilled	80%

A 266 000 tonne per annum Energy from Waste (EfW) incinerator came on-line in 1998. The gate fee is less than the current cost of disposal to landfill. The municipality has secured long term disposal contracts with surrounding communities to ensure adequate availability of feedstock material. The province of Brescia has 1,040,000 inhabitants living in 206 communities and of the MSW generated in the whole province, 387,000 tonnes of material were sent to landfill in 1995. There is the opportunity for the operation of the EfW incinerator to significantly reduce this figure. The emission standards adopted for this state-of-the-art facility are more stringent than current European or Italian emission standards, therefore public acceptance of the facility has been high.

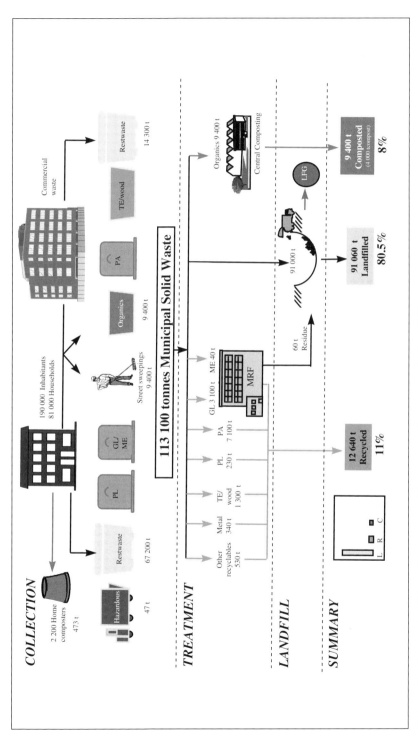

Figure 3.4 Integrated Waste Management in Brescia Commune, Italy, 1996.

Collection

All waste collection and recyclable material collection in Brescia has been developed around kerb-side 'environmental points'. These areas contain both restwaste and recycling containers (either 2400 litre bins or 2–3 m³ containers). Restwaste is collected by automatic side-loading vehicles for the 2400 litre kerbside containers 6 days a week. In the old centre of Brescia, smaller vehicles are used as the streets are too narrow to accommodate a side loader. Household organic material collection (kitchen and garden waste) takes place three times per week from brown 2400 litre bins, in a pilot area that covers 50,000 inhabitants. This programme was expanded in 1998 to cover a total of 110,000 residents. Collected materials are either transported to a transfer station out of town at ASM headquarters or directly to the landfill site, which is 25 km from Brescia.

Treatment

ASM manages the treatment of all waste arising in the city of Brescia. There is one transfer station, at ASM headquarters, where recyclable material/waste is collected for transportation to materials reprocessors/landfill. There are no MRF facilities required to carry out central sorting due to the separate collection scheme operated by ASM in the Brescia commune. However, plastics and glass/metal streams are sent to privately operated sorting facilities. Additionally, a private facility sorts and recovers all materials that were placed in the 'miscellaneous' bins at the recycling centres. Most of the materials recycling occurs by transportation of the collected materials directly to the reprocessors. This makes the contamination rate difficult to assess. Organic materials were delivered to one of two privately owned central composting plants, but in 1997 ASM opened its own composting plant, which accepts all organic wastes collected.

In 1998 a new 266,000 tonne/year thermal treatment plant opened in Brescia. The new facility produces 200 GWh of electricity per year and the equivalent of 350 GWh of heat per year. This is enough energy and heat to supply 25% of all energy used in Brescia. The emission standards that have been adopted for the plant are more stringent than European, Italian and regional limits (see Table 3.6).

Compound	European Union	Italy	Brescia region limit	ASM planned limit
Particulates	30	30	10	5
SO₂	300	300	150	100
NOx	–	650	200	<100
HCl	50	30	30	20
HF	2	2	1	1
CO	100	100	100	50
Pb Cr Cu Mn	5	5	2	0.5
Ni As	1	–	2	0.5
Cd	0.2	0.2	0.1	0.05
Hg	0.2	0.2	0.1	0.05
Dioxin (ng/Nm³)	–	4000*	0.1	0.1

Table 3.6 Limits for energy from waste facility air emissions. All units mg/Nm³ except dioxins

*Dioxins + Furans

Because of these stringent standards adopted, and due to ASM's favourable reputation with the local citizens, public resistance to the project was minimal, although an extensive communication campaign explaining the need for the facility was run from 1991 to 1995. This positive public attitude is a fundamental cornerstone for waste management in the Brescia region.

The project also benefits from favourable electricity prices for the first 8 years of operation. The subsidised price for non-fossil fuel projects is about three to four times the national average price of energy. The cost per tonne of the thermal treatment is estimated to be lower than the cost of disposal to landfill. ASM is in the process of securing longer term disposal contracts from surrounding communes to ensure an adequate flow of material.

Landfill

All final restwaste was deposited in the sanitary landfill in 1996. This landfill site, which was opened in 1989, is located about 25 km from Brescia in Calcinato (which had a 1,400,000 tonne capacity) and closed in March 1998. From April 1998, all landfilled materials are deposited in the new Montichiari landfill. Currently all leachate is collected and treated (there are also several monitoring wells to check for landfill liner leakage) and a system to collect landfill gas is installed and operating. Four new generators have been installed and the gas is being used to generate electricity, with any excess gas being flared.

Additional information

Complete operational flexibility of the Integrated Waste Management system in Brescia is now possible as the EfW incinerator is on-line. Waste material within the system can be directed via different treatments as economic or environmental conditions change.

ASM have expanded the horizons of Brescia's waste management system by expanding to manage MSW arisings from the whole province rather than just Brescia commune. This has enabled (both financially and with respect to the total amount of MSW available for treatment) the development of a truly integrated system, which over a period of time will be able to be further optimised with regard to both economics and the environment.

Hampshire, England, 1996/97

Key IWM Characteristics: Economy of scale, flexibility and public acceptability.

In the late 1980s Hampshire's landfill sites were rapidly filling up and the five incinerators (without energy recovery) in the county were reaching the end of their working lives. The incinerators ceased operation in 1996 as they did not meet new European standards on emissions. A private waste management company, Hampshire Waste Services Ltd (HWS) was selected as the contractor for the provision of disposal services for the County of Hampshire. They put forward a detailed proposal utilising a number of treatment facilities that would provide a *long-term service* for the treatment of waste in the county (see Figure 3.5).

Close co-operation between the 13 local collection authorities and the county disposal authority in Hampshire and HWS has resulted in the development and implementation of Project Integra, the Integrated Waste Management strategy for Hampshire County. Public consultation has been extensive (see Petts, 1995), and the *views of the public* have been reflected in the strategy that evolved over a period of years.

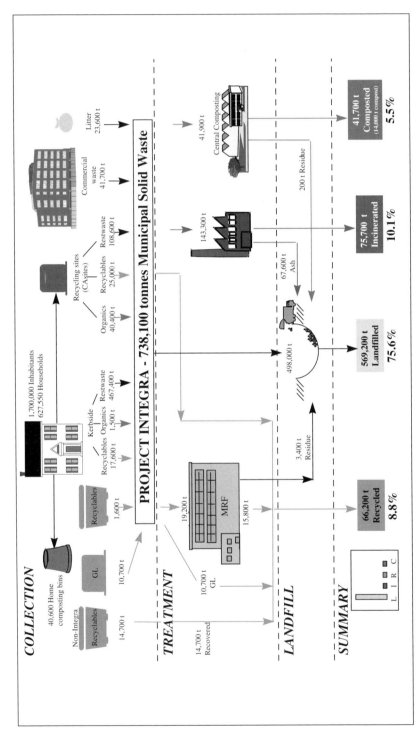

Figure 3.5 Integrated Waste Management in Hampshire, England, 1996/97.

Summary – Hampshire

Population	1,600,000
Total tonnes of material managed	738,100
Total operating costs	£30 million per year (for the next 25 years)
	(45.8 million euros, 47.5 million US dollars)
Collection	42% (estimate)
Treatment and Landfill	58%
Collection method	Kerbside collection (co-mingled collection of
	Recyclables, which are then sorted at MRF)
	Source separation of Organic material, Restwaste
	and Recyclables (PA, PL, GL and ME)
Total composted	6%
Total recycled	9%
Total incinerated	10%
Total landfilled	75%

This case study was carried out during a transitional period in the development of Hampshire's Integrated Waste Management strategy. The figure presented for incineration refers to the throughput of the last of five old incinerators, which were closed during the necessary changes to the waste management infrastructure.

Collection

Waste collection practices are still being optimised by each of the 13 collection authorities; the numbers of households with some form of kerbside collection of recyclable material are increasing on a monthly basis. Each collection authority has implemented the type of collection system that best meets the local circumstances and financial constraints. As a result of this flexibility, there is a range of different kerbside collection schemes in place in the county.

One of the fundamental features of Project Integra is waste delivery points. HWS is required to provide waste delivery points throughout the county. Each of the 10 waste delivery points accepts a range of wastes, according to local demand, such as recyclables, compostables, residual waste, etc. They provide a 'One Stop Shop' for municipal waste deliveries. In practice, the waste delivery points are either landfill sites or transfer stations with the facilities to handle a range of wastes. They are crucial to this Integrated Waste Management approach because they provide a strong interface between waste collection and waste disposal and rationalise transport networks.

Treatment

Waste recovery is currently based on recycling and composting: three interim MRFs and three composting facilities are currently operational. Planning permission for three Energy from Waste (EfW) incinerators has been submitted and an anaerobic digestion facility is under consideration. These facilities will provide the flexibility necessary to optimise the treatment and disposal of Hampshire's waste with respect to economic and environmental concerns.

Landfill

Five landfills currently operate within Hampshire but they only have sufficient capacity to service Project Integra for about 10 years. Within this period the three county EfW incinerators are expected to be fully operational, significantly reducing the amount of material requiring final disposal and thus further extending the lifespan of the landfills. It is also expected that during this period recycling rates and composting rates will increase across Hampshire, further alleviating this problem. Unfortunately, however successful these efforts are at prolonging the lifespan of the five landfills, new landfill space will still be required in Hampshire.

Additional information

The integrated approach that has been adopted by Hampshire has ensured that *as the infrastructure develops the amount of waste requiring final disposal is further reduced* and more time becomes available to select the best sites (based on economic, environmental and social considerations) and for the construction of new facilities.

Helsinki, Finland, 1997

Key IWM Characteristics: Economy of scale and control of all waste arisings.
The Helsinki Metropolitan Area Council (YTV) was formed by the *amalgamation of three municipalities* in 1983/4 to manage waste treatment and disposal for the Helsinki metropolitan area (see Figure 3.6). YTV's main activity is landfilling as they own the Ämmässuo landfill, which serves the whole of this area. They *control* household and commercial waste collection by awarding contracts for specific locations to private waste collection companies. YTV's functions also include regional transport, air-quality monitoring and land use planning. Helsinki metropolitan area took part in 'The Urban CO_2 Reduction Project' and afterwards committed itself to develop a long-term action plan towards sustainability.

Summary – Helsinki

Population	905,800
Total tonnes of material managed	856,000
Total operating costs	195 million FIM
	(32.8 million euros, 34 million US dollars)
Collection method	Bring system that requires:
	Source separation of PA, Cardboard and Organic material, Restwaste, Hazardous material and Scrap metal
	Deposit system for glass
Total composted	16%
Total recycled	26%
Total landfilled	58%

In the Helsinki Metropolitan area source separation is required depending on the amount of waste generated. Therefore people living in large apartment blocks (75% of the population) must separate paper, cardboard and biowaste, while people living in buildings containing less

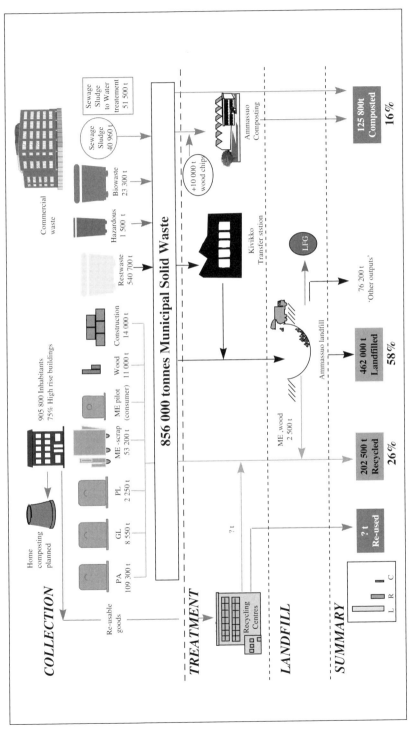

Figure 3.6 Integrated Waste Management in Helsinki, Finland, 1997.

than five apartments must only separate paper from their waste stream. This system allows the collection of the easily accessible majority of source-separated paper, cardboard and biowaste but avoids trying to collect the remaining small amounts of this material, as this has been found to be uneconomic. Finland as a whole has a well functioning deposit system for refillable glass bottles. Approximately 80% of all glass bottles used are refilled.

Collection

The total cost of waste transport and treatment is covered by consumer fees paid by households; the fee depends on the type of household and amount of waste generated. YTV has plans to introduce selective fees to favour those who sort waste in households to encourage recycling. YTV arranges waste transport for 80% of the population. The rest, central urban areas and major institutions, deal directly with private waste hauliers. Waste transport is based on competitive bidding among private companies. Uniform planning and competitive bidding for collection contracts has resulted in a reduction in transport costs by one-third over a 10-year period.

Building residents are obliged to participate in on-site source separation as follows: five apartments or more must collect paper; 10 apartments or more must collect paper and biowaste; while in larger apartment blocks where production of materials is greater than 50 kg per week, paper, cardboard and biowaste must be collected.

YTV is also responsible for the collection, treatment and disposal of hazardous waste from households and many of the small and medium-sized enterprises in its operating area.

Treatment

As recyclables are collected separately no sorting facilities are necessary; however, some additional sorting of the collected paper and glass does occur prior to reprocessing. This is carried out by private contractors.

The collected organic material is transported to a central facility where it is composted using a closed process, which allows closer temperature, moisture and aeration control than the conventional windrow technique (see Chapter 11).

YTV has organised five fixed, single operator scrap metal collection points in the metropolitan area. In spring and autumn scrap metal collection vehicles operate in the YTV area collecting scrap metal from over 400 sites in the metropolitan region. The largest scrap metal items are also collected for reuse from mixed waste loads at the Ämmässuo landfill.

The metropolitan region's waste management regulations include an obligation to construction sites to make separate collections of construction and demolition waste that is suitable for reuse: untreated construction lumber waste, scrap metal and collectable cardboard waste. These regulations only apply if a construction site produces over 50 kg of material weekly. This has resulted in a significant decrease in the amount of construction waste being sent to landfill.

Landfill

As YTV effectively controls the issue of collection contracts and the acceptance of material for final disposal, they have complete control over the quality and amount of material entering the Ämmässuo landfill. The site is currently expected to be operational until 2030. Leachate is collected and treated off site and all of the site's landfill gas will be collected and used for electricity generation by 2002, even though the separation and composting of organic material is expected to increase during this period.

Additional information

As YTV is a non-profit organisation its aim is not to maximise income by accepting large amounts of waste material at their landfill site but to focus on *developing systems to effectively recover and recycle appropriate materials.* In Helsinki it is clear that private recycling (in this case paper) becomes more economically viable as the cost of final disposal increases.

Lahn-Dill-Kreis, Germany, 1996

Key IWM Characteristics: Control of all waste arisings (except packaging) and long-term planning.

The county of Lahn-Dill-Kreis (LDK) manages almost all of the collection, and transport of household and some assimilated industrial waste and the treatment and disposal of most solid waste occurring within the county. LDK *manages the collection, treatment and disposal of all waste arisings,* except for packaging waste, which is managed by the DSD programme (see Figure 3.7).

Summary – Lahn-Dill-Kreis

Population	260,000
Total tonnes of material managed	268,000
Total operating costs	Not available
Collection method	Bring system for glass
	Source separation of PA, Organics, Packaging material and Restwaste.
	Kerbside collection of PA, Organics and Restwaste.
	Kerbside collection (DSD) of Packaging material
Total composted	11%
Total recycled	42%
Total landfilled	47%

A breakdown of the 42% MSW recycled in Lahn-Dill-Kreis is as follows: 8.3% Paper, 2.1% Glass, 1.4% Packaging, 9.9% Construction and demolition waste and 19.8% Soil reclamation.

The German Packaging Ordinance came into effect on 12 June 1991. This law makes manufacturers responsible for their product's packaging 'from the cradle to the grave'. Manufacturers and retailers could be exempted from their obligation to accept returned packaging if they joined a privately organised system for the collection and recycling of packaging waste. To reach the recycling targets specified by the German Packaging Ordinance, which are significantly higher than those specified by European Union Legislation, a national packaging recovery scheme was established by Duales System Deutschland (DSD). Now DSD collects packaging material (in a yellow bag) throughout Germany as a separate stream and recycles this packaging material as a separate stream.

Collection

Household waste is collected door to door every 2 weeks. All residents receive a grey bin for household restwaste, a brown bio-bin; a blue bin for mixed paper (this collection system is

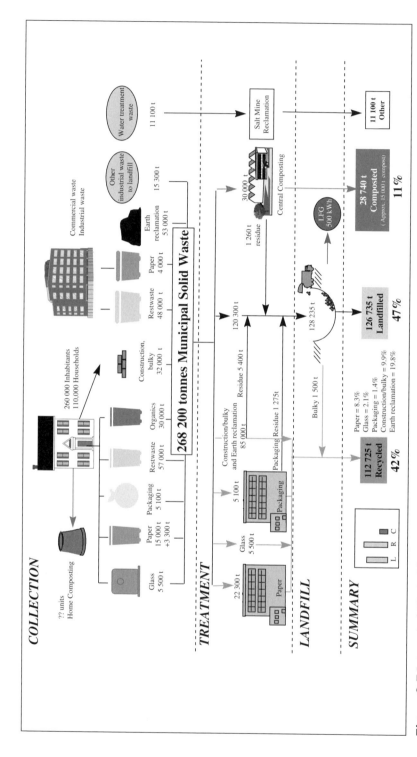

Figure 3.7 Integrated Waste Management in Lahn-Dill-Kreis, Germany, 1996, a non-integrated system (note: the the two Material Recovery Facilities).

financed 25% by Duales System Deutschland – DSD, the so-called Green Dot system) and a yellow bag for packaging materials (funded entirely by DSD). This type of collection is characteristic of the DSD system throughout Germany. Colour-separated glass is collected in containers at kerbside collection points (this collection is also financed by DSD and run under the auspices of the county system).

Treatment

Private, DSD-run central sorting exists for the yellow bag materials as well as the paper collected. Materials are sorted into paper, composite containers, mixed plastics, plastic film and aluminium and ferrous fractions. There is approximately 25% residue from the sorting process. The sorting plant for paper separates the paper into mixed, cardboard, carton, printed and newspaper fractions. Materials recycling is managed under the auspices of DSD.

All collected bio-waste is taken to one of the three composting sites in LDK, located at the LDK landfills in Asslar, Oberseheld and Beilstein. The composting takes place in containers (rot-boxes), and after the initial biodegradation, windrows are created outside for final maturation. The entire process takes between 2 and 4 months. Material sales are managed by a private firm, with some compost bagged and some sold directly for agricultural use. The composting process is rigorously controlled, and the produced compost is of very high quality.

No Energy from Waste incinerator currently exists within LDK County.

Landfill

In 1996, all household waste collected in the grey restwaste bins was sent to landfill. The landfill is controlled, with gas and leachate collection, leachate transfer and treatment. About 500 kWh of electricity is produced from two gas engines utilising the landfill gas. Construction waste is also recycled at the landfill, with a large processing machine arriving every 6 weeks to crush the concrete and brick material down to gravel-size pieces. This material is then sold.

Additional information – how to move towards Integrated Waste Management

This system, which initially may appear to be integrated, is not, because of the parallel waste management system that exists for packaging materials. Therefore, the system is operating at sub-optimal efficiency at both an economic and an environmental level. This is highlighted by the average price for recycled plastic, which was 2500 Deutschmarks (1278 euros, 1325 US dollars) per tonne in Germany (EP&WL, 1999) compared with the price for virgin plastic, which was 1400–1800 Deutschmarks (715–920 euros, 765–985 US dollars) at the time of writing. Collection and recycling of packaging materials is carried out by a separate waste management system rather than being integrated into and treated as part of a single solid waste stream. This has resulted in the duplication of collection rounds and even in the duplication of Materials Recovery Facilities (one for packaging and one for other recyclable material). This parallel packaging collection system has been shown to be very expensive (Staudt, 1997), but even if costs are driven down, it is unlikely that they will ever be as low as would be possible if a fully Integrated Waste Management system was implemented. Although the current waste management system has seen the volume of materials recycled steadily increase over the past 5 years, a steady increase in the amount of material being recycled annually was the trend in Germany before the Packaging Ordinance came into force.

A waste stream analysis of the grey (restwaste) bins by LDK revealed that the DSD programme is only moderately effective at collecting of packaging materials. LDK discovered that the restwaste contained a high proportion of non-recovered DSD materials and had an organic waste content of approximately 40%. With a ban on organic materials entering landfills in the year 2005 (in German legislation) and no desire to build an Energy from Waste incinerator in the region, the county waste managers collaborated with the University of Kassel to develop *a more environmentally and economically feasible solution for future waste management in LDK.*

The system that has been developed seeks to stabilise, through a composting process, the rest waste, and then remove metals, batteries and inert materials (such as sand and gravel, which would be used as a construction material). The resulting product (the Dry Stabilate) has a calorific value similar to high-grade brown coal or low-grade black coal and can be used directly as a fuel in industrial processes. When burned this product yields less ash than conventional coal incineration. The expectation is that this stabilised product will replace coal imports and reduce overall greenhouse emissions, as the original waste material will not be landfilled and will be valorised, replacing coal.

The composting and recovery process is operational and the stabilised fuel is currently being baled and stored while emissions tests are being carried out. Plans are being developed for the construction of an incinerator to burn the Dry Stabilate and recover both heat and power (see Chapter 11, Biological treatment).

Changes in the 1996 collection system are envisioned by the LDK management to complement their new waste management system. In this new system, the yellow sack of the DSD system will be eliminated. Households will still have grey bins, which will now include plastic and metal packaging material and metals, and bio-bins for organic waste. Glass containers will still accept glass, with the addition of another container for large plastic bottles. Blue bins will only accept high-quality paper, and a blue bag will accept packaging paper and composite packaging materials (which are currently placed in the yellow sack).

As indicated above, the existing waste management system in LDK is not fully integrated as the parallel collection and treatment of packaging material occurs. The proposed new waste management system is integrated as packaging material is included in a single solid waste stream. From this solid waste stream paper, glass and plastic bottles are recycled, the separated organic fraction is composted and all remaining restwaste will enter the new Dry Stabilisation process that is linked to EfW incineration. The small percentage of residue remaining (mainly ash) will be landfilled. This integrated system is expected to reduce both the cost and the environmental burden of waste management in Lahn-Dill County *in the long term.*

The anticipated alteration of the waste management approach in LDK has led to a legal dispute between the LDK Local Authorities and the Dual System company. At the time of writing the final outcome of the related court case is still to be decided. A preliminary injunction rules that LDK must not be a shareholder in a waste management company competing with DSD. The legal reasoning behind this decision touches on fundamental aspects of the German Packaging Ordinance. An analysis might suggest that the text of the German Ordinance currently in force could possibly hinder the development of creative and innovative approaches towards Sustainable Waste Management.

Vienna, Austria, 1996

Key IWM Characteristics: Control of all waste arisings (except packaging), public acceptability and continual technology improvements.

Waste collection, street sweeping, recycling, composting and disposal of the municipal waste arising within Vienna municipality *are all operated* by Magistratsabteilung 48 (MA48); this department is directly controlled by the City Council of Vienna. Energy from Waste (EfW) incineration (district heating and electricity generation) is managed by another company (Fernwärme Wien), and the slag is returned to MA48 for disposal (see Figure 3.8).

The first comprehensive waste strategy including separate collection in Vienna was introduced in 1985, but separate collection of recyclable material (mainly glass) has been ongoing since the late 1970s.

Summary – Vienna

Population	1,640,000
Total tonnes of material managed	855,000
Total operating costs	2.4 billion ATS
	(174.4 million euros, 180.1 million US dollars)
Collection	62%
Recycling and Composting	8%
Incineration	24%
Landfill	6%
Collection method	Kerbside collection of PA, Cardboard, Restwaste
	and Packaging materials (ARA collection)
	Bring system for Glass
Total composted	11%
Total recycled	27%
Total incinerated	31%
Total landfilled	31%

On February 5th 1993, Austrian industry founded Altstoff Recycling Austria AG (ARA) for the implementation of the country's Packaging Ordinance. The company's brief was to organise a nation-wide system for the collection and recovery or recycling of packaging waste. Unlike DSD, ARA is responsible for sales packaging as well as for secondary and transport packaging.

Since the first recycling programme for paper was established in 1974, *communication with the public has been regular and extensive*. The current waste management system enjoys wide public support. After a brief and unsuccessful trial with a 'dirty MRF' in the early 1970s, Magistratsabteilung 48 now believes that the separate collection of recyclables is key to ensuring a good quality product.

Collection

Household waste and paper and cardboard collection in Vienna takes place door to door. Collection of clear glass, coloured glass, metals, plastics and composite cartons takes place from kerbside collection points. Biowaste bins are located throughout the city at the kerbside

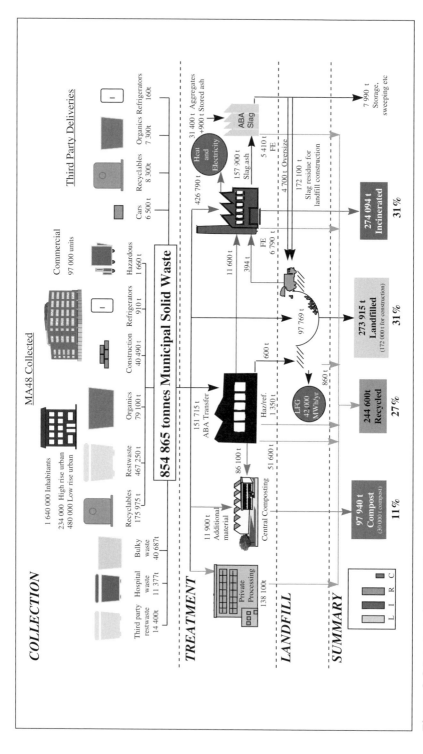

Figure 3.8 Integrated Waste Management in Vienna, Austria, 1996.

collection points. In the town outskirts, biowaste bins are also located in gardens of one or two family houses with sufficient space. Restwaste is collected, on average, once or twice per week from individual households and delivered to one of the two EfW incinerators.

Treatment

In Vienna the materials that are separately collected are transported directly to reprocessors, with the exception of plastics which, are sorted by resin type.

There is one transfer station, where recyclable materials and waste are collected for transport to materials reprocessors or landfill. The station acts not only as a transfer station but also as a sorting and processing plant. Organic material is 'prepared', sorted and homogenised and mixed with additional material to improve its structure. The output of the transfer station is transported to the composting plant, to recycling firms, to incinerators and to the landfill.

The 'pre-compost' is transported to the Lobau composting plant. This plant is equipped with leachate collection and the resulting leachate is recirculated through windrows (see Chapter 11). The entire process takes about 3–4 months.

The composting process undergoes rigorous quality controls every 3–4 weeks and processes are monitored for pH change, water loss, nitrate, nitrite and total nitrogen. Additionally heavy metal content is regularly sampled. All final compost is tested by an outside laboratory and must meet Austrian Standards S2200 and S2203. Each batch is recorded with its relevant data. The resulting compost is then transported to the Schafflerhof composting plant where it is screened and marketed. It is then spread on the farmland of Vienna, distributed free of charge at the recycling centres and given to gardeners.

The majority of restwaste is taken to one of the two EfW incinerators (which are managed by a separate department) – Flötzersteig (in the west, 200,000 tonnes/year) and Spittelau (in the centre, 250,000 tonnes/year) – operating in the Vienna Region. Along with household waste, sewage sludge is also treated. The Spittelau facility provides both electricity and district heating, while Flötzersteig only supplies district heating. Although both incinerators were built in the 1970s, both have been upgraded and are equipped with air filters and de-NOx devices (see Chapter 12). The slag from the incineration process is taken back by MA48 and after metal is removed, slag and ash are mixed with concrete and used in the construction of the landfill site infrastructure.

Landfill

The Rautenweg landfill is the only landfill in Vienna, located to the north-east of the city in an old gravel pit. All restwaste is sent directly to this landfill. Landfill gas is collected, and since 1994, electricity has been generated from 12 gas engines, supplying 42,000 MWh in 1996. All electricity generated is supplied to the Vienna grid.

Additional information

For all packaging materials collected, the municipality receives a refund from ARA. In 1996, 39% of all paper collected qualified for a refund, along with 100% of glass, 65% of all metals, and 100% of plastics. In 1996, this refund to MA48 amounted to 11% of the organisation's total revenue. Again, as in the Lahn-Dill-Kreis case study, the inefficiencies created due to the existence of a parallel collection, transport and recovery scheme for packaging ensure that the waste management system as a whole is operating at sub-optimum levels both environmentally and economically.

In Vienna the separate collection of recyclables is an integral part of the management scenario, although questions on system cost are being raised (Brunner, 1998). Vienna's waste management strategy currently requires more capacity for either EfW incineration or landfill as waste generation is increasing. This extra capacity is unlikely to become available through the building of a third incinerator due to the current political climate. A possible solution may be to attempt to increase composting and recycling levels to compensate for the extra waste being generated. As Vienna is so incinerator dependent, innovative developments with respect to the incineration of selected waste fractions offer possible solutions.

Malmö Region, Sweden, 1996

IWM Key Characteristics: Economy of scale, control of all waste arisings and continuous technology improvements.

In the Malmö Region, household and industrial waste collection has remained a responsibility of each of the nine municipalities (either through private companies or municipally owned ones). *Waste treatment and disposal are managed by SYSAV*, a public/private company that was set up by the municipalities. Waste treatment and disposal is managed using an integrated approach that includes composting, recycling, EfW incineration and landfill. SYSAV has served to centralise and co-ordinate waste treatment and disposal for the whole region (see Figure 3.9).

Since the establishment of SYSAV and the construction of the first incineration plant in Malmö in 1974, *continuous facility improvements* coupled with more stringent pollution control limits have allowed incineration to play a significant role in the management of solid waste in the region.

Summary – Malmö Region

Population	500,000
Total tonnes of material managed	554,000
Total operating costs	Not available
Collection	70–80% (estimated)
Treatment and Landfill	20–30% (estimated)
Collection method	Kerbside collection of Restwaste and Organic material
	Bring system for GL, PA, PL, ME and Hazardous
	Deposit system for Glass bottles.
Total composted	5%
Total recycled	38%
Total incinerated	29%
Total landfilled	28%

Solid waste is managed within the guidelines of the 1990 Regional Waste Management Plan and the following 'Eco-cycling plan'. The principle of eco-cycling is stated as 'that which is removed from nature shall, in an endurable way, be used, recovered, recycled or disposed of in such a way as to occasion the least possible exploitation of resources and without causing damage to the environment' (SYSAV, 1996). Within this concept is the acceptance of market forces. A market for recovered materials is understood to be essential for the collection of any material. If no viable market exists, no recycling occurs.

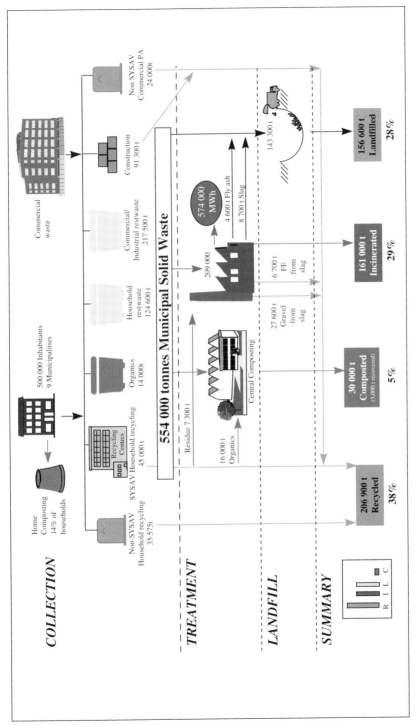

Figure 3.9 Integrated Waste Management in Malmö Region, Sweden, 1996.

Currently Producer Responsibility legislation, which requires the separate collection of packaging materials, a future ban on landfilling combustible material (2002) and landfilling organic material (2005) is threatening to undermine the efficiently operating Integrated Waste Management system of the Malmö Region.

Collection

Restwaste is collected by each municipality separately. Due to the demographic variation and as each of the nine municipalities is responsible for their own solid waste collection, different types of collection systems exist in each municipality. Six of the municipalities use private companies for waste collection, while three rely on municipally run departments.

Treatment

All treatment of solid waste arising in the region, both municipal and industrial is managed by SYSAV. Recycling is managed either through municipal activities, Producer Responsibility initiatives or one of SYSAV's nine recycling centres. No central sorting of recovered materials exists as it is believed that working conditions in Materials Recovery Facilities are unacceptable. Two transfer stations, one in Trelleborg and one in Lund, serve to help optimise transport of household waste to the Malmö EfW incinerator.

Two source-separated composting sites are currently being operated by SYSAV, one at the Trelleborg landfill and the other at the Malmö landfill. Organic materials are deposited in unused landfill cells and 'mattress' composted. This Danish system deposits organic materials in layers and turns them regularly. The total process takes about 3 years from start to finish. The advantages of this type of system are multiple: no new site need be constructed, leachate collection is already established, the location is central, and costs are low. As only garden waste is accepted for composting, the compost produced in the Malmö Region has qualified for KRAV eco-production status, and is of very good quality. Compost is sold at the recycling centres, for soil improvement or for use on golf courses.

The EfW incinerator at the Malmö site has the capacity to accept about 220,000 tonnes per annum and in 1996 treated 209,000 tonnes, of which about 59% was household restwaste. Fly ash is deposited in a special cell at the landfill. Slag is further sorted, and metals are removed. The remaining slag is graded and sold as aggregate. The aggregate can be sold; since over the last decade extensive tests have been performed on the environmental performance of the aggregate and it has been found to have very low leaching properties.

In addition to the aggregate, the EfW incinerator also provides heat to the district heating scheme, supplying some 20–25% of Burlöv and Malmö's needs (543,000 MWh sold in 1996). Both dry scrubbing and fabric filters are used in the flue gas cleaning.

Landfill

The area has two landfills, one near Malmö and the other near Trelleborg. As most of the household restwaste passes through the incinerator first, only 6% of household restwaste was directly landfilled. Of the household restwaste that is landfilled, much of it is deposited in biocells, where special landfill gas collection and energy generation takes place. Additionally, all leachate collected is filtered on site, and several drainage ponds using reed bed filters are being tested. The Malmö landfill is built out into the sea. If the landfill leaked, sea water would flow in, rather than leachate flowing out.

Additional information

Future developments include the possible construction of a new EfW incinerator with heat recovery and electricity generation. If planning permission is approved then this facility would accept wastes from neighbouring regions. This may counter some of the effects of the parallel system for packaging materials and the inevitable erosion of available waste arisings. An anaerobic digestion plant is also planned, which will accept organic material from food processors and agricultural wastes.

The existing Integrated Waste Management system in Malmö is the result of *a continual development process*. The combination of *effective management* and *control of all waste arisings* have allowed the creation of an effective Integrated Waste Management system. This integrated system is now threatened by the establishment of a parallel system for the collection and treatment of packaging waste, which will result in the duplication in some of the operations that are already being carried out effectively. The environmental and economic benefits of such parallel systems have yet to be proven, and in Malmö's case it is difficult to see where any such benefits will occur.

Zürich, Switzerland, 1997

IWM Key Characteristics: Economy of scale, stability, control of all waste arisings and continuous technology improvements.

Abfuhrwesen Zürich (AWZ) is the municipal department of the City of Zürich, which operates the regional Integrated Waste Management system. *AWZ manages the collection, recycling, composting, incineration and disposal of the municipal waste arising within the city of Zürich* and the surrounding 54 communes (see Figure 3.10).

Summary – Zurich

Population	360,000
Total tonnes of material managed	239,000
Total operational costs	140 million CHF
	(87.6 million euros, 90.8 million US dollars)
Collection method	Kerbside collection of Restwaste (Zuri-Sack) and Paper
	Bring system for Glass and Organic material
Total composted	6%
Total recycled	19%
Total incinerated	56%
Total landfilled	19%

The first incineration plant was built in Zürich in 1904, and since then, continual facility improvements coupled with increased pollution controls have meant that incineration (at a relatively high fixed cost) has played a central role in the MSW management in Zürich.

In 1993 AWZ instituted a polluter pays scheme, known as the Züri-Sack. The extra income raised from these fees was supposed to help offset higher waste management costs, affecting the people who produce the waste directly. The introduction of the Züri-Sack has seen an increase in

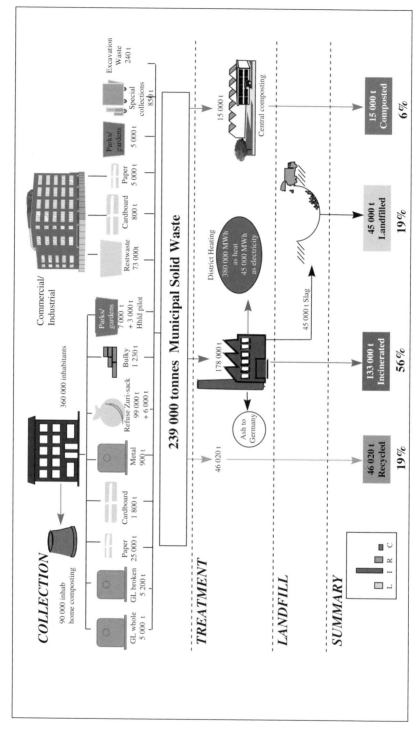

Figure 3.10 Integrated Waste Management in Zürich, Switzerland, 1997.

recycling activities and composting, and a reduction of approximately 30% in the amount of waste available in the restwaste stream. Zuri-Sack charges are fixed at the original 1993 price.

Collection

All household restwaste must be in a Züri-sack and is placed either in front of the building, or in special waste collection containers. This is collected, on average, twice per week from households and delivered to one of the two Energy from Waste (EfW) incinerators. Due to the decrease in the amount of waste collected over the past 5 years, waste will be collected on a weekly basis. This change will take place gradually over the years 1997–1998. A network of community composting centres exists and collection of organic material can also be arranged with AWZ.

Glass is collected in clear, green and brown fractions from 164 streetside collection points, which are serviced once every week or once every 2 weeks. Whole wine and champagne bottles are collected in separate containers. Bundled paper is collected from households and offices once per month. Cardboard collection takes place every 2 months from households and offices. Composite packaging and laminated paper are not included. Tin cans, metal foils, wires, cables, beverage cans, pipes and tops of jars are among the metal items targeted for the metal container collection. Additionally, special pick-up of bulky metal items can be arranged.

Treatment

As the two EfW incinerators are relatively near the city centre, there are no transfer stations. No central sorting of recovered materials exists. The majority of materials recycling occurs by direct transport of the collected materials to reprocessors. This makes the quality of the collected material difficult to assess, but a very impressive purity rate of 98% for glass has been measured.

At a central composting facility all organic materials are shredded and composted in open windrows (see Chapter 11). They are turned regularly and heavy metal concentrations are measured six times a year. The compost produced is of a high quality, but no market development has occurred and compost is given away free to farmers (AWZ pays the farmers to cover haulage costs).

Two EfW plants, Josefstrasse (two incineration lines – one rebuilt in 1975, the other built in 1996) and Hagenholz (two incineration lines – built in 1968, refurbished in 1989, 1993) operate in Zürich city. Household restwaste, assimilated waste and some hospital waste are incinerated. Both facilities provide electricity and district heating. *Both have been continually upgraded and rebuilt, with extensive emission control technology being added in the early 1990s.* As the composition of waste entering the incinerator has changed, the calorific value has also steadily risen. Current estimates set the figure at about 12 MJ/kg. As the facilities were not built for such high energy inputs, certain difficulties have been experienced and additional facility maintenance has been required.

Pre-1997, the incinerator slag was used for road construction, however, a recent ordinance now forbids this. Currently, all slag and filter dust are taken to special landfills outside of the Zürich area. It is estimated that the slag contains approximately 9% metal, and although no metal separation currently takes place, it is possible that it will be added in the future.

Landfill

Only slag from the incineration plant is landfilled. The ash is exported to Germany for disposal in hazardous waste landfills. The slag is deposited in one of three controlled landfills (one in St Gallen, and two in Bern). All are special reactive landfills with leachate collection and treatment. As none accepts any biodegradable material, no methane is generated from these facilities.

Additional information

AWZ in Zürich can be regarded as a victim of its own success. AWZ and city politicians have *successfully educated the public* in how to manage their solid waste effectively. However, decreasing tonnages entering the incinerators, exacerbated by an economic recession, have meant that only 50% of the available capacity is used. Logically, the per-tonne-costs of the incinerators (already operating with high fixed costs) have further increased.

Other factors also contribute to this situation of escalating costs. From 1997, industries and businesses within Zürich city were no longer obliged to use AWZ's collection or treatment services. This liberalisation of the market has meant that waste arisings in Zürich need not be treated or disposed of in Zürich (several of the regions surrounding Zürich have incineration facilities with significantly lower gate fees). Additionally, as of 1999, the 54 surrounding communities in the Zürich canton will no longer be forced to use the AWZ facilities.

Despite the fact that AWZ has operated an effective Integrated Waste Management system and has developed initiatives that have resulted in a significant reduction in the amount of rest-waste generated by the general public (corresponding to an increase in both recycling and composting) the entire system is under severe pressure.

The loss of control of all waste arisings will result in a further loss of material and the efficiency of the IWM system, which is already operating sub optimally, will drop further still. AWZ incinerator costs will continue to rise to compensate for the lack of available feedstock. In an attempt to break this vicious circle (that has been imposed upon them by a combination of their own successful waste minimisation scheme and political forces beyond their control), AWZ are proposing a series of actions. Investigation into the possible incineration of 10,000 tonnes of sewage sludge are underway. The price of the Zuri-Sacks was increased in August 1998 (interestingly this may result in a further reduction in the amount of restwaste that is available), and the frequency of collection may be decreased to once per week to reduce collection costs. AWZ will also actively try to win back industrial clients by adopting a more flexible approach to contracts and offer a more transparent pricing structure.

AWZ accepts that it is essential for them to become more competitive compared to private waste contractors (see O'Brien, 1998) and surrounding regions' waste management operations *to secure adequate waste arisings to operate their Integrated Waste Management system under optimum conditions*. This rapid management response to a very serious crisis should result in this IWM system *evolving once more to add economic sustainability* to a system that has already focused on environmental effectiveness and social acceptability.

Copenhagen, Denmark, 1996

IWM Key Characteristics: Economy of scale, stability, control of all waste arisings, continuous technology improvements, market-orientated and enabling legislation.

The source separation of recyclable material, incineration of 'clean' residue, utilisation of incinerator ash and recycling of construction waste results in the recovery of approximately 97% of all waste arisings in Copenhagen (see Figure 3.11). Copenhagen's Municipal Solid Waste stream includes construction and demolition waste, which is 48% of all waste arisings (90% of the construction and demolition waste is recyclable). If this material is subtracted from the waste management system, to allow a comparison with the other European case studies a slightly different, but by no means less impressive, picture of Copenhagen's integrated system can be seen. The total amount of material recycled falls to 40%, which is still very high compared to the figures from other systems, composting increases to 4%, incineration increases dramatically to 53% and the total amount of material landfilled is 3%. The figures given below include construction and demolition waste in the figure for total MSW generated.

Summary – Copenhagen

Population	555,000
Total tonnes of material managed	867,000 (including construction and demolition waste)
Total operating costs	Not available
Collection method	Kerbside collection of Organic material and Restwaste
	Bring system for GL, PA, Hazardous waste
Total composted	2%
Total recycled	64%
Total incinerated	27%
Total landfilled	4%
Special treatment	3%

Copenhagen operates a truly Integrated Waste Management system. All waste streams are source separated (no materials recycling facilities are required). Recyclable material is delivered directly to reprocessors. Organic material is composted. Non-recyclable material is incinerated with energy recovery (supplying the district heating scheme) and the resulting ash contains little or no hazardous compounds and is used as a raw material in road building. All construction and demolition waste is recycled as several different grades of aggregates. Less than 4% of the total waste arisings in the Copenhagen area are disposed of to landfill.

Collection

In the municipality of Copenhagen and Frederiksberg householders *are obliged by law to source separate their waste*. Organic waste and residual restwaste are collected by the public/private company Kjøbenhavns Grundejeres Renholdelsesselskab (R'98), while all other recyclable material, such as glass, paper and magazines, are taken to neighbourhood kerbside collection containers. These kerbside collection areas also have hazardous waste collection containers. All recyclable (wood, metal, plastic, etc.) and non-recyclable materials are also accepted at large district recycling stations. Commercial and industrial waste (which must also be source separated) is collected by private waste transport companies.

Concepts and Case Studies

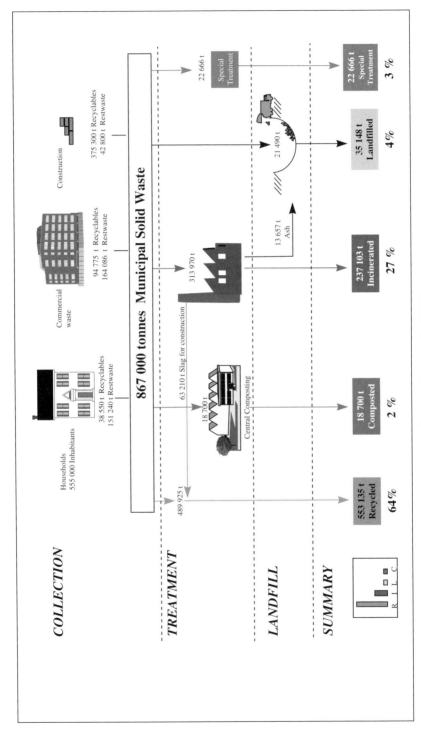

Figure 3.11 Integrated Waste Management in Copenhagen, Denmark, 1996.

Treatment

All source-separated material is delivered directly to reprocessors. The collected organic material is composted; no organic material enters the landfill. All household restwaste, commercial restwaste and industrial restwaste is burnt in EfW incinerators. This restwaste material contains extremely low levels of heavy metals and other hazardous compounds, therefore the ash from the incinerators is considered inert enough to be used as a building material. Both of Copenhagen's EfW incinerators are connected to the district heating scheme.

Construction waste is recycled at a single large facility close to the centre of Copenhagen. The majority of the construction waste is recycled as aggregate for road building and is sold at 15% of the price of virgin aggregate.

Landfill

A sanitary landfill, opened in 1989 and at the time of writing having a remaining capacity of 2,000,000 m^3 (an estimated lifespan of 8 years) is available for final disposal of inert non-recyclable material.

The overall result of this well-observed and well-enforced source separation of recyclable material, incineration of 'clean' residue, utilisation of incinerator ash and recycling of construction waste is that approximately 96% of all waste arisings in Copenhagen are recycled or recovered (i.e. incinerated with energy recovery). Only 35,147 tonnes of inert non-recyclable material were sent to landfill in Copenhagen in 1996.

Additional information

This is yet another example of an efficiently operating Integrated Waste Management system. Again this system has *developed over a long period of time*. R'98 was initially granted the monopoly on waste collection in 1964. Several waste management problems such as a lack of landfill space and groundwater pollution (both 1983–85) have driven the development of the integrated system forwards, as solutions were required within short time-scales.

The *continual improvement* of the EfW incinerators' gas cleaning technology has ensured that these incinerators are viewed as a source of clean renewable energy by the people of Copenhagen (WMIC, 1995), and the district heating scheme results in a major reduction in Copenhagen's CO_2 balance, and may further help in the public acceptance of the EfW incinerators. Good source separation of household waste enables the ash remaining from the incineration of this waste to be used as a raw material in road building.

Public support for the waste management system in Copenhagen has allowed the development of an integrated system that can recover very high percentages of all waste arisings (96%). To achieve this figure it is essential that the recycling of construction waste take place as this significantly reduces the amount of material requiring final disposal to landfill.

The public/private waste management company, R'98, has used its *control* on material entering the landfill to encourage recycling, and the well-informed people of Copenhagen separate recyclable materials with high efficiency.

Enabling legislation exists in Copenhagen that provides a framework within which an integrated approach to waste management can operate. For example this legislation allows for the prosecution of individuals and companies who do not separate their waste as well as collection companies that accept unsorted waste material, thus ensuring the smooth operation of the collection and recycling operations.

Concepts and Case Studies

Seattle, USA, 1998

Key IWM Characteristics: Control of all waste arisings and public support.

In 1987 waste management in Seattle faced a crisis: the city's last two landfills had closed and waste was being sent 32 km south-east of Seattle to the Cedar Hills Regional Landfill in King County. This resulted in an increase in the cost of waste management to Seattle's public of 82%. The waste management contract with King County required Seattle to find an alternative disposal site by 1993 or be locked into the existing contract for the next 40 years. The City initially considered waste incineration as a solution to this problem but the public were totally opposed to this approach. The City therefore decided that the only acceptable approach was to maximise the amount of material that could be diverted from final disposal to landfill by recycling and composting. Waste minimisation was encouraged and recycling was promoted by the provision of kerbside collection of recyclable material and garden waste. A variable rate pricing system was developed based on the volume of garbage (restwaste) bins required by each household.

Up until 1997, the Seattle Engineering Department Solid Waste Utility (SWU) was responsible for all solid waste planning and management. In January 1997, SWU became Seattle Public Utilities (SPU), which now provides stormwater, water, drainage and wastewater services as well as solid waste services (see Figure 3.12).

Summary – Seattle

Population	533,660
Total tonnes of material managed	725,107
Total operating costs	82.1 million US dollars in 1998
Collection method	Kerbside collection system that requires:
	source separation of Newspaper, Mixed paper, Glass, Ferrous and Non-ferrous metals, PET and HDPE plastic bottles
	Separate kerbside collection for Garden waste.
Total composted	7%
Total recycled	36%
Total landfilled	57%

In 1989, the document *On the Road to Recovery: Seattle's Integrated Solid Waste Management Plan* was published and this outlined how Seattle would achieve recycling or composting of 60% of all wastes by 1998. In 1998, the document *On the Path to Sustainability* was published: this is Seattle's new solid waste management plan and it includes environmental, economic and social considerations within its stated goals, policies and programmes.

Collection

State law (RCW 35.21, 1962) gives Seattle express authority to control all waste collection and recycling through exclusive contracts. Residential waste collection (restwaste, garden waste and recyclables) and Commercial waste collection (restwaste and recyclables) are currently contracted out to private companies.

Two different kerbside collection systems for recyclables operate in Seattle. The North system requires source separation into 'three stackable containers' and collection is weekly. The South system is a co-mingled collection and occurs on a monthly basis: all recyclables

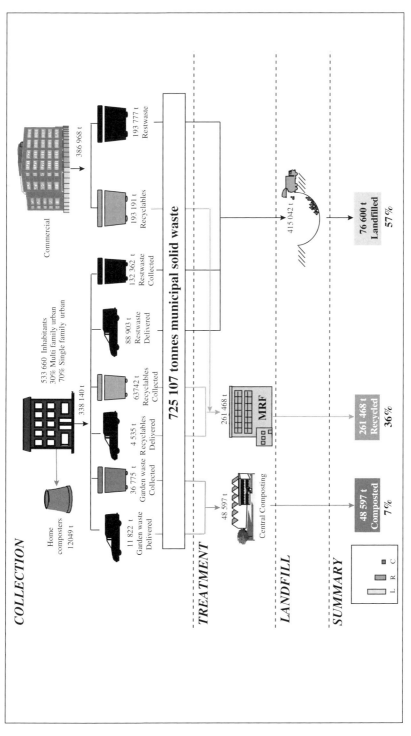

Figure 3.12 Integrated Waste Management in Seattle, USA, 1998.

except glass are put into a 120 litre green bin, and glass is placed into a basket that 'nests' inside the co-mingled bin to prevent broken glass from contaminating the other recyclables. The frequency of garden waste collection depends on the time of year; more collections are made during Autumn (for leaves) and Spring (for grass cuttings). In 1988 the City adopted an ordinance requiring the separation of garden waste from restwaste.

SPU owns and operates two of the four transfer stations (known as Recycling and Disposal stations) in Seattle. These facilities accept delivered restwaste (a fee applies) recyclables (no fee) and garden waste (small fee applies). Some 333,400 vehicles used the two facilities in 1997. Two privately owned transfer stations accept commercial restwaste and sorted construction and demolition waste.

All restwaste picked up by the collection vehicles is delivered to the transfer stations and this material plus all of the delivered restwaste is compacted (approximately 25 tonnes) into 12 metre sealed shipping containers. The containers are delivered by truck to a railway station in the City, where they are loaded onto railcars for onward transportation to Oregon.

Treatment

Collected garden waste is delivered to a privately owned composting facility (Cedar Grove) where nutrient-rich compost is produced. The majority of this compost is sold in bulk as a fertiliser/soil improver and is purchased by landscapers and homeowners. Cedar Grove also bag (two cubic foot bags) some of the compost and this is sold at grocery stores, hardware stores and other retail outlets.

Two private facilities sort and process the recyclable materials collected in Seattle. At the time of writing, both of the kerbside recycling contractors sort and sell the materials they col-lect. New recycling contracts started in April 2000, when all of the collected recyclable materi-als went to a single private facility in south Seattle, which will be responsible for both sorting and selling the material. Currently the majority of the recycled materials are processed within North America. Only 25% of mixed paper and newspaper is exported to Asia, as well as 40% of HDPE and 50% of PET. All of the cardboard, glass and metal collected in Seattle is recycled within North America.

Landfill

SPU has a contract with Washington Waste Systems (WWS) for rail haulage and disposal of the restwaste from Seattle. The contract runs until 2028, with the possibility of termination in either 2001 or 2014 if SPU chooses. WWS is responsible for transporting the waste containers by rail from Seattle to 10 miles south of Arlington, Oregon, to a siding at the Columbia Ridge Landfill, where they are unloaded and hauled by truck a short distance to the working face of the landfill, a total journey of well over 260 km. Each waste train is made up of approximately 100 waste containers and leaves Seattle three times a week.

The Columbia Ridge Landfill and Recycling Centre is located on a 800 hectare site in Gilliam County, Oregon. The actual disposal cells (of which there are 20) of the landfill occupy 260 hectares. The area offers a good degree of natural environmental protection. It is geologically stable, receives less than 23 centimetres of rain per year and the groundwater is 150 metres below the bottom of the landfill protected by naturally occurring, low-permeability soils. The landfill includes a composite liner system and collection systems to control both leachate and gas. In the event of a leak, an extensive monitoring system of gas monitoring probes, and vadose zone (the unsaturated zone between the water table and the surface of the ground) monitoring will allow early detection and repair. As one cell is being filled, the next is being excavated and when each cell is full, it is capped with 1.2 metres of clay and topsoil and planted for grazing. The site has been operational since 1990 and is permitted and regulated by the Oregon Department of Environmental Quality. In order to use the landfill, SPU has to satisfy strict recycling requirements established by Oregon law.

Additional information

The schematic diagram presented in Figure 3.12 shows the material flow of the basic waste management system in Seattle and shows some details of the amount of material collected, treated and disposed of by the overall system. Contamination rates of the material collected for recycling and composting are not available, therefore the amount of residue from these processes cannot be calculated. As sorting, processing and composting are carried out by private companies, the waste they produce is included in the commercial waste streams, therefore the final amounts of material being sent to landfill are correct.

Seattle's waste management operations are paid for by a self-supporting solid waste fund that receives no subsidies. The revenue from the variable rate system pays for all collection, landfill, as well as closure costs of Seattle's old landfills together with litter and graffiti clean up. All residents within the City of Seattle are required by law to have waste containers and pay for the collection service. The costs of different containers are presented in Table 3.7.

In 1998 the Solid Waste Fund expenses were $82.1 million. This is broken down as follows:

- Garbage, yard waste and recycling collection – 31%
- Transfer station, hauling and disposal – 29%
- Household hazardous waste – 1%
- Litter control – 3%
- General administrative – 7%
- Customer billing and collection – 5%
- Taxes – 14%
- Depreciation – 6%
- Interest expense and amortisation – 4%

Public support for the existing waste management system is high: 90% of the people interviewed in a city-wide residential survey said that they were very satisfied with the system.

Bin size	Single-family residential rate	Multi-family residential rate
Micro-bin (45 litre)	$10.55	$9.75
Mini-bin (72 litre)	$12.35	$12.05
One bin (120 litre)	$16.10	$15.80
Two bins	$32.15	$31.85
Additional (per 120 litre bin)	$16.10	$16.10
Garden waste (a maximum of 20 × 30 kg bundles per month)	$4.25	$4.25

Table 3.7 Monthly kerbside collection rates

Case study analysis – conclusions

1. The variation between each of the case study programmes was seen to be extensive but common 'Drivers', such as closure of incinerators and lack of landfill space (often a crisis forces a change) were identified.

2. Contrary to the belief that considers environmental issues ahead of social and economic issues, there was common agreement amongst the majority of the waste managers from the case study programmes that in reality their priority order for sustainable waste management has generally been:

   2.1 economic viability
   2.2 social pressure
   2.3 environmental benefits

   Environmental benefits cannot be engineered into the development of a waste management system unless that system is economically viable and socially acceptable, hence all three areas must be addressed simultaneously.

3. An integrated approach to waste management is being adopted at a local level throughout Europe.

4. Enabling legislation can have a positive impact on the development of effective waste management systems whereas prescriptive legislation can have a negative influence on both existing and developing waste management systems.

5. The waste hierarchy was only considered to be useful as a list of possible treatment options.

6. More flexibility and data-based decisions are necessary at the level of waste management operations.

7. There is also evidence for system evolution, from waste management practices, to Integrated Waste Management, to resources management and on towards sustainability.

What the case studies show is that although waste management systems are complex, the adoption of an integrated approach can be characterised by a series of clear steps that lead to the implementation of the general principles of IWM as described in Chapter 2. These steps do not occur in a rigid order, but as would be expected in these diverse systems, they occur as and when they become necessary for the further development (or evolution) of the system.

Madras, India, 1999 – a case study from a country with a developing economy

The previous case studies all relate to countries with developed economies. In these scenarios the majority of residents can afford to pay taxes or fees, which in turn pay for the maintenance and development of waste management services. In countries with developing economies, those who can afford to pay taxes are often in the minority. The proportion of tax revenue allocated to waste management, although high (Bartone, 1999), is often insufficient to fund either effective collection or disposal due to the poor existing municipal waste management infrastructure and the huge number of people the system must serve (see Chapter 2, Integrated Waste Management in countries with developing economies).

A partial solution to India's increasing waste management problems has been developed by a community group in Madras that helps local people collect, compost and recycle a significant fraction of the waste generated in their neighbourhood. Although the system does require residents to contribute a separate (small) waste management fee, this is clearly considered acceptable with respect to the significant improvement that they see in their immediate environment. This is an example of a small-scale, community-based Integrated Waste Management system.

Introduction

In India the problems of collection, transport, treatment and disposal of Municipal Solid Wastes (MSW) are straining both the financial resources of the local authorities and their physical capabilities. The lack of adequate resources, both financial and human often, leads to the problem of waste management being neglected; few of the waste management laws are actually enforced and there is an acute lack of suitable waste disposal sites.

The population of the four metropolises of India (Bombay, Delhi, Madras and Calcutta) produce 0.5 kg/person/day of waste. Approximately 60% of the dry recyclable material in MSW in India is recycled by scavengers (known as rag pickers), thus little real economic advantage exists for the municipal authorities to attempt to recover further materials or energy from the MSW. The Indian life style results in the waste containing a large organic fraction and therefore its moisture content is as high as 50% by weight. (Due to the Indian diet, this organic material is virtually meat free, making it an ideal material to compost.) This organic fraction is not recovered by the rag pickers. Consequently, a large fraction of the total MSW generated accumulates on the side of the streets, where it is dumped indiscriminately by residents and small commercial operations.

The city of Madras (174 km^2) is the capital of the state of Tamil Nadu located on the southeastern coast of India on the Bay of Bengal, which had a population of 5,900,000 in 1995, and a growth rate of 2.34% per annum. The population density is approximately 22,000 per km^2 with 39% of the population living in slums. Currently, 68% of the MSW in Madras is of residential origin, 14% is generated from commercial establishments, schools and institutions generate 12% of the MSW arisings, while industry accounts for just 2%; the remainder comes from hospitals and clinics (which is also disposed of with the MSW). The cost of MSW management in Madras is between US$40 and 45 per tonne.

Residents' associations in the city of Madras met with local authorities to voice their concern over the lack of an effective waste management system in the area. In the absence of any sustained effort or resources being provided by the local authority, the lead was taken by a non-governmental organisation, EXNORA, which utilised scavengers and the unemployed to

Cost/revenue	Amount	
	Rupees	US dollars
EXNORA signpost	1000 per unit	25
Tricycles for waste collection	7500 per unit	188
Tricycle maintenance	100–150 per year	2.5–4.0
Uniforms	350–500 per year	8.8–13.0
Street Beautifiers		
– salary	1000–1500 per month	25–40
– salary increments and bonus	50 per year, 85 per year	1.25, 2.25
Revenue – collection charges		
from residents	10–50 per month	0.25–1.25
Revenue – from recyclables	10–15 per day	0.25–0.40

Table 3.8 Cost of operations of Civic EXNORA

manage wastes without requiring funding from the local authorities. Since 1989 over 5000 dirty streets of Madras in South India have undergone total transformation as residents manage their own solid wastes by employing rag pickers to clean streets and collect wastes from homes. EXNORA has succeeded in motivating residents to improve their own environment, and has gained national and international recognition as the concept expands to other Indian cities. The group was featured in the 100 best practices selected globally by the UN Council for Human Settlements (Habitat).

The development of EXNORA

In September 1988 a bank officer in Madras, concerned about the build-up of municipal waste in the streets, concluded that the solution involved the participation of the local people who generated the waste. This led to the development of an organisation that planned to formulate and practice EXcellent, NOvel and RAdical ideas to solve this waste management problem. EXNORA was first introduced in the Adyar locality of Madras in October 1988. The residents took collective responsibility to ensure that the waste generated in their homes was collected and transported to identified dump-sites by a local rag picker now employed as a 'street beautifier'. The scheme is funded by small subscriptions (US$3.00 per year) from the residents themselves.

EXNORA organises local neighbourhoods typically consisting of 80–150 households, each of which is called a 'Civic EXNORA'. They employ a 'street beautifier', who is provided with a uniform and a tricycle cart and training. Each 'street beautifier' collects waste from every household in their neighbourhood. Currently this waste is not separated, but home sorting is planned in the future by starting a public education programme. The collected waste is sorted into a biodegradable fraction and a recyclable fraction. The 'street beautifier' sells the recyclable fraction; this revenue is in addition to a basic salary of approximately US$25–40 per month, which is collected from the residents. The organic fraction and any residue from the recyclable fraction is taken to the local municipal waste collection point. The 'street beautifier' must also sweep the streets in his area. Before and after photographs show remarkable results.

Loading capacity per tricycle: 100–150 kg, 2 × trips per day = 200–300 kg/day.

Total waste collection by EXNORA in Madras: 3000 units × 250 kg
= 750 tons per day.

Total waste generated in Madras is 2600 tonnes/day, therefore approximately 29% of the MSW is collected by EXNORAs:

- 39% of this collected material is organic and therefore can be composted,
- 4.5% of paper, and
- 4.5% of rags, as well as
- 3.5% of a mixture of plastic, glass, rubber, wood, leather, etc. can also be recycled.

This results in EXNORA diverting 386 tonnes (or 51.5%) of collected waste from final disposal.

EXNORA currently diverts almost 15% of the total waste arising in Madras from final disposal.

Box 3.1 Waste collected in Madras by EXNORA.

The system is based entirely on public participation through neighbourhood committees. The role of the EXNORA Central office is basically to initiate and catalyse the process, and provide the practical expertise. Initially, residents had to be persuaded to address the issue. Now that their success is well known, neighbourhoods ask for assistance to start their own Civic EXNORA. Depending on the community, EXNORA either pays the initial start-up costs of approximately US$250, or asks the neighbourhood to provide it. A breakdown of the costs of a typical EXNORA is presented in Table 3.8. A major advantage of the EXNORA approach is the low capital and operational costs required to manage MSW. This is primarily because of the simplicity of the scheme and the efficiency of the small-scale units that are strongly supported by the residents of each area (as they experience the improvement in their immediate environment).

EXNORAs are in the process of scaling up activities to larger communities. This involves organising 'street beautifiers', rag pickers and local unemployed young people to form a micro-enterprise (waste collection, sorting and composting) with the support of the community. Each micro-enterprise produces and markets compost made from the organic fraction of the waste from approximately 4000 families.

The scale of EXNORA's success

Since the first EXNORA project began in 1988, the concept has spread rapidly throughout India. There are now approximately 5500 schemes in operation serving some 2,750,000 people. These figures are based upon the assumption that the average number of people in a typical Indian household is five and 100 households make up a typical Civic EXNORA unit.

There were over 3000 Civic EXNORAs in Madras and neighbouring Kancheepuram and Tiruvallur districts in 1998. They collect approximately 30% of the total waste generated in Madras (about 2600 tons per day, excluding 500 tons of construction debris generated per day). Previously, the collected waste was managed by the municipality, which simply transported all of the material to the city's dumps. Civic EXNORAs now support home composting and 'zero waste' (to the municipality) with initiatives in 30 localities. Residents are being encouraged to separate their waste at source into organic and recyclable fractions. Systems such as these are to be introduced in all Civic EXNORAs as the next development in the residents' waste management skills.

The waste generated in Madras is largely (60%) biodegradable and with a significant inert fraction (27%), together accounting for about 90% of the material. The waste has a high organic content (39%) coupled with high moisture levels (28%), and a carbon:nitrogen ratio of 31; thus it is suitable for biological treatment. The major sources of the waste that is generated in Madras are the residential areas, contributing about 60% of the total waste stream. With the help of EXNORA approximately 29% of the total Madras waste stream is collected and 15% is diverted from final disposal (see Box 3.1 for the calculation).

Future plans for EXNORA

The work of EXNORA is to continue, as outlined in Table 3.9.

Conclusions

Stakeholder participation has been shown to be essential in Madras to support an integrated approach to waste management, which includes an *innovative but appropriate* (low technology) collection system (the employment of a scavenger), materials recycling and composting. The EXNORA scheme has been *successfully reapplied* in and around Madras resulting in 29% of total MSW arisings being collected, 11% of total MSW arisings being composted and 3.6% being recycled. A further 8% of the collected material (the inert fraction) has been identified as suitable for use as a substrate material for road building; this option is currently under investigation by EXNORA.

Optimisation of Integrated Waste Management systems

The principles of an integrated approach to solid waste management demonstrated in the case studies can be seen to enable an increase in the levels of recycling and energy recovery. This results in a decrease in the amount of material requiring final disposal to landfill. In the past, achievements such as these would be enough to satisfy most waste managers. But now, with environmental issues such as 'Global Warming', 'Eutrophication' and 'Acid rain' receiving considerable media attention, waste managers need more data on the environmental burdens associated with each stage of the waste management system, from collection to final disposal, to help support their decisions with respect to each of these environmental concerns.

The tool of Life Cycle Inventory, described in the next chapter, can be applied to Integrated Solid Waste Management systems. It can provide waste managers with the data they require to make informed decisions relating to their obligations to their customers (those people whose waste they manage) to their environment at a local level, and to the environment at a global level.

Item	Description	Further details
Primary collection	Source segregation by households Volume reduction of bulky garden cuttings Separate handling of biomedical wastes	Households will collect waste in two separate containers for organic and inorganic wastes. Street beautifier to be provided with a manual shredder and cutter to minimise the volume of the garden cuttings. Separate collection vehicles for medical waste.
Secondary collection	Need for transfer stations may be reduced in view of source separation and volume reduction schemes	Considerable reduction in transportation costs. Redeployment of some of the collection fleet to areas unable to finance a Civic EXNORA. Landfills to be used for disposal of inert wastes only; smaller volumes, less leachate production and less landfill gas production.
Disposal	Organic waste from households will be composted locally	Inorganic material to be sorted by street beautifier, cleaned and sold to recyclers. Bulky garden cuttings to be shredded and composted at street corners/housing complexes.
Zero waste schemes	Community-based schemes	Wastes from market places, restaurants, abattoirs, etc. to be composted in two-stage process by municipality with help from EXNORA.

Table 3.9 Waste management initiatives proposed by EXNORA

CHAPTER 4

Life Cycle Assessment

Concepts and Case Studies

Summary

The technique of Life Cycle Assessment (LCA) is introduced. It is a tool that takes into account all of the operations involved in providing a product or service. It consists of four stages: Goal Definition and Scope, Inventory Analysis, Impact Assessment and Interpretation, of which the first two are well developed and the latter two have been developed more recently. In view of this, the analysis of solid waste management will be limited to a Life Cycle Inventory (LCI) (comprising Goal Definition and Scope and Inventory Analysis stages). The analysis will not involve a detailed Life Cycle Impact Assessment or Life Cycle Interpretation, although the following text does describe these phases of LCA for completeness. Other assessments for economics, safety, and site-specific environmental impact, although not part of an LCA, are equally important but separate parts of an overall assessment of waste management systems.

> **Definition**: Life Cycle Assessment (LCA) is a compilation and evaluation of the inputs, outputs and the potential environmental impacts of a product system throughout its Life Cycle. ISO 14040: 1997 Environmental Management – Life Cycle Assessment – Principles and Framework.

What is Life Cycle Assessment?

Life Cycle Assessment (LCA) is an environmental management tool increasingly used to understand and compare how a product or service is provided 'from cradle to grave'. The technique examines every stage of the Life Cycle, from raw materials acquisition, through manufacture, distribution, use, possible reuse/recycling and then final disposal. In addition, every operation or unit process within a stage is included. For each operation within a stage, the inputs (raw materials, resources and energy) and outputs (emissions to air, water and solid waste) are calculated. These inputs and outputs are then aggregated over the Life Cycle. The environmental issues associated with these inputs and outputs are then evaluated in the Life Cycle Impact Assessment. This provides a general overview of the product system and other assessment tools can then be combined with this information to evaluate the product or service over the entire Life Cycle. Conducting LCAs for alternative products or services thus allows for improved understanding and comparisons to be made. An LCA will not necessarily guarantee that one can choose which option is 'environmentally superior' or better than another, but it will allow the trade-offs associated with each option to be assessed.

An explanation of these terms and their important differences is given in Table 4.1.

Life Cycle Assessment	A process to analyse the materials, energy, emissions, and wastes of a product or service system, over the whole Life Cycle 'from cradle to grave', i.e. from raw material mining to final disposal. Currently considered to consist of four stages: **Goal Definition**, **Inventory Analysis**, **Life Cycle Impact Assessment** and **Life Cycle Interpretation.**
Goal Definition and Scope	Stage at which the **functional unit** for comparison is defined (normally per equivalent use), as well as the study purpose, system boundaries, Life Cycle stages, unit processes and scope of the assessment.
Life Cycle Inventory Analysis	Process of accounting for all the inputs and outputs of the product system over the Life Cycle. Will result in a list of raw material and energy inputs, and of individual emissions to air, water and as solid waste.
Life Cycle Impact Assessment	Associates the inputs and outputs with particular environmental issues, e.g. ozone depletion, and converts the inventory of materials, energy, and emissions into representative indicators, e.g. an aggregate loading of ozone-depleting chemicals.
Life Cycle Interpretation	Evaluation of the significance of the inputs, outputs, and indicators of the system Life Cycle. This stage is the least well accepted or defined.

Table 4.1 Life Cycle terminology

Life Cycle Assessment requirements have also been included in legislation. The European Community (EC) Ecolabelling Regulation (1992) requires that the whole Life Cycle be considered when setting labelling criteria. Provision for Life Cycle Assessment is also included in the EC Packaging and Packaging Waste Directive (1994), which states that 'Life Cycle Assessments should be completed as soon as possible to justify a clear hierarchy between reusable, recyclable and recoverable packaging.' Practically, this will have to be carried out on a case-by-case basis. As it is developed further, it is likely that Life Cycle Assessment will find many additional applications.

Benefits of the Life Cycle Approach

Life Cycle Assessment is an inclusive tool. The Life Cycle Inventory phase is essentially an accounting process or mass balance for a system. All necessary inputs and emissions in many stages and operations of the Life Cycle are considered to be within the system boundaries. This includes not only direct inputs and emissions for production, distribution, use and disposal, but also indirect inputs and emissions, such as from the initial production of the energy used. It is essential that all of the processes are included in the boundaries to conduct a fair and transparent analysis. The reader should recognise that the analysis aggregates over time, i.e. all inputs and emissions over the whole Life Cycle, are included regardless of when they occur, and aggregates over space, i.e. all of the sites are included, regardless of where they are located. If real environmental improvements are to be made, it is important to use LCA so that any system changes do not cause greater environmental deteriorations at another time or another location in the Life Cycle.

LCA offers the prospect of mapping the energy and material flows as well as the resources, solid wastes, and emissions of the total system, i.e. it provides a system 'map' that sets the stage for a holistic approach. Comparing such system maps for different options, whether for different products or waste management systems, allows the identification of areas where environmental improvement can be made.

Concern over the environment is sometimes expressed in terms of individual issues, such as acidification. Concentrating on one issue alone, however, ignores and may even worsen the system with respect to other environmental issues. The power of LCA is that it expands the debate on environmental concerns beyond a single issue, and attempts to address a broad range of environmental issues. By using a quantitative methodology, at least for the inventory inputs and outputs, it also gives an objective basis for decision making. The system map also allows other environmental information and assessment tools to be incorporated to be used in conjunction with LCA. This helps to take some of the emotional element out of environmental debates.

Limitations of the Life Cycle Approach

The seemingly all encompassing nature of LCA has proved very attractive. It may appear to new users that it is a single tool that can accomplish 'everything' with regard to environmental assessment. Many people have viewed LCA as being able to give a comprehensive, overall assessment of a product, service or package. As a result, there have been ill-advised efforts to use LCA as the only measurement tool when developing product labelling systems and during policy making.

Unfortunately, there is a dilemma at the heart of LCA. As LCA employs an overall system balance and functional unit to aggregate resource use, solid waste and emissions over time and space, it is **not** able to assess the actual environmental effects of the product, package or service system. The International Standards Organisation (ISO) Life Cycle Impact Assessment document (ISO/FDIS, 1999) specifically cautions that LCA does not predict actual impacts or assess safety, risks, or whether thresholds are exceeded. The actual environmental effects of emissions and wastes will depend on when, where and how they are released into the environment, and other assessment tools and information must be utilised.

For example, an aggregated emission, if released in one event from a point source such as a refinery, will have a very different environmental effect than releasing it continuously over years from many diffuse sources. In addition to this, the inventory will allocate the inputs and outputs of a refinery to many products, i.e. different product systems. Recalling that LCA deals with

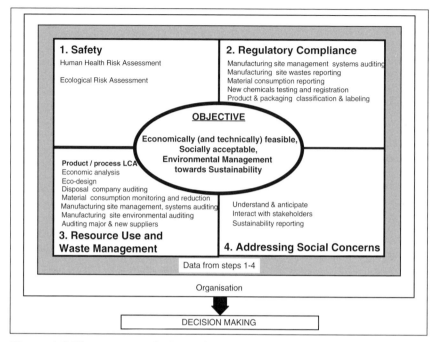

Figure 4.1 The position of LCA within an environmental management framework. Source: White *et al.* (1995b).

only the inputs to a single system, only a small percentage of the total activity will be considered for a single product. This problem is generally acknowledged, and the term 'indicator' is now used to show that LCA does not predict actual environmental effects (ISO, 1999b; Owens, 1999).

The dilemma, therefore, is that LCA is the only tool that attempts to include the whole Life Cycle, and all environmental issues associated with a product, package or service system, and the only one that relates this to the functional unit, yet it cannot predict the actual environmental effects that are likely to occur. Other tools, such as risk assessment, are able to predict the actual effects likely to occur, but they do not cover all environmental issues in the Life Cycle, neither do they link the effects to the functional unit.

Clearly no single tool can do everything – a combination of tools with complementary strengths is needed for overall environmental management. Figure 4.1 presents a generic environmental management framework and shows the position of LCA with respect to other environmental tools.

International Standards Organisation (ISO) – The ISO 14040 series

Based on the work carried out by the Society for Environmental Toxicology and Chemistry (SETAC), the ISO has further developed, and has managed to reach agreement among its global membership on a series of standards: the ISO 14040 series on Life Cycle Assessment (see Figure 4.2).

ISO 14040 Environmental Management – Life Cycle Assessment – Principles and Framework (ISO, 1997).

ISO 14041 Environmental Management – Life Cycle Assessment – Goal and Scope Definition and Life Cycle Inventory Analysis (ISO, 1998).

ISO 14042 Environmental Management – Life Cycle Assessment – Life Cycle Impact Assessment (ISO/FDIS, 1999).

ISO 14043 Environmental Management – Life Cycle Assessment – Life Cycle Interpretation (ISO/FDIS, 1999).

Figure 4.2 ISO 14040 series on LCA.

Within the 14040 series ISO is trying to establish a flexible framework under which LCAs can be carried out in a technically credible and practical manner. The ISO 14040 series is intended as a non-prescriptive guide. There is no single method or end use for conducting LCA studies, so organisations should have the flexibility to implement LCA practically, based upon the specific application and the requirements of the user. There are some mandatory steps within the methodology that must be completed in order for the end results to be described as 'fit for use'.

Structure of a Life Cycle Assessment

As a result of intensive recent efforts to define LCA structure and harmonise the various methods used, LCA is now considered to consist of the following four distinct phases (ISO, 1997) (see Figure 4.3 and 4.4).

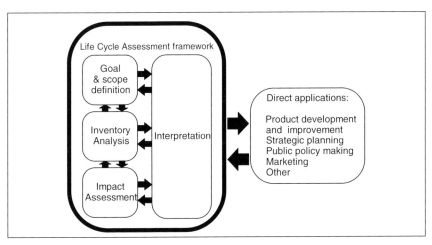

Figure 4.3 Phases of an LCA. Source: ISO 14040 (1997).

1. Goal definition
   Define:
   – options to be compared
   – intended use of results
   – the functional unit
   – the system boundaries.

2. Life Cycle Inventory Analysis (LCI)
   Account for:
   – all materials and energy, both inputs and outputs across the whole Life Cycle.

3. Life Cycle Impact Assessment (LCIA)
   – organises or classifies the LCI inputs and outputs into specific issues or categories
   – models the inputs and outputs for each category into an aggregate indicator.

4. Life Cycle Interpretation
   – the process of balancing the importance of different effects
   – no agreed scientific method
   – requires public debate.

Note: A Life Cycle Inventory (LCI) – the basis of this book – includes the goal definition and inventory stages.

Figure 4.4 The phases of a Life Cycle Assessment.

Goal and scope definition

Defining the Goal of the study

The Goal of an LCA study shall unambiguously state the intended application, the reasons for carrying out the study and the intended audience – ISO 14041.

The Goal Definition component states the reason for performing a specific study, defines the options that will be compared and the intended use of the results. The intended use of the LCA will influence the type of study carried out and the type of data required. This stage also involves identifying the system boundaries (technical, geographical and time) and the procedures for handling the data. Rules and assumptions must be documented, especially with respect to allocation rules for co-products and open-loop recycling and aggregation. This is fundamental to any LCA investigation as LCA is not a precise science. Every stage will involve the need for choices and value judgements. All of the judgements that need to be taken while carrying out an LCA should be made in the light of the purpose of the study. A clearly defined goal will partly address the need for transparency and will help ensure the end result is fit for use. Transparency is essential throughout all stages of the Life Cycle procedure; this allows data tracking and calculation verification to be carried out if necessary. A complete and transparent record of a study is compiled in a final study report.

Defining the Scope of the study

> The Scope should be sufficiently well defined to ensure that the breadth, the depth and the details of the study are compatible and sufficient to address the stated Goal – ISO 14040.

The Scope of a study basically outlines the parameters within which the study will be carried out. These need to be compatible with the Goals of the study. In defining the Scope of an LCA study, ISO 14040 requires that the following are clearly described: the functions of the product system(s); the functional unit; the product system to be studied; the product system boundaries; allocation procedures; the types of indicators and the methodology of Life Cycle Impact Assessment and subsequent Life Cycle Interpretation to be used; data requirements; assumptions; limitations; the initial data quality requirements; the type of critical review (if any) and the type and format of the report required for the study. As LCA is an iterative process and therefore as more is learnt about the system being studied, it is likely that amendments to the scope will have to be made.

Product System

> A Product System is a collection of operations connected by flows of intermediate products, which perform one or more defined functions. The system should be defined in sufficient detail and clarity to allow another practitioner to duplicate the Life Cycle Inventory Analysis – ISO 14041.

Product Systems are subdivided into unit processes.

> Each unit process encompasses the activities of a single operation or a group of operations. Unit processes are linked to one another, by flows of intermediate products and/or waste for treatment and to other Product Systems by product flows – ISO 14041.

The Product System is the series of interconnected operations that occur during the Life Cycle of a product or the delivery of a defined service. The system itself lies within a System Boundary.

Functional unit

> The Scope of an LCA study shall clearly specify the functions of the system being studied. A functional unit is a measure of the performance of the functional outputs of the Product System – ISO 14040.

The functional unit is the basis on which the products or services will be compared. The importance of defining the most appropriate Functional Unit cannot be over-emphasised. The functional unit is the cornerstone of an LCA study, providing the reference point to which both inputs and outputs are related and allowing clear comparison of LCA results. The stages and unit processes connected to the functional unit are known as the Product System.

The functional unit is often expressed in terms of amount of product (e.g. per kg or litre), but should be related to the function served by a product or service, i.e. per equivalent use, for example, the packaging used to deliver a given volume of milk. As all of the inputs and outputs are calculated per functional unit, any alteration in the size of the functional unit, e.g. to allow for higher performance of one of the products compared, will have a major effect on the outcome of the assessment. It is essential that various performance attributes of each system be considered and incorporated into the functional unit, otherwise any comparison will not be made on a fair and equivalent basis.

System Boundaries

The System Boundary defines the unit processes that will be included in the system to be modelled. Ideally, the Product System should be modelled in such a manner that inputs and outputs at its boundary are elementary flows – ISO 14041.

Note: Elementary flow: (1) material or energy entering the system being studied, which has been drawn from the environment without previous human transformation; (2) material or energy leaving the system being studied, which is discarded into the environment without subsequent human transformation.

The System Boundaries, i.e. what stages, operations, and inputs and outputs are included within the assessment, and what is omitted, need to be defined. Again, these boundary decisions on inclusion and omission must be balanced between different systems for any comparison to be fair and equivalent.

The inclusion involves defining exactly where the boundaries or 'cradle and grave' of the Life Cycle lie. Should the mining of the raw materials be included in the 'cradle'? Similarly should the 'grave' include the emissions from used materials after they have been buried in the ground, i.e. landfilled?

The Scope must also determine how much detail will be included at each stage of the Life Cycle. Should the unit operations of making the factory and equipment that make the product also be included? In this instance, previous studies have shown that such 'second level' inputs, when divided among the number of units that the factory produces, are insignificant, so can be omitted from most analyses. There is also the question of which inputs and outputs to include in the inventory, and this is typically determined by the environmental issues that a study wishes to address. Thus, Life Cycle Inventory and Life Cycle Impact Assessment must be co-ordinated in the study goal and scope planning.

It should be stressed that a Life Cycle Assessment can be done on any system, whatever the boundaries defined. There are no right or wrong boundaries to choose, but some are more appropriate to the defined goal than others. If meaningful comparisons are to be made of products or services, then the same or equivalent boundaries must be used in each system in the study.

In any Life Cycle Assessment, defining the Functional Unit and the boundaries of the system being assessed are both important steps. Reporting the Functional Unit and the boundaries used needs to be clear and detailed with any omissions stated and justified. This is as important as giving the actual results.

Life Cycle Inventory Analysis (LCI)

> A Life Cycle Inventory Analysis is concerned with the data collection and calculation procedures necessary to complete the inventory – ISO 14041.

This stage consists of accounting for all of the material and energy inputs and outputs over the whole Life Cycle of the product or service. The operational steps are presented in Figure 4.5.

The procedure entails describing the Life Cycle as a series of steps, and then calculating the inputs and outputs for each of these steps (see Figure 4.6). This amounts to constructing a materials and energy balance for each step in the Life Cycle. The analysis of all inputs and outputs for each stage in the Life Cycle can then be combined to give the overall Life Cycle Inventory.

The methodology for the inventory stage is now accepted and well-used. The large amount of data generated makes decision making challenging. There is a need to find methods of aggregating data but this is full of dangers (see page 95, paragraph 3 of the section on Sensitivity and uncertainty analysis). Given the numerous possible choices, it is unlikely that allocation will be standardised. As with functional unit and boundary decisions, the allocation choices must be equivalent in their effect on different systems for a fair comparison to be made.

Notwithstanding these limitations, an LCI provides enormous knowledge about and insights into the operations of a given system, which can provide the basis for applying and integrating other environmental information and assessment tools into a system comparison.

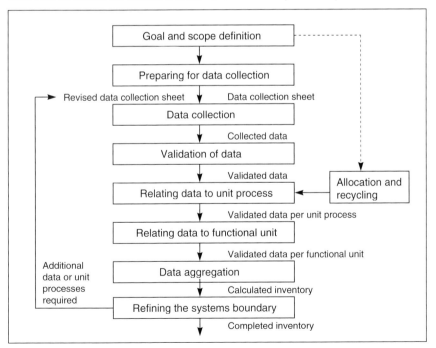

Figure 4.5 Procedures for Life Cycle Inventory analysis. Source: ISO 14041 (1998).

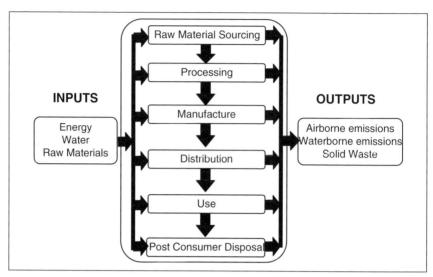

Figure 4.6 The stages of a product's Life Cycle.

Data quality requirements
There are two main categories of data used:

1. Specific data, for production, distribution and waste management.
2. Generic data, for energy production, raw material extraction and transportation.

Data quality requirements should address: time-related, geographical and technology issues; the precision, completeness and representativeness of the data; the consistency and reproducibility of the methods used throughout the LCA; the sources of the data and their representativeness, and the uncertainty of the information.

There is an increasing amount of generic data becoming available on basic and commonly used processes and products. The Society for the Promotion of Life Cycle Development (SPOLD) has prepared a directory of these sources of LCI data (SPOLD, 1995). However, there can often be difficulties in understanding the underlying assumptions and areas of applicability of these data. For this reason, SPOLD has developed a common electronic file format for LCI data. This provides a format, common to all processes, in which existing LCI data can be made available and additional data collected. The SPOLD format for LCI data allows different databases and software to communicate in a common LCI 'language'. Several database owners and software producers have already announced that they will use the SPOLD format in their applications. More information and a copy of the SPOLD file format are available for downloading from the Internet at www.spold.org.

Sensitivity and uncertainty analysis
According to ISO 14043, LCIs should also undergo sensitivity and uncertainty analysis. The data and results should not be used without understanding their quality and limitations. These processes also reflect the fact that LCI involves data uncertainties and value judgements.

Sensitivity analysis can help to identify whether any of the assumptions made, for example, about missing data, have a significant influence on the final outcome of the LCI, and if so which assumptions have the greatest influence. It can therefore provide information about the robustness of the LCI results, and about where there is the greatest need for more, or more precise data to improve the inventory. Ideally sensitivity analysis should be carried out on every parameter within the LCI study, but in practice this is often limited to a selected number of parameters. Particularly important parameters for sensitivity analysis are those that are being omitted either on purpose, because of simplification steps, or accidentally because of lack of data (SETAC, 1997).

Uncertainty analysis is also essential. Generic emission data may involve a wide range in the level of emissions from one or more unit operations or these numbers may have changed since the emissions were measured. This injects a degree of uncertainty into almost every number within the inventory that must be considered. As the objective is to make comparisons, it is essential the decision makers understand when apparent differences are real or uncertain. One of the systematic methods for performing sensitivity analysis, developed by Heijungs (1996), uses confidence limits for all input parameters and identifies those parameters for which the margins of uncertainty have large influences on the final result. Subsequently, the margins of uncertainty for these parameters can be improved.

Transparency

As LCA is subjective, the rationale behind all stages of the study should be clear. This will give context and meaning to the analysis. A transparent approach is essential in helping 'users' understand the approach used and any assumptions made.

Critical review

Guidelines for critical or peer review are included in ISO 14040. An LCI study is likely to affect interested parties who have been external to the study, such as consumers, non-government organisations, local authorities and industry. The opinions of stakeholders need to be taken into account and a critical review is where this is possible. A critical review ensures that the methods used to carry out the LCI are scientifically and technically valid, and that the data used is appropriate and reasonable in relation to the goal of the study. There are three possible forms of critical review.

1. Internal expert review. This type of review should be carried out by an LCI expert within the same organisation who is independent from the study, but has a good knowledge of the system being investigated. Internal expert review should be used when minor changes have been made to a design or initial comparisons are being made between a variety of options using generic data, e.g. proprietary LCI software.
2. External expert review. External expert reviews are again carried out by a proficient LCI practitioner, but this time they should be independent from the study and the organisation carrying out the LCI work. This often means using an LCI practitioner from a relevant industry body or an LCI consultant company. Going external with LCI data can sometimes require the use of confidentiality agreements to protect proprietary information. External reviews of LCI studies should be carried out if the results of the study are going to be used as part of a major decision, i.e. a product launch or a major upgrade or process change.

3. Review by a panel of interested parties. Some LCI studies involve product, package or process changes that can have a perceived impact on people or organisations outside of the company commissioning the LCI study. LCI studies are a good tool for communicating environmental burden information in this type of circumstance; because they take the 'cradle-to-grave' approach they include the manufacturing facilities and the waste disposal phase of the products Life Cycle.

In situations like this, an external review panel with representatives from these interested parties should be set up. An external, independent expert is selected by the original study commissioner to act as chairperson of the review panel. Based on the goal, scope and budget available the chairperson selects other independent, qualified reviewers. This panel should include other interested parties affected by the conclusions drawn from the LCI study, such as government agencies, non-government groups and possibly competitors. The panel will produce a statement and report on the comments and recommendations made during the review process. This should be included in the final LCI study report.

Life Cycle Impact Assessment (LCIA)

> The LCIA phase aims to examine the product system from an environmental perspective using category indicators, derived from the LCI results. The LCIA phase also provides information for the interpretation phase – ISO 14042.

The LCIA phase of an LCA study provides a system-wide perspective of environmental and resource issues for product or service systems. To achieve this, LCIA assigns LCI results to specific, selected impact categories (an impact category is used to group certain LCI results that are associated with a particular environmental issue). For each impact category, appropriate indicators are selected and a characterisation model is used to calculate indicator results. The collection of indicator results – the LCIA profile – provides an environmental context for the emissions and resource use associated with the product or service system.

LCIA is composed of several mandatory elements that convert LCI results to indicator results. There are also optional elements for normalisation, grouping or weighting of the indicator results and data quality analysis techniques. The framework presented in Figure 4.7 includes both the mandatory and optional elements of LCIA.

It is now generally accepted that the most meaningful aggregation of LCI results in an impact category will conform to a scientifically based environmental mechanism (SETAC, 1992). Thus all releases that combine together in a chemical mechanism to deplete ozone will be modelled and aggregated together.

It should be recognised that some impact categories may not be scientifically based, but may involve aggregation among independent mechanisms that cannot be scientifically combined (Owens, 1998). This information can still be useful in pointing to particular issues for further evaluation and analysis, but will be less objective than if they were based on an environmental mechanism. An example is given in Figure 4.9 between a global warming indicator and a toxicity indicator that combines different effects. However, ISO requires that indicators be scientifically and technically valid for claims or comparative assertions.

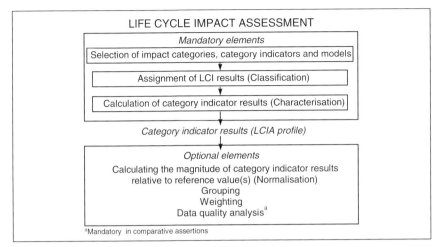

Figure 4.7 Elements of LCIA. Source: ISO/FD15 14042 (1999).

Each indicator should be assessed for environmental relevance as shown in Figure 4.8. Most current impact category models simply aggregate the emissions released and, as noted, do not analyse the actual environmental impact. As the time and space of releases are involved in environmental impacts, many current indicators lack environmental relevance. However, more complex and environmentally relevant models and indicators are being developed. ISO 14042 sets out a series of criteria on which users may judge the environmental relevance of different models and indicators, and requires environmental relevance for making claims or comparative assertions.

The result of the impact assessment will be the contribution of the Life Cycle to the selected environmental issues. Figure 4.8 presents a schematic diagram that outlines the procedure for converting LCI results to category indicator results.

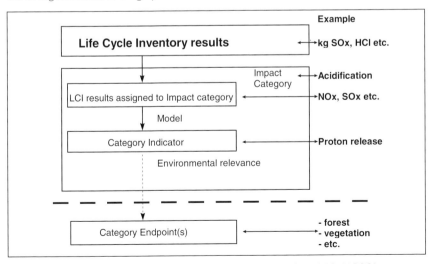

Figure 4.8 Concept of indicators. Source: ISO/FD15 14042 (1999).

Issue	Endpoint	Indicator	Inventory result
Global climate change	Increase in average global temperature	Aggregate loading of greenhouse gases expressed as CO_2 equivalents	Emission across system boundary of greenhouse gases, e.g. CO_2, CH_4, CFCs and HCFCs, etc.
Acidification	Loss of aquatic life as pH of receiving waters decreases	Aggregate loading of all atmospheric and aquatic emissions expressed as acidification potential (proton or H^+ equivalents)	Emission across system boundary of acids and substances possibly converted to acids. e.g. HCl, SO_2, NOx, etc.
Eutrophication	Loss of aquatic life as dissolved oxygen levels in receiving waters decrease	Loading of nutrient substances, which may cause eutrophication and organic substances, which may decrease dissolved oxygen during mineralisation	Emission across system boundaries of major nutrients, such as phosphorous and nitrogen, and readily metabolised organic materials, e.g. BOD.
Photochemical smog	Human health effects. e.g. increased severity of asthma or breathing difficulties	Loading of VOC emissions expressed as photochemical ozone formation potentials and other key reactant, nitrogen oxide. No common reference conditions exist for combining these	Emission across system boundaries of VOCs and nitrogen oxides (note: as reactivity and mass emitted combine to give total formation potential, careful review of inventory is necessary).
Human toxicity	Adverse effects on human health, ranging from the occurrence of cancer to irritation of the skin and eyes	No agreed indicator. Some practitioners attempt a total toxicity aggregate, others use subscores for cancer, reproduction, etc. Use of concurrent exposure assumptions and subjectivity in combining different effects	Emission across system boundaries of various toxic substances. Approximate toxic potency and effect must be known.
Resources	Depletion of resources	Depletion rate of each mineral resource, subjectivity is involved in combining different resources	Input across system boundaries of various resources.

Table 4.2 Examples of issues, categories and indicators

Classification

The classification stage requires the identification of inventory data relevant to each specific impact category and assignment of the appropriate LCI results to each category. Data may belong to more than one category, e.g. NOx has a global warming and an acidifying effect.

Selection of impact categories

Table 4.2 presents the type of impact categories that are commonly examined. The categories should be selected based on the goals and scope of the LCA study.

Characterisation

The aim of characterisation is to provide a basis for the aggregation of inventory results into an indicator for each category. Each impact category requires a specific model to convert the inventory results into the indicator – ISO 14042.

The characterisation or modelling stage requires calculations to be made to evaluate the relative significance of each contributor to the overall impact of the system or operation being studied, by converting these to a common indicator. For example, in the case of global warming the most common indicator used is Global Warming Potential (GWP) in CO_2 equivalents. Basically, there are two steps in the calculation. Each greenhouse gas is first converted into carbon dioxide equivalents based on a particular characterisation factor. The individual carbon dioxide equivalents are then added into a total indicator.

Normalisation

If this is undertaken, it involves relating the characterised data to a broader data set or situation, for example, relating SOx emissions to a country's total SOx emissions. Normalisation can provide insights but should be treated with caution as results can differ significantly if different data sets are used. Normalisation is often omitted from LCA studies.

Weighting

Weighting is the process of converting indicator results of different impact categories into scores by using numerical factors based on values. Weighting may include aggregation of the weighted results into an overall score. This is the most subjective stage of an LCA and is based on value judgements and is not scientific. ISO 14042 cautions that different individuals or organisations may have different preferences or values. Therefore, different parties are likely to reach different weighting results based on the same indicator results. Figure 4.9 shows a schematic representation of the decreasing objectivity across an LCA.

Whilst there are significantly fewer impact categories than inventory categories, there are still many environmental issues to be considered. Comparing one Life Cycle option with another will not normally show which is 'environmentally superior' (except in the case where one has a lower impact in all categories), but will demonstrate the trade-offs between the two options.

For the foreseeable future, decision making on the basis of the impact assessment results should be done by open public debate as part of the democratic process. Any scheme that weights or aggregates the impact categories to a single score may appear to make decision making easier. However, in the process, the assumptions and priorities upon which the decision will be based are obscured and only the values or opinions of one group are used. Thus, broad

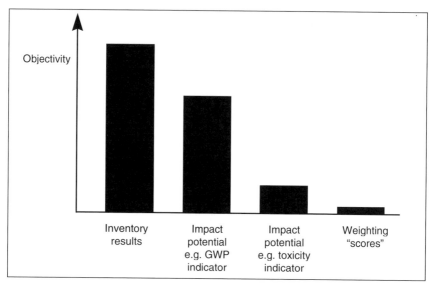

Figure 4.9 Decreasing objectivity and reliability across an LCA. Source: IEM (1998), with additions.

acceptance of the outcome of the decision is uncertain. That verdict will only be passed when the results become evident to those affected.

ISO 14042 states that 'weighting across impact categories shall not be used for comparative assertions disclosed to the public'. ISO 14042 also makes it clear that impact assessment is designed to 'support better decision making'. Life Cycle Impact Assessment does not replace the decision-making process.

Life Cycle Interpretation

Life Cycle Interpretation is a systematic technique to identify, qualify, check, and evaluate information from the results of the Life Cycle Inventory (LCI) analysis and/or LCIA of a product system, and present them in order to meet the requirements of the application as described in the goal and scope of the study – ISO 14043.

The interpretation stage of the LCA process is closely linked to the iterative nature of the processes of scope definition, inventory analysis and impact assessment. Interpretation involves a review of all of the stages in the LCA process and a check that all assumptions are consistent. Data quality should be checked and a sensitivity analysis performed to establish the significance of data uncertainty on the results of the study. This may result in revisiting and revising certain areas of the study. Links between LCA and any other environmental management tools are necessary to highlight the strengths and limits of the LCA in relation to its goal and scope definition phase. This procedure, including transparent reporting, is necessary, as these results can be used as a basis for conclusions, recommendations and decision making in accordance with the goal and scope definition phase.

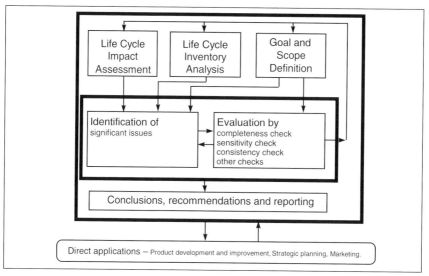

Figure 4.10 Relationships of the elements within the interpretation phase with the other phases of LCA. Source: ISO 14043 (1998).

There are three elements in the Life Cycle Interpretation phase of an LCA (see Figure 4.10).

1. Identify the significant issues based on the LCI and LCIA phases of the LCA.
2. Evaluate the significant issues based upon completeness, sensitivity and consistency checks.
3. Draw conclusions, make recommendations and report the significant issues.

Identification of significant issues

The objective of this element is to structure the results from the LCI or LCIA phases in such a way that it is possible to determine the significant issues. This should include any implications of the particular method used and any assumptions made. Allocation rules, cut-off decisions, choice of indicators and characterisation methods must all be addressed.

The determination of the significant issues of a product system may be simple or complex. ISO standard 14043 does not provide guidance on why an issue may or may not be relevant in a study, or why an issue may or may not be significant for a product system.

Evaluation

> The objectives of the evaluation element are to establish and enhance the confidence in and the reliability of the result of the study – ISO 14043.

The results of the evaluation should be presented in such a way as to allow the reader a clear and understandable view of the outcome of the study. To achieve this, a completeness check (to ensure all relevant information for interpretation is available and complete), a sensitivity check (to assess the reliability of the results by assessing the uncertainty of the significant issues affecting the conclusion) and a consistency check (to determine whether the assumptions, methods and data are consistent with the goal and scope) should be carried out.

Conclusions, recommendations and reporting

> Drawing conclusions from a study should be done interactively with the other elements in the Life Cycle interpretation phase – ISO 14043.

Preliminary conclusions must be checked to ensure that they are consistent with the goal and scope of the study. If this is the case then they may be reported as full conclusions. Recommendations should be based on the final conclusions of the study and the final report shall present a complete, unbiased and transparent account of the whole study.

Life Cycle Inventory of solid waste

A Life Cycle Inventory (and assessment) may be used with other environmental information and assessment tools to improve the environmental performance of a product or service. Given that the product will be used and disposed of in a fixed disposal system (i.e. relative levels of recycling/incineration/landfill) it is possible to determine how changes in the product will alter the various aspects of the environmental performance of the system, e.g. changes in energy consumption and emissions on a Life Cycle basis. However, it is possible to conduct 'Life Cycle (Inventory) in reverse' (White et al., 1993), by keeping the product constant and changing the disposal conditions to see how this affects the overall environmental impact.

This is essentially what is involved in a Life Cycle Inventory of solid waste. Assuming that the composition of waste produced is fixed, and details of its composition are known, at least for household waste, it is possible to determine how the use of different options for waste management affect the environmental performance of waste disposal. This technique will be used throughout this book, and the general format of an LCI for waste will be explored in the next chapter.

CHAPTER 5

A Life Cycle Inventory of Solid Waste

Summary

The Life Cycle Inventory (LCI) technique described in Chapter 4 is applied to waste management. The possible uses for an LCI of different waste management options are discussed. The functional unit for the comparison is defined, as are the system boundaries. This includes defining the 'cradle' and 'grave' for waste. The general structure of waste management systems, which forms the basis of the LCI model, is mapped out, and the computer model developed to conduct the LCI is introduced.

> **Definition**. The goal of a Life Cycle Inventory for solid waste is to be able to, as accurately as possible, predict the environmental burdens of an Integrated Waste Management system.

Integrated Waste Management and Life Cycle Inventory

The objective of Integrated Waste Management (IWM) is to deal with society's waste in a way that is environmentally and economically sustainable and socially acceptable (Chapter 2). To assess such sustainability, tools that can predict the environmental burdens and likely overall cost of any system are needed. Life Cycle assessment is an environmental management tool that allows prediction of the environmental burdens associated with a product or service over the whole Life Cycle, from 'cradle to grave'. This technique can be applied to waste management to assess environmental sustainability. At the same time, a parallel economic assessment can determine the economic sustainability of waste management systems – a criterion crucial to their continued operation.

As described in Chapter 4, a Life Cycle Assessment consists of four stages: Goal and Scope definition, Life Cycle Inventory, Life Cycle Impact Assessment and Life Cycle Interpretation (Table 4.1). Currently the Goal and Scope definition and Life Cycle Inventory stages (which together comprise a Life Cycle Inventory study) are routinely carried out in a variety of applications; Life Cycle Impact Assessment and Life Cycle Interpretation still require value judgements and therefore present significant challenges. This book and the associated computer model are based on a Life Cycle Inventory of solid waste, although the model also aggregates LCI data for carbon dioxide, methane and nitrogen dioxide based on their Global Warming Potential, using the weighting proposed by the Intergovernmental Panel on Climatic Change (IPCC, 1996) over a 100-year time span (see Table 4.2 and Figure 4.9). This chapter addresses the concept and practicalities of using Life Cycle Assessment to compare waste management systems.

The Life Cycle technique has been used to compare specific options for waste management (e.g. Kirkpatrick, 1992; Denison, 1996; Finnveden and Ekvall, 1997) and has now also been used to assess complete IWM systems (Wilson, 1997, 1998; Thurgood, 1998). In spite of these and other similar studies it remains necessary for reasons of transparency to first address such basic questions as where is the 'cradle' of waste, and where is its 'grave'?

A Life Cycle Inventory of waste

Goal definition

Here three major questions are addressed:

1. What is the purpose of the study?
2. What will be compared, i.e. what is the functional unit for comparison?
3. What are the boundaries of the system (see Table 5.1)?

This last question defines what will be included in the study and what will be omitted, and specifies the 'length' and 'breadth' of the study.

1. Options to be compared:	Different systems for managing solid waste
2. Purposes:	To predict environmental performance (emissions and resource consumption) of IWM systems To allow 'What if... ?' calculations To support achieving environmental sustainability To demonstrate interactions within IWM systems To supply waste management data for use in individual product LCIs
3. Functional unit:	The management of the household and similar commercial waste arisings from a given geographical area in a given time period (e.g. 1 year)
4. System Boundaries:	Cradle (for waste): when material ceases to have value and becomes waste (e.g. the household dustbin) Grave: when waste becomes inert landfill material or is converted to air and/or water emissions or assumes a value (intrinsic or economic) Breadth: 'second level' effects such as building of capital equipment ignored. Indirect effects of energy consumption included.

Table 5.1 A Life Cycle Inventory of waste: Goal definition

What are the purposes of the LCI?

1. To predict the environmental performance of an IWM system. Because specific data for all parts of the Life Cycle are not available, generic (typically averaged) data will be frequently used, and the result of the inventory will not be 100% accurate. However, it will provide a 'first cut' and will provide rough comparisons between different system options.

   The objective of predicting environmental performance of waste management systems can be met in two ways. Detailed LCI studies can be run for several individual waste management systems, and general conclusions extrapolated from the results. The alternative is to construct a generic, flexible tool that can be applied to any waste management system to assess the overall environmental performance. Constructing a model that is flexible enough to describe all possible waste management scenarios is a very challenging task, but is the option attempted in this book. A general model such as this will rely on generic data, so will not give such accurate results as specific studies, which describe particular waste management systems. However, the flexibility to apply the same model to a range of waste management systems, both existing or planned, is considered to be more useful than a very accurate model of a single waste management system.

2. To demonstrate the interactions that occur within waste management. As it attempts to model the whole waste system, a Life Cycle model will show how different parts of the system are inter-connected and will help improve understanding of the system's behaviour.

3. To clarify the objectives of the waste management system. It has been argued above that the objective of waste management is environmental and economic sustainability, i.e. minimising the environmental burdens for an acceptable cost. Because it specifically calculates both the cost and individual environmental burdens (i.e. emissions due to energy consumption, emissions to air, emissions to water, landfill requirements, etc.) it focuses attention on which parameters need to be maximised or minimised. It is important that potential users understand that the tool of Life Cycle Assessment will not in itself decide this, but it will provide the data on which these societal/political decisions can be based.

4. To allow for 'What if... ?' calculations. The use of a computer model for the LCI allows the user to compare a number of hypothetical waste management systems, their environmental burdens and economic costs.

5. To provide data on waste management methods, which can be used in LCI studies of individual products and packages.

6. To provide an economic assessment of the IWM system using the same system boundaries as the LCI, ensuring that the two data sets may be analysed in parallel.

Defining the functional unit

The functional unit is the unit of comparison in a Life Cycle Inventory. Historically LCI studies have been related to products, such as washing machines (DoE/DTI, 1991) or detergents (Stalmans, 1992), or packages (White et al., 1993; Kuta et al., 1995; Smith and White, 1999). The functional unit in such cases relates to the product/package made, and the comparisons are made on the basis of per amount or per equivalent use of the product. The functional unit is therefore expressed in terms of the system's output (see Figure 5.1). Such studies are usually run to see how changes to the product will affect their overall environmental burdens.

1. Functional unit

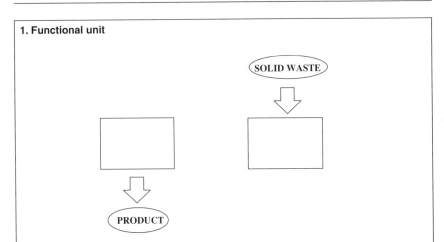

In a product LCI the functional unit is defined in terms of the system's output, i.e. the product, for example, per kg of product made, or per number of laundry loads washed.

In the LCI of waste the functional unit is defined in terms of the system's input, i.e. the waste. The functional unit would be the management of the waste of one household or the total waste of a defined geographical region in a given time (e.g. 1 year).

2. Boundaries – definition of cradle and grave

Product LCI studies consider the whole Life Cycle of one particular product, from raw material extraction, through manufacture, distribution and use, to final disposal. The last part of the Life Cycle will be spent as waste in a waste management system (Figure 5.2). This book considers the Life Cycle of waste, from the moment it becomes waste by losing value, to the moment it regains value or leaves the waste management system as an emission.

A Product LCI considers the whole Life Cycle of a single product; an LCI of waste includes part of the Life Cycles of all products.

Using 'What if ... ?' scenarios

Product LCIs are normally used to determine the environmental effect of changes to the product. This LCI of waste can be used to determine the environmental effect of changes to the waste management system.

Figure 5.1 Key differences between LCI studies for products and an LCI study of waste.

The function of a waste management system, in contrast, is not to produce anything, but to deal with the waste of a given area. Therefore, the functional unit in an LCI of waste is the waste of the geographical area under study. In this study this functional unit is refined further into the household and similar commercial waste of the specified geographical area. This is a key difference in approach: in a product LCI the functional unit is defined by the output (product) of the system; in an LCI of waste the functional unit is defined in terms of the system's input, i.e. the waste.

Using this LCI method, different systems for dealing with the solid waste of a given area can be compared. The geographical area under study, and the waste produced by this area are defined by the user, and the LCI model will calculate the environmental burdens and overall cost of different options chosen for dealing with this waste. The waste input from any given area will be constant, and the LCI can be used to assess the overall performance, both environmental and economic, of different waste management systems.

The environmental burdens and costs for the whole system can be broken down further, however. For economic costs, it is useful to have the cost attributed per household, since revenue is usually collected in this way for domestic solid waste. Alternatively, costs can be attributed per tonne of waste collected, which is especially relevant to commercial/industrial waste where charges are levied in this way. As well as overall system costs and environmental burdens, costs and environmental burdens per tonne and per household will also be calculated, but neither alone is satisfactory.

A figure that is often quoted in reports of recycling schemes is the cost per tonne of recovered or recycled material (usually high!). This will not be used as the functional unit in this assessment, since it is not the function of an Integrated Waste Management scheme to produce recycled material. The objective is to deal with waste in an environmentally and economically sustainable way; recovering and recycling materials is a means to this end but not the end in itself.

System boundaries

Where is the cradle of waste and where is the grave?

All Life Cycles run from cradle to grave. When considering the Life Cycle of products, inventories usually go back to the source of the raw materials, by mining for example, to define the product's 'cradle'. The 'grave' is the final disposal of the product, often back into the earth as landfill. Whilst it shares the same grave as individual products, the Life Cycle of 'waste' does not share the same cradle (see Figure 5.2). Waste only becomes waste at the point at which it is thrown away, i.e. ceases to have any value to the owner. Thus the 'cradle' of waste, in households at least, is usually the dustbin. This is another key difference between an LCI for a product and an LCI for waste. Every product spends part of its Life Cycle as waste, or conversely a Life Cycle study of waste includes part (but only part) of the Life Cycle of every product or package.

Clearly there are overlaps between the two approaches, since the LCI for a product includes the time the particular product spends in the waste management system, and the LCI for waste includes the waste management stages of all products and packages. However, there are fundamental differences, since the two applications have different functional units and therefore different uses (and potential users). A product LCI can be used to optimise a specific product Life Cycle, normally within a given infrastructure system (energy generation system, transport system, solid waste management system, etc.). A solid waste LCI, in contrast, aims to optimise the infrastructure system for managing a given amount and composition of waste. Hence product LCIs are of use to those who can control product design and manufacture; solid waste LCIs are of use to those who plan or manage solid waste management systems. They represent two different tools for two different user groups. It is important that this distinction is appreciated, since different questions relating to product and waste management will require the choice of the appropriate tool. For example, a solid waste LCI (the horizontal approach in Figure 5.2)

Concepts and Case Studies

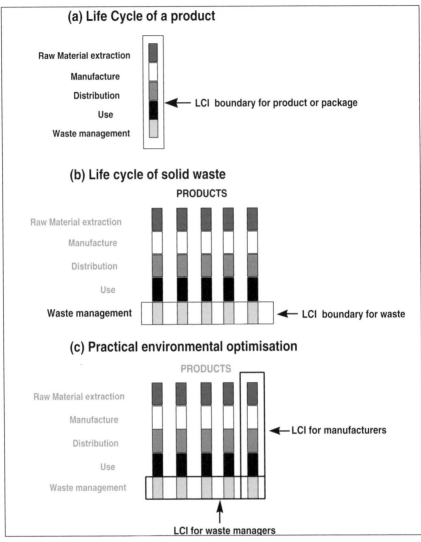

Figure 5.2 The Life Cycle of a product (a), the Life Cycle of waste (b), and a practical approach to environmental optimisation (c). The LCI for solid waste considers part of the Life Cycle of all products and packages. Designers and manufacturers optimise performance of products and packages (vertical analysis), whilst waste managers, municipalities and policy makers optimise Integrated Waste Management systems (horizontal analysis).

attempts to assess the environmental burdens of the waste, once produced. Since it takes the solid waste as a given (the zero burden approach; see definition below), this method cannot be used to assess how waste prevention can best be achieved, since this occurs prior to the creation of waste.

> **Definition**: The *zero burden approach* takes the waste to be managed by the system as a given. No credit or burden is allocated to the calorific value or chemical composition of the waste entering the system. None of the burdens associated with the previous use of the products or materials, prior to them becoming waste, are included either.
>
> A strict thermodynamic mass and energy balance approach would require all inputs and outputs to the system to be equal. However, this is not useful for waste management purposes. What are required are the net energy and emissions burdens associated with managing the waste in question.
>
> For example, the calorific content of plastic waste is considered as neither a credit nor a debit to the system, although incineration of plastic will result in the production of energy, which will be considered as a net output. Similarly the carbon content of waste paper is taken as a given, but burdens associated with incineration, landfilling or recycling paper plus possible avoided burdens of displacing virgin paper are included in the output.

LCI can look at the consequences on a waste management system of changes in waste composition, which may arise through waste prevention measures, but cannot identify how and where waste prevention should occur. Since each product system will be different, the opportunities for waste prevention must be identified on a product by product basis, through the use of product LCIs (the vertical approach). Therefore, comparisons of product systems, such as reusable versus one-way packaging systems, need to be done using a product LCI, on a product by product basis. In contrast, comparisons between treating a given waste by recycling, composting, incineration or landfilling, and how to achieve the optimal combination of such options in an IWM system, can be achieved using a solid waste LCI (horizontal approach).

The cradle

Household and similar commercial waste can be either collected or delivered (e.g. to bottle banks) in a variety of ways, so for a practical boundary, the 'cradle' of such waste in this study is taken as the point at which it leaves the household or commercial property.

Prior to this stage, individual items in the waste stream will be affected by source reduction, waste minimisation and other processes for environmental improvement. Whilst these are valuable contributions, they occur during product manufacture, distribution or use and are thus upstream of waste management. As a result they will not be included here. Other waste treatment methods can also be used to reduce the amount of solid waste within households, prior to collection. Home composting of organic material is a good example of this. Such treatment methods will similarly not be considered within the boundaries of this LCI, since they occur on-site, and prior to the materials leaving the household.

The grave

The 'grave' of the waste Life Cycle is its final disposal back into the environment. Incineration and landfilling are often described as 'final disposal' options, but neither represents the true end of the Life Cycle for the materials involved. Incineration produces ash, for example, which then needs disposal, often by landfilling. Similarly, landfills are not the final resting place for some of the materials contained, since they can in turn release gas emissions and leachate. It has been suggested that landfills should not be considered as environmental burdens that should be measured, but as waste management processes, which in turn produce gas and water emissions with measurable burdens (Finnveden, 1992, 1995).

The 'grave' or end of the waste Life Cycle used here, is considered to be when the waste becomes inert landfill material, or is converted into air or water emissions. Alternatively, the waste can regain some value (as compost, secondary material or fuel) and thus ceases to be waste. 'Value' normally implies positive economic value.

Defining the exact point at which waste acquires value and ceases to be waste has important practical consequences, as well as being essential in this analysis. Different regulations generally apply to handling and transporting raw materials rather than waste, and different emission levels are allowed when burning fuels rather than burning waste. Thus a clear definition of exactly when waste becomes a secondary raw material or a fuel is essential.

One area where difficulties arise is the recovery of materials for recycling. Once materials are collected and sorted, they regain value as a secondary raw material. This value equates to positive economic value in many cases, when sorted materials are sold from materials collection banks or from Materials Recovery Facilities (MRFs). Consequently, this point has been chosen as the boundary of the waste management system for such materials. In some schemes, however, where collected material supply greatly exceeds recycling capacity, the collected material may still have a negative economic value at this point. Such material would not then regain value until after it had been converted into recycled resin pellets. For example, choice of this point as the boundary would mean including the whole recycling industry within the waste management system. Since many recycling processes, e.g. for steel, aluminium, paper, etc., are integrated within the virgin material industries, this would necessitate taking most of industry into the waste management system, which would make system modelling unwieldy, if not impossible.

Since all industries are interrelated, defining hard and fast boundaries is not easy, but some boundaries must be chosen. For this study, recovered materials are considered to leave the waste management system when they leave the MRF or are collected from materials banks. They then enter the plastics/paper/metal/glass industry system. These reclaimed materials will eventually replace other virgin materials in the manufacture of new products. In many cases this will lead to environmental improvements via energy savings, and reduced raw material consumption. These potential savings (or costs) are not included within the boundaries of this LCI study, but are addressed and calculated in Chapter 14, Materials Recycling (and in the model) so that their significance can be assessed.

The general system boundaries for an Integrated Waste Management system are shown schematically in Figure 5.3. Along with the waste itself, there are energy and other raw material inputs (e.g. petrol, diesel) to the system. The outputs from the system are useful products in

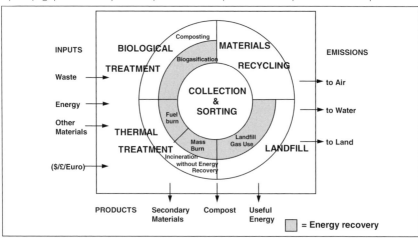

Figure 5.3 System boundaries for the Life Cycle Inventory of solid waste.

the form of reclaimed materials and compost, emissions to air and to water and inert landfill material. Energy will also be produced in Energy from Waste options (which also includes use of recovered landfill gas); combining this with the energy inputs to the system gives a net value for energy consumption/generation. The exact boundaries at which materials and energy enter or leave the waste management system studied here are defined in Table 5.2.

For the energy inputs into the systems as fuels (e.g. petrol, diesel, gas or as electricity), in addition to the actual energy content delivered, energy will have been expended in drilling, mining, and during production. These processes would also have generated emissions to air and water, and solid waste. Consequently these environmental burdens associated with energy production will also be included whenever energy is consumed within this Life Cycle Inventory. Effectively, the energy production industries are included within the system boundaries of this LCI study (although they are not depicted in Figures 5.5–5.7).

It is assumed that any energy recovered by the waste management system is converted into electrical power. Subtracting the amount of electrical energy produced from that consumed gives the net amount of electrical energy consumed by the waste management system.

	System boundary	Units
Inputs: Waste	Point where the waste leaves the household or commercial unit	tonnes
Inputs: Energy	Extraction of fuel resources	GJ thermal energy
Outputs: Energy	Electric power cable leaving Energy from Waste facility. (The electrical energy generated is subtracted from the energy consumed, i.e. is effectively used within the system, and not exported)	GJ thermal energy
Outputs: Recovered Materials	Material collection bank or Exit of Material Recovery Facility or Exit of Refuse-Derived Fuel plant or Exit of biological treatment plant	tonnes
Outputs: Compost	Exit of biological treatment plant	tonnes
Outputs: Air Emissions	Exhaust of transport vehicles or stack of thermal treatment plant, i.e. after emission controls or stack of power station (for electricity generation) or landfill lining/cap	kilograms
Outputs: Water Emissions	Outlet of biological treatment plant or outlet of thermal treatment plant or outlet of power station (electricity) or outlet of landfill leachate treatment plant	kilograms
Outputs: Final Solid Waste	Content of landfill at end of biologically active period	tonnes or cubic meters of material

Table 5.2 Boundary definitions for the LCI of waste

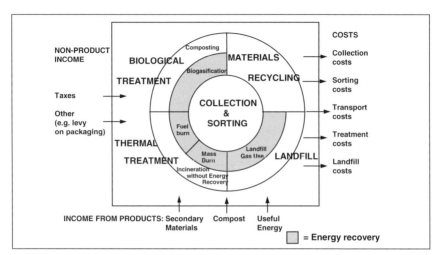

Figure 5.4 System boundaries for an economic assessment of solid waste.

The system boundaries for the parallel economic assessment are shown in Figure 5.4. Economic inputs to a waste management system include costs for collection, sorting, various forms of treatment, transport and for final disposal to landfill. Revenues produced by the system come from sale of reclaimed materials, compost and energy. Subtracting the revenues from the costs will give the net cost of operating the system.

As discussed above, reclaimed materials will displace virgin materials, so if the reclaimed materials are produced for less than virgin materials, further economic savings occur. As with the corresponding environmental savings, these cost savings are not included within the boundaries of the basic LCI system defined in Figure 5.4, but they are calculated in Chapter 14 so that their significance can be assessed.

What level of detail?

What level of detail should be included in the study? Should the burdens of manufacturing the trucks that carry the waste be included in the overall assessment? What about the burdens of constructing incinerators?

In most Life Cycle Inventories to date, such 'second level' burdens are considered to be insignificant, when spread over the Life Cycle of the equipment or facility, and so are normally omitted. This practice will be followed in this study for the environmental burdens of waste management. Second level burdens have been estimated to be up to 10% of the total environmental burden of waste management (Chem Systems, 1997; Schwing, 1999), although, in future studies these second level burdens are likely to become more significant as emission standards become more stringent. As standards are raised, facilities require more or new infrastructure to house more sophisticated equipment. Second level burdens are generally of increasing interest with respect to LCA/LCI studies and the amount of research in this area is increasing (see, for example, Frischknecht et al., 1995, 1996).

The approach described above cannot be justified, however, when considering the economic costs of waste management. The capital cost of installing plant and vehicles is clearly significant in comparison to running costs, since it must be financed and overheads must be paid. When considering the economic costs, therefore, the full inclusive cost of waste management will be addressed, including purchase of capital equipment, depreciated over a suitable time span. These full and inclusive costs must be entered into the model under the heading of 'Processing costs/tonne'.

The Inventory stage

The Inventory stage looks at all of the inputs and outputs in the Life Cycle of waste. The first step, however, is to define the Life Cycle. Since the objective of the LCI is to be able to describe the majority of waste management systems, existing or planned, all possible processes and technically feasible combinations of processes need to be possible. The main stages, and their interconnections in the Life Cycle of solid waste are shown schematically in Figure 5.5,

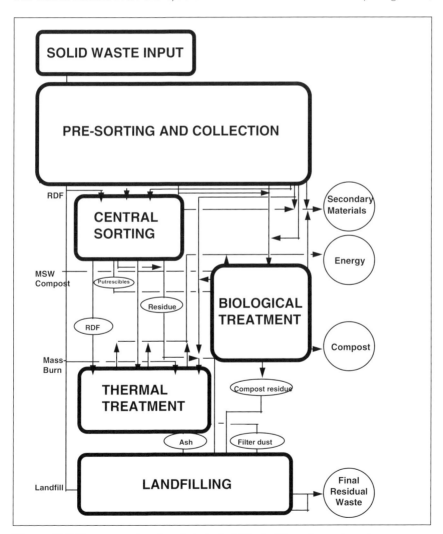

Figure 5.5 Components of an Integrated Waste Management system.

Concepts and Case Studies

comprising pre-sorting and collection, sorting, biological treatment, thermal treatment and landfilling. It is then necessary to consider the processes within each stage (Figure 5.6), and list all materials and energy entering and leaving each process. By linking up all processes within each stage and then all stages in the Life Cycle, it is possible to define the overall waste management system. This is shown in Figure 5.7 and Table 5.3, which also defines the boundaries

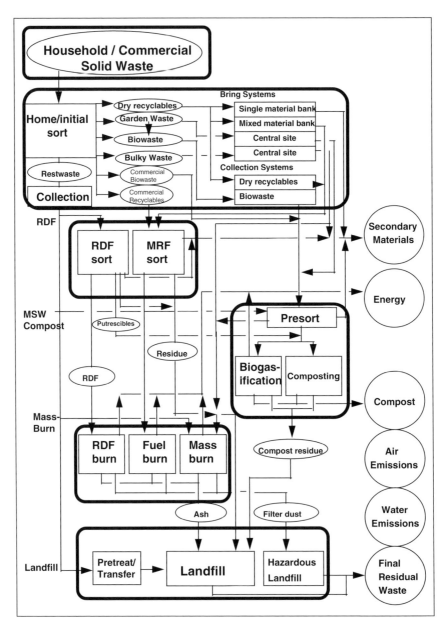

Figure 5.6 Detailed structure of an Integrated Waste Management system.

of the system, and is the basis on which the Life Cycle Inventory is performed. Materials enter the system primarily as waste, and leave the system when they are converted into recovered materials or compost, emitted to air or water, or are deposited as solid waste in landfills.

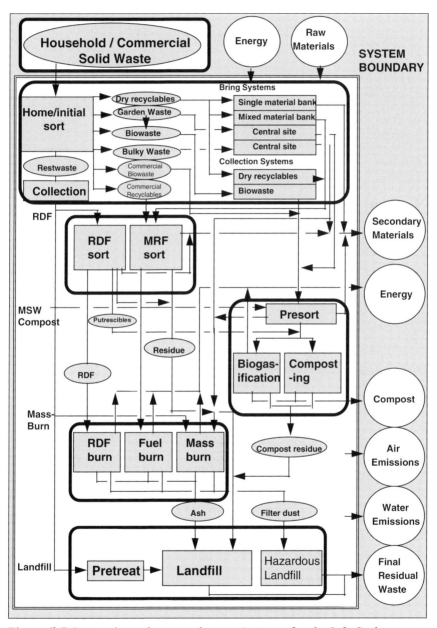

Figure 5.7 System boundaries and inputs/outputs for the Life Cycle Inventory of solid waste.

1. System studied.	See Figure 5.7
2. Data inputs.	
(a) Types of data:	Variable data (inserted by user), which define the waste management system considered. Fixed data (embedded in model), which defines inputs and outputs for unit waste management processes.
(b) Sources of data:	Waste composition and amounts: international agencies, technical literature. Collection and sorting: ERRA and other local schemes. Treatment processes: technical literature, company reports.
(c) Data quality:	Data available for treatment of mixed waste stream. Little/no data on treatment of individual fractions of waste. Overall poor quality, especially for waste quantity and composition.
3. Data outputs	
(a) System inputs:	Net energy consumption.
(b) System outputs:	Amounts of recovered materials.
– Products:	Amount of compost produced.
– Emissions:	Emissions to air. Emissions to water. Emissions to land.

Table 5.3 Life Cycle Inventory of waste: Inventory stage

A Life Cycle Inventory for solid waste management consists of two main steps. Firstly, the waste management system to be considered must be described. This involves choosing between the different possible waste treatment options. A large amount of variable data are needed to show how the waste is treated in any given system, and what route materials take through the system, e.g. how much of the household waste is separated at the kerbside, how many fractions it is sorted into, whether organic material is composted separately or left with the residue, what proportion is incinerated as opposed to directly landfilled and so on. Answers to these questions must be chosen by the user, and can be altered to carry out 'What if ... ?' calculations.

Secondly, the inputs and outputs of the chosen processes must be calculated, using fixed data for each process. Such fixed data are dependent on the performance of the equipment and technologies involved and are expressed relative to the amount of material treated. For example, the energy generation and emissions per tonne of waste incinerated in an Energy from Waste facility.

Fixed data are generally lacking for most routine manufacturing processes, let alone for waste management. The lack of quality data is a serious problem in any LCI of waste, and was probably one of the main reasons why this approach has take such a long time to be both accepted and now developed. Suitable data sources are now becoming available, have been identified, and will be used in this book.

There is one area in which current fixed data sources are generally still scarce, however: the allocation of burdens to the different fractions of the waste stream. Data are available for most waste management treatment processes, such as composting or incineration, in terms of inputs and outputs per tonne of waste treated, but solid waste varies not only in quantity, but also in composition.

In an Integrated Waste Management system, different materials within the waste stream are separated and treated in different ways. Thus the residual waste that enters a mixed waste treatment method, e.g. incinerator or landfill, will vary in composition depending on what has been removed for other treatment methods. Without data allocating the inputs and outputs to the particular fractions of waste, it is not possible to reliably predict the inputs and outputs of any mixed waste treatment methods. Where process inputs and outputs can be allocated to individual materials in this LCI this will be applied, otherwise average data for mixed waste streams will be used.

The important distinction between generic and specific data should also be borne in mind. Generic data are generally averages, and although useful for giving an approximate measure of the environmental burdens associated with a system, can never be more specific than that. To assess the actual burdens of a given system requires measurement of all the processes in that actual system. Clearly this is costly both in terms of money and time; Life Cycle Inventories collecting and using specific data have taken several years to complete. Most of the data used in this LCA model for waste will be generic data, as this is often the best available. In the model, the user can overwrite any of the generic data provided with updated or system-specific data as it becomes available.

Results of the Life Cycle Inventory model: system inputs and outputs

Net energy consumption

All energy inputs to the system and the energy produced during certain treatment processes must be considered. The inherent energy of the waste is not included since it is common to all possible options for dealing with the same amount of waste. The energy delivered by fuels and electricity is included, plus the indirect energy consumption during fuel and electricity production (also known as Primary Energy).

Air and water emissions

Since there are a large number of different emissions from many single processes, let alone complete waste management systems, it is necessary to define which emissions will be considered. Not surprisingly, previous studies have varied in the parameters that they have considered. The list of emissions to air and water considered most important by the authors, and thus included in this LCI study, are given in Table 5.4. This list of emissions is used for each of the processes included within the waste system boundaries, but since a range of data sources is used, not all data sets contain all the required information, nor use the same parameters. Therefore it has sometimes been necessary to convert data from other sources into the parameters required.

Emissions to air:	Emissions to water:
Particulates	BOD/COD
CO	Suspended solids
CO_2	Total organic compounds
CH_4	AOX (Adsorbable Organic Halides)
N_2O	Chlorinated HCs
NOx	Dioxins/furans (TEQ)
SOx	Phenols
HCl	Aluminium
HF	Ammonium
H_2S	Arsenic
Chlorinated hydrocarbons	Barium
Dioxins/furans	Cadmium
Ammonia	Chloride
Arsenic	Chromium
Cadmium	Copper
Chromium	Cyanide
Copper	Fluoride
Lead	Iron
Manganese	Lead
Mercury	Mercury
Nickel	Nickel
Zinc	Nitrate
	Phosphate
	Sulphate
	Sulphide
	Zinc

Table 5.4 Emission categories used in the LCI of solid waste

Landfill volume

Since landfills fill up rather than get too heavy, volume rather than weight is the key measure. This volume needs to reflect the level of compaction that occurs naturally in the landfill.

Recovered materials and compost

Recovered materials and compost are products, and therefore outputs of the system. It is important to predict the types and amounts of materials that are likely to be produced by any system, as both reprocessing capacity and markets will be needed.

Other statistics

Although descriptors of the system rather than inputs or outputs, it is useful to use the data from the model to generate certain statistics since Government targets are usually set in these terms:

- **Material recovery rate.** The percentage of the waste stream recovered as usable secondary materials.

$$\text{Material recovery rate \%} = 100\% \times \frac{\text{Amount of recovered recyclable materials leaving the system}}{\text{Total amount of waste entering waste management system}}$$

- **Overall material recovery rate.** This would include both 'dry recyclables' and compost.

$$\text{Overall material recovery rate \%} = 100\% \times \frac{\text{Total amount of recovered recyclable materials and compost produced}}{\text{Total amount of waste entering waste management system}}$$

- **Landfill diversion rate.** The percentage of the waste stream that is diverted away from final disposal in a landfill. This is not the same as the material recycling rate, as diverted material may be released as emissions, e.g. during compost production.

$$\text{Landfill diversion rate} = 100\% \times \frac{(1 - \text{amount of waste entering landfill})}{\text{Total amount of waste entering system}}$$

Fuel and electricity consumption in the Life Cycle of solid waste

Wherever fuels or electricity are used within the waste management system, there will be environmental burdens not only due to their actual use, but also due to the mining, drilling transport and production of the fuels and electricity. By using generic data throughout the Life Cycle Inventory, every time that fuel or electricity are consumed, the relevant consumption of thermal energy (including 'pre-combustion' energy consumption), emissions to air and water and production of solid waste are added to the overall Inventory totals.

Electricity consumption

The overall thermal energy consumption, emissions and solid waste generated by electricity production will vary according to the method used, and its efficiency. Ideally, the energy mix for the electricity grid of the country under study should be used to give the most accurate results (see Table 5.5). Although the model provides data on country-specific mixtures of electricity generation grids (Table 5.5) and their efficiencies, the initial default values (which can be changed by the user) are based on the UCPTE (Union for the Connection, Production and Transport of Electricity) 1994 model.

Clearly, the environmental burdens of generating 1 kWh of electricity from fossil fuels will differ markedly from hydro-electric generation of the same amount. The burdens associated with consuming 1 kWh of electricity from different fuel sources are presented in Table 5.6.

Producer	Hard coal	Brown coal	Oil	Natural gas	Nuclear	Hydro	Efficiency
Austria	7.1	8.8	2.8	9.9	6.4	65	48.1
Belgium	22.8	0	2.2	14.2	58.4	2.4	28.7
Brazil	3	0	0	2	0	95	25
Canada	25.9	0	15.4	2.2	8.6	47.9	51.2
Denmark	73.8	0.1	3.7	4.5	5.2	12.7	31.5
Finland	21.1	0	14.4	9.1	31.2	24.2	33.2
France	7.3	0.4	2.1	1.5	72.9	15.8	30.4
Germany (W)	28.9	18.8	2.2	4.5	39.3	6.3	28.6
Iceland	0	0	0	5.5	0	94.5	76.4
Italy	9.6	0.5	43.4	14.5	8.5	23.5	33.6
Japan	10	5	30	16	30	9	25
Luxembourg	25.7	14.7	1.7	10.3	37.9	9.7	29.7
Mexico	10	0	50	10	5	25	25
Netherlands	30.7	3.1	3.6	48.9	12.8	0.9	30.9
Norway	0.1	0	0.2	0.2	0.3	99.2	75.6
Portugal	36.3	0.7	43.4	0.4	2.6	16.6	31
Slovenia	16.3	15.7	3	3.2	28.6	33.2	34.5
Spain	30.5	9.9	9.2	1.7	35.4	13.3	30
Sweden	0.9	0	5.9	0.1	39.5	53.6	41.6
Switzerland	4.7	1.4	2.5	1.2	50	40.2	36.8
UK	59	0	8.2	3.5	26.4	2.9	28.7
USA	56.7	0	2.9	9.8	22	8.6	32.9
Venezuela	7	0	7	11	0	75	25
UCPTE (1994) (W. Europe excluding UK)	17.4	7.8	10.7	7.4	40.3	16.4	31

Table 5.5 Life Cycle data for national electricity generating grids. Source: BUWAL 250 (1998) with additions. Note: all figures are percentages and UCPTE (1994) is the default setting in the computer model

The amount of electrical power generated by Energy from Waste methods within the waste management system is subtracted from the electrical energy used, to give the overall net electrical energy consumption. This amount is then used to calculate the thermal energy consumption, emissions and solid waste that are due to electricity generation, using a displaced energy grid, the default setting of which also uses the UCPTE (1994) data, unless this has been changed by the user. If the amount of electrical energy recovered from waste should exceed the amount consumed by the system, there will be a net export of electrical energy from the waste management system, and a net saving of the emissions and solid waste that would have been associated with the production of this amount of electricity using the conventional generation methods.

	Units	Hard coal	Brown coal	Oil	Natural gas	Nuclear	Hydro
Efficiency (generation and supply)	%	28.5	24.8	27.1	34.2	27.2	76.5
Emissions to air							
Particulates	g	1.73	2.02	0.376	0.0652	0.0179	0
CO	g	0.125	0.133	0.223	0.267	0.00565	0
CO_2	g	979	1350	880	767	5.71	0
Methane	g	4.26	0.262	1.1	1.76	0.0138	0
NOx (as NO_2)	g	2.52	1.97	1.96	1.49	0.0199	0
N_2O	g	0.00606	0.00686	0.0194	0.00558	0.000152	0
SOx (as SO_2)	g	4.02	6.86	9.30	0.265	0.0214	0
HCl	g	0.3	0.291	0.00969	0.000619	0.000361	0
HF	g	0.0319	0.0228	0.000973	0.0000528	0.000108	0
H_2S							
Chlorinated CH	g	0.0000000106	0.0000000104	0.00000000155	0.00000000155	0.000000761	0
Dioxins/furans							
Ammonia	g	0.00584	0.000505	0.000756	0.000194	0.000146	0
Cadmium	g	0.00000466	0.0000216	0.0000596	0.000000229	0.000000123	0
Chromium							
Copper							
Lead	g	0.000183	0.0000494	0.000518	0.00000361	0.000000795	0
Manganese	g	0.000111	0.0000383	0.000158	0.00000296	0.000000194	0
Mercury	g	0.0000385	0.0000503	0.00000279	0.0000146	0.000000146	0
Nickel	g	0.000464	0.0000663	0.00452	0.00000488	0.00000398	0
Zinc	g	0.000350	0.000335	0.000379	0.00000582	0.00000948	0
Emissions to water							
BOD	g	0.000139	0.00000659	0.000601	0.0000195	0.0000245	0
COD	g	0.00435	0.000161	0.0107	0.000249	0.000167	0
Suspended solids	g	0.0363	0.00678	0.786	0.220	0.0802	0
TOC	g	0.00494	0.00164	0.107	0.239	0.00312	0
AOX	g	0.00000152	0.000000302	0.0000476	0.000000118	0.0000000869	0
Chlorinated HC	g	0.000000492	0.0000000877	0.0000121	0.00000338	0.000000046	0
Dioxins/furans(TEQ)							
Phenols	g	0.0000662	0.0000119	0.00198	0.0000386	0.00000361	0
Aluminium	g	0.975	0.00715	0.00399	0.0890	0.0102	0
Ammonium	g	0.00142	0.000186	0.0138	0.000664	0.00426	0
Arsenic	g	0.00197	0.0000144	0.0000190	0.000178	0.00000807	0
Barium	g	0.0791	0.000774	0.0348	0.00719	0.000108	0
Cadmium	g	0.0000507	0.000000486	0.0000150	0.00000449	0.00000249	0
Chloride	g	6.30	0.125	7.38	0.631	0.136	0
Chromium	g	0.00977	0.0000716	0.000156	0.000911	0.0000348	0
Copper	g	0.00489	0.0000356	0.0000441	0.000445	0.00000100	0
Cyanide	g	0.00000528	0.000000598	0.0000528	0.0000685	0.000000161	0
Iron	g	0.303	2.37	0.00827	0.0282	0.00559	0
Lead	g	0.00490	0.0000434	0.0000413	0.000448	0.000502	0
Mercury	g	0.00000139	0.0000000175	0.000000165	0.000000782	0.0000000105	0
Nickel	g	0.00493	0.0000360	0.0000583	0.000448	0.0000167	0
Nitrate	g	0.0235	0.000222	0.00927	0.0000527	0.000531	0
Phosphate	g	0.0585	0.000422	0.000369	0.00533	0.0000633	0
Sulphate	g	4.35	5.02	0.274	0.462	2.07	0
Sulphide	g	0.0000140	0.00000283	0.000421	0.0000648	0.00000263	0
Zinc	g	0.00982	0.0000723	0.000159	0.000894	0.0000572	0
Solid waste	g	219.9	108.7	16.5	27.1	4	0

Table 5.6 Life Cycle Inventory data for electrical energy generation (per kWh). Source: BUWAL 250 (1998) Part 2, Table 16.7

Petrol and diesel consumption

The delivered energy contents of petrol and diesel are 34.35 and 38.14 MJ/litre, respectively (BUWAL, 1998). Additional energy has been expended, however, to drill, transport and process the crude oil to produce these fuels. Similarly, air and water emissions and solid waste will have been generated. For example, the energy efficiency of the diesel production industry in Europe is estimated as 76% (BUWAL, 1998). Thus to supply 1 MJ of energy as diesel requires a gross energy input of 1.32 MJ. Such 'pre-combustion' burdens of petrol and diesel production are included in the data presented in Table 5.7 (litres are used as the unit since this is how fuel consumption is totalled over the Life Cycle). The use of petrol and diesel will generate further emissions to air. The actual amounts generated, per litre of fuel used, will vary according the vehicle or machine involved. Since most of the petrol and diesel consumption in the waste management system will be by private cars and heavy goods vehicles, respectively, data for these vehicles are used as the basis for calculating the emissions per litre as fuel used. (Note: by calculating emissions in this way, the fuel consumption of different vehicle types will be accounted for, since this approach determines the total amount of fuel consumed.) Air emission data for fuel usage are included in Table 5.7 so the data represents the total burdens due to fuel production and use.

Natural gas consumption

Data for production and use of natural gas are also presented in Table 5.7. Although the energy content of natural gas will vary along with its composition, the data are presented per cubic metre.

The economic assessment

The parallel economic assessment of the overall costs of the waste management system also requires both variable and fixed data. The variable data will be the same as for the environmental LCI, but details for the cost of each process in the Life Cycle are required. These also need to be entered by the user, since generic data on costs are of little use. Costs for waste management processes are extremely variable, both between and even within countries, and reflect availability of local facilities, salaries and land prices (see Chapter 4). In the following chapters examples of average costs for a range of countries are given, but local cost estimates are needed for any reliability in the result. For consistency, all costs in this book will be quoted in euros and US dollars, using the exchange rates current at the time of writing (September 1999). These conversion rates are given at the end of the book in Appendix 1.

The methods used for costing waste management processes vary. Collection costs for domestic waste depend on the number of households visited, and so are often calculated per household per year (IGD, 1992). Collection costs for commercial and industrial waste, and processing costs for waste treatment and disposal relate to the amount of waste handled, so are calculated per tonne. As the aim of this model is to compare the costs of the complete waste management system, the cost will be calculated as the overall cost per year. For comparison with other literature, this can be broken down into cost per household per year and cost per tonne of throughput per year.

	Units	Natural gas (m^3)	Petrol (l)	Diesel (l)
Energy thermal	GJ	0.04020	0.03435	0.038136
Efficiency (production and supply)	%	80.2	63.6	75.9
Emissions to air				
Particulates	g	0.123	0.498	1.2432
CO	g	0.97	65.025	16.548
CO_2	g	2290	2985	3015.6
CH_4	g	6.46	3.9375	3.6708
NOx (as NO_2)	g	2.34	28.875	54.264
N_2O	g	0.0247	0.23175	0.072828
SOx (as SO_2)	g	1.29	4.8975	4.5444
HCl	g	0.0143	0.01335	0.0061656
HF	g	0.00151	0.001395	0.00064428
H_2S				
Chlorinated HC	g	5.04E-08	3.375E-08	1.7304E-08
Dioxins/furans				
Ammonia	g	0.000228	0.000156	0.000081984
Arsenic				
Cadmium	g	0.00000172	0.00012075	0.000029316
Chromium				
Copper				
Lead	g	0.0000166	0.003105	0.00016128
Manganese	g	0.00000718	4.8075E-06	2.4696E-06
Mercury	g	0.0000725	0.000008475	3.0156E-06
Nickel	g	0.0000994	0.0026025	0.0014532
Zinc	g	0.0000343	0.000825	0.000966
Emissions to water				
BOD	g	0.000111	0.003915	0.0041328
COD	g	0.0015	0.1275	0.13524
Suspended solids	g	1.13	2.565	2.6208
TOC	g	1.11	0.4275	0.42252
AOX	g	0.00000131	0.000171	0.0001806
Chlorinated HC	g	0.0000173	0.000039375	0.000040236
Dioxins/furans (TEQ)				
Phenols	g	0.000232	0.005895	0.0060816
Aluminium	g	0.0346	0.0234	0.012096
Ammonium	g	0.000811	0.09525	0.1008
Arsenic	g	0.0000696	0.000081	0.000060144
Barium	g	0.00363	0.114	0.11592
Cadmium	g	0.0000023	0.0000492	0.000050988
Chloride	g	0.0437	23.85	24.528
Chromium	g	0.000455	0.00060825	0.00050736
Copper	g	0.000171	0.00019425	0.00014196
Cyanide	g	0.00000274	0.00017175	0.00018144
Iron	g	0.056	0.0438	0.025788
Lead	g	0.000204	0.00018825	0.00012432
Mercury	g	0.00000367	0.00000049725	0.0000004536
Nickel	g	0.000174	0.00024	0.00018816
Nitrate	g	0.00105	0.028875	0.03024
Phosphate	g	0.00206	0.00183	0.0011928
Sulphate	g	0.0391	1.05	0.8652
Sulphide	g	0.0000125	0.001365	0.0014448
Zinc	g	0.000349	0.00062625	0.00053676
Solid waste	g	3	5.3	5.7

Table 5.7 Life Cycle Inventory data for fuel (including production and use). Source: BUWAL 250 (1998) Part 2, Table 16.9

The main differences between IWM-1 and IWM-2 Life Cycle Inventory models

The first version of the Life Cycle Inventory model for Integrated Waste Management – IWM-1 – was published in 1995. Since then the model has been used by waste managers and planners, consultants, academics and students. IWM-2 was developed to improve certain aspects of the original model and to include more recent global (rather than solely European) data. Many of the improvements to the new model were based on feedback from actual users. Such dialogue is very valuable during the planning stages of any development programme. Table 5.8 summarises the main differences between IWM-1 and IWM-2.

IWM-1	IWM-2
Excel spreadsheet.	Stand-alone programme, Windows style interface.
Although in spreadsheet format, difficult to track calculations.	All calculations are fully transparent, due to a 'drill down' capability.
Old data.	Updated data.
No help system.	Help system and Glossary included in the model.
Rigid waste collection section.	More flexible waste collection section (up to four kerbside collection systems and four material bank collection systems).
Fixed contamination rates.	User can edit default contamination rates.
Biological treatment – the user must select either composting or biogasification, not both.	Biological treatment, the user can select to use both composting and biogasification.
Thermal treatment section is a simple model based on equal allocation of emissions depending upon waste input composition.	Thermal treatment, carbon dioxide and flue gas emissions are based upon a stoichiometric approach. The non-metal emissions CO, SO_2, NOx, HCl, particulates and dioxins/furans are calculated using a stoichiometric approach. Metal emissions are based on the metals composition of individual waste components.
Results presented on single page, a large, somewhat overwhelming, amount of data.	Results presented in tab note book style, more manageable amounts of data.
Scenario comparisons not possible.	The output of up to eight scenarios can be compared and the results displayed graphically.

Table 5.8 Summary of the main differences between model IWM-1 and model IWM-2

The details of the developments in IWM-2 are fully described in Chapters 15–28.

Other LCI models for waste management

US Environmental Protection Agency Life Cycle model for waste management

The US Environmental Protection Agency is currently working to apply recent Life Cycle Assessment methods to develop tools for evaluating Integrated Waste Management. The research began in August 1994 and is expected to be completed in 2001. The outputs from this research will include (1) a database of Life Cycle Inventory data on the various solid waste components (i.e. glass, metals, plastic, and paper) and the different waste management activities (i.e. transportation, recycling/composting, landfilling, and combustion); (2) a decision support tool for applying Life Cycle Assessment tools on a site-specific basis to evaluate different Integrated Waste Management strategies; and (3) case studies of several state and local governments applying the decision support tool. The stakeholders group for the research includes state and local governments, trade associations, industry, academia, and environmental interest groups. This research will develop data and information for a variety of technologies and multimedia pollutant data.

The goal of the project is to develop information and tools that enable users to evaluate the cost and environmental trade-offs associated with different integrated solid waste management strategies. These are intended for use in determining the baseline conditions for a particular solid waste management system strategy or operation.

The decision support tool being developed through this research integrates default data from the database, system materials flow equations, LCI and cost methodologies, and an optimisation routine in a user-friendly interface. This tool is being designed to allow MSW planners to enter their site-specific data (or rely on the default data) to compare alternative MSW management strategies for their community's waste quantity and composition and other constraints. The final beta version of the tool was completed in December 1999. To test the LCI and cost methodologies and the overall decision support tool, a number of community case studies are being carried out. At the time of writing, initial case studies are underway with communities in Ohio and Iowa and these are designed to test the methodologies developed for individual operations. Future case studies will test the prototype decision support tool.

The ultimate products of this research will include a database and decision support tool that will enable users to perform an LCI and cost analyses based on locality-specific data on MSW generation and management. The decision support tool will be supported by the database, which will contain data on LCI parameters for individual solid waste management unit operations.

The UK Environment Agency model

The UK Environment Agency has been undertaking a long-term research programme the aim of which is not only to provide a basis for making more sound and more sustainable decisions in waste management, but also to put the ability to do this in the hands of waste managers. To do this they have set out to produce a software package that can be used to aid decision makers in both the waste management industry and the regulatory and local authorities. Although the ability to trace data is important in LCA, it is doubly so when it is used to aid decisions in waste management, where confidence needs to be built in both the method and the results. What the Environment Agency has tried to do is to develop a basis for making strategic decisions on the way in which particular wastes should be dealt with, in order to develop a truly sustainable strategy for the management of waste. To achieve this, the Environment Agency has developed software (WISARD), which can be used by waste managers, in the widest sense of the word, to examine different strategies for dealing with municipal waste, produce a realistic comparison of different options and assess the results in terms of their overall costs and benefits.

The Agency envisage the tool being used at three levels:

1. For strategic waste planning to develop overall strategies for waste at the regional and county level.
2. To improve waste management systems by waste collection authorities, for example, examining where the main burdens from particular systems arise and looking at what improvements they might achieve.
3. At the facility level by the manager of say, a landfill site, to examine where changes would impact most significantly on the environment.

Ultimately, the quality of the decisions made will depend on the relevance and quality of the data used. Obtaining the amount of detailed data required to use Life Cycle Assessment has been no easy task. However, before WISARD is used to make real decisions, the Agency must ensure the data are sufficiently sound and identify any further gaps that need to be filled. To this end, the Agency will work closely with a number of local authorities in the UK using WISARD to ensure the results help provide environmentally sound solutions to real problems.

CSR/EPIC model

Two Canadian industry associations, CSR: Corporations Supporting Recycling (CSR) and the Environment and Plastics Industry Council (EPIC), whose mandates are to assist municipalities to develop Sustainable Waste Management systems, commissioned the development of a model that could be used by municipal waste managers to quantify the Life Cycle environmental effects of waste management processes. The project is being carried out under the direction of a steering committee of industry, municipal government and federal government representatives. The City of London in Ontario was a co-participant in the development of the model. The City was used as the 'test case' for the model's initial development and as the first case study for the model. The objective of the project is to provide Canadian municipalities with a tool that will provide municipalities with information on the environmental performance of the various elements of their existing or proposed waste management systems. It is envisaged that this information will, in conjunction with economic, social and political considerations, provide input into decisions on the selection of system elements for Integrated Waste Management systems.

The relationship between a Life Cycle Inventory for waste and product or packaging Life Cycle Inventories

This chapter has emphasised the clear difference between the Life Cycle of a product (or package) and the Life Cycle of solid waste (see Figure 5.2). Similarly, LCI studies of solid waste systems and LCIs of products or packages fulfil different functions. An LCI of solid waste aims to optimise the waste treatment system for a given input of waste. This will be of use to waste managers, whether in national or local governments or in private waste management companies. It will not, however, predict whether one form of a product or package is better or worse for the environment than another. A product or package only spends a part of its Life Cycle in the waste management system, and its compatibility with waste management processes may be offset by environmental burdens in raw material sourcing, manufacture, distribution or use. Any comparison should be on a cradle-to-grave basis for the product or package Life Cycle, i.e. a product-specific LCI. Similarly, this model is not designed to answer questions such as whether one-way packaging or returnable packaging is preferable from an environmental point of view. An example often used is the comparison between returnable bottles and single-use cartons for milk packaging. This comparison needs to look at the packaging LCIs for the two options, including the initial manufacture of the bottles and cartons, use, and subsequent refilling, recycling or disposal processes as appropriate.

Product LCIs and Waste LCIs are complementary – the waste LCI can be used to optimise the overall waste management system for all waste materials; product-specific LCIs can optimise the individual items that end up within the waste stream. Although there is an area of overlap in the processes, the two different objectives are distinct.

Concepts and Case Studies

CHAPTER 6

LCI Case Studies

Summary

The following chapter demonstrates some of the practical applications of LCI for waste management systems. Each case study shows how the results of the LCI can provide data that can be used as part of the decision-making process, either for future improvements to the system or for validation of previous changes to the system. The range of different applications of the LCI model shows the suitability of the Life Cycle approach to the analysis of waste management systems and the geographical spread of the case studies indicates that the approach is now being accepted on a more global basis.

Introduction

The following case studies provide details of some of the possible applications of LCI to the field of waste management. A summary of the case studies is presented in Table 6.1. In each case, the results of the LCI modelling procedure did not provide a definitive answer. System A is rarely better than system B for all parameters, but the procedure provides data upon which decisions can be made. In certain cases (for example, the retrospective LCI from Canada) the procedure can provide data that supports decisions made in the past.

Caracas, Venezuela – LCI scenarios for the recovery of recyclable material

A preliminary exercise to investigate the benefits of adopting an integrated approach to waste management in Caracas was carried out by the Asociación para la Defensa del Ambiente y la Naturaleza (ADAN), a Venezuelan non-governmental organisation charged with promoting Integrated Waste Management and the Instituto de Estudios Superiores de Administración (IESA), a Venezuelan graduate business school. It was decided that LCI was an appropriate tool for evaluating the environmental and economic advantages and disadvantages associated with waste management.

LCI tool
The LCI tool used to model the current waste management system in Caracas and the addition of a possible materials recycling scenario was IWM-1 developed by White et al. (1995).

Case study	Description
Caracas, Venezuela	A comparative LCI, which investigated the burdens associated with the establishment of a materials recycling programme.
Pamplona, Spain	A comparative LCI, which investigated the burdens associated with the composting of the organic fraction of MSW with the composting of the organic fraction plus the paper fraction.
Gloucestershire, UK	An example of 'What if . . .?' modelling, where the environmental burdens of a number of very different waste management options were compared to help to identify the most suitable system for a particular area.
Barcelona, Spain	The use of LCI in the development of a long-term waste management strategy.
London, Ontario, Canada	The details of a retrospective LCI are presented and an LCI used to compare the environmental burdens of different tenders for a collection contract is described. The retrospective LCI used data sets gathered before and after changes to a waste management system were made, and the comparison established whether the predicted environmental benefits due to the system change actually occurred. The use of LCI to compare tenders for waste management contracts allowed managers to award contracts based on both economic (lowest cost) and environmental (lowest burdens) data.
US EPA case study summary	A number of different applications of the US EPA's Life Cycle model are described, including: the effects of changing recycling rates, collection systems, garden waste collection and landfill fees.
UK EA case study summary	A review of 11 local authorities, which used the UK Environment Agency model, WISARD, to compare a number of proposed waste management scenarios is presented.

Table 6.1 LCI case study summary

Baseline scenario

The baseline scenario was as follows:

- four people per home (the average for Caracas)
- collection three times a week from every home
- 100 km total distance for waste transfer to landfill (with liner, no gas collection)
- transport and landfilling costs of $5 per tonne
- no recovery of recyclable material
- no recovery or treatment of landfill leachate or gas.

Recycling scenario

The scenario compared against the baseline scenario was the same as above, but included the recovery and sale of recyclable materials by scavengers on the streets and at the landfills:

- glass: 35% of all glass in the waste stream at 0.07 $/kg
- aluminium: 80% of all aluminium in the waste stream at 0.70 $/kg
- paper: 13% of all paper in the waste stream at 0.07 $/kg
- plastic film: 5% of all the plastic film in the waste stream at 0.07 $/kg.

Comparison

The IWM-1 model was used to assess the environmental and economic burdens of including materials recycling as part of a new waste management strategy for the city. Using the energetic values for Europe that were contained in the software, but using the composition of domestic solid waste as was measured for Caracas, two scenarios were developed. The results are summarised in Table 6.2.

Parameter	Baseline scenario	Recycling scenario
Total cost ($)	55,000,000	27,000,000
Cost per tonne ($)	58.4	28.47
Cost per home ($)	72.42	35.30
Savings due to recycling ($)	0	13988.54
Total energy use (GJ)	288,464	–2,195,611 (savings)
Energy savings (GJ)	0	2,714,936
Total waste to landfill (tonnes)	959,000	885,000
Total CH_4 (kg)	3.84×10^7	3.66×10^7
Total CO_2 (kg)	2.29×10^8	2.33×10^8
Total CO (kg)	1.66×10^5	-2.78×10^4 (savings)
Total BOD (kg)	3.10×10^5	-2.13×10^5 (savings)
Total COD (kg)	3.10×10^5	-6.08×10^5 (savings)

Table 6.2 Comparison between the baseline and recycling scenarios for the city of Caracas

Conclusions

The results presented in Table 6.2 show that there would be significant financial savings and considerable environmental advantages associated with the formal establishment of a multi-materials recycling programme in Caracus. The environmental benefits include an 8% reduction in the amount of material requiring final disposal to landfill, reductions in the emission to air of methane and a saving in the emission to air of carbon monoxide and a saving in the emissions to water of both Biological Oxygen Demand (BOD) and Chemical Oxygen Demand (COD).

The results of this preliminary study allowed ADAN and IESA to conclude that 'Life Cycle Inventory represents a valuable tool for environmental management. The tool allows optimisation in the use of natural resources and energy, and the minimisation of waste generation, discharges and emissions from processes, products or services. It does clarify the factors to be weighted in the evaluation of the environmental performance of a product, process or service. LCI should be used jointly with other environmental management tools, such as risk analysis, cost-benefit analysis, and toxicity tests. This will provide the means to make decisions relating to waste management that reduce the environmental burden most relevant to that region, while including both economic and sustainability considerations' (Cardinale, 1998).

From this statement it is clear that ADAN and IESA, while recognising the potential benefits of the application of LCI to waste management, also recognise the limitations of the tool and the possibility that misuse could lead to an erosion of credibility in the tool itself.

Acknowledgement

This case study is based on original research by Pablo Cardinale of Instituto de Estudios Superiores de Administración (IESA), Venezuela.

Pamplona, Spain – LCI scenarios for separate collection of organic material

The Mancomunidad de la Comarca de Pamplona (Association of Municipalities of the Pamplona District) is committed to the efficient management of Municipal Solid Waste, to involve the local population, to promote state-of-the-art management techniques and to protect the environment. A multi-materials recycling scheme has been set up in Pamplona in order to address these issues. With the possibility of organic material being banned from landfill by the impending EU Landfill Directive and the fact that the people of Pamplona want to divert more material from final disposal to landfill, the Mancomunidad evaluated the possible advantages of selective collection and composting of the organic fraction of the waste stream.

LCI tool

The LCI tool used to model the different scenarios possible when adding the selective collection of organic material for composting to the waste management infrastructure that existed in Pamplona was IWM-1 by White et al. (1995).

Baseline scenario

The baseline scenario was the existing waste management strategy operating in Pamplona and this was used as the reference against which all new scenarios were measured. In Pamplona a close to home bring system operates for recyclables (paper, plastic, metal and glass) and for restwaste. The recyclable material is sent to a Materials Recovery Facility (MRF). The residue from the MRF and the restwaste are sent to landfill with gas collection and energy recovery.

Pamplona scenarios

The scenarios modelled for Pamplona are presented in Figure 6.1 and were selected to allow the evaluation of the advantages and disadvantages of adding the following treatment options to the existing waste management infrastructure:

- composting of the organic fraction
- composting plus an extra kerbside collection round
- composting of the organic and paper fractions
- composting of the organic and paper fractions plus an extra kerbside collection round.

In addition to the scenarios described above a worst case scenario was also considered. All waste arisings being sent to landfill without any treatment is the worst case waste management scenario for most developed countries. This scenario is often modelled in order to provide perspective, with respect to the environmental burdens associated with waste management and also to ensure that none of the subsequently run scenarios have significantly worse burdens. Modelling this scenario is also often a useful first step in the data-gathering process.

Results

Increased recycling results in more of the municipal waste stream being diverted from landfill than the existing waste management system in Pamplona. Even more material is diverted by a combination of recycling and composting. Obviously, the more material that is recycled and composted, the less waste remains for final disposal to landfill. This is demonstrated in Figure 6.1, with the worst case scenario representing all waste arisings in Pamplona being disposed of to landfill.

A comparison of the net energy use of each of the six scenarios revealed that if the benefits of recycling are not included in the model, then the collection of landfill gas from the anaerobic breakdown of organic material, and the subsequent generation of electricity plays a major role in the energy balance of the whole waste management system in Pamplona.

The worst case scenario where all MSW (including paper and organic material) is disposed of to landfill produces 62,000 GJ of electricity from landfill gas, while the baseline scenario only produces 35,000 GJ of electricity as the paper fraction is recycled leaving less biodegradable material in the landfill.

The scenarios that model composting of organic material and composting of organic material with an extra kerbside collection use 9000 and 21,000 GJ of energy, respectively. This is because of (1) the energy required to operate the composting process; (2) the organic material not being disposed of to landfill, therefore in the LCI model producing no biogas; and (3) the extra collection requiring more fuel for the collection vehicles.

The scenario where the organic material is composted with paper shows an increase in the amount of energy required by the overall system to 12,000 GJ, because now more material is being composted (more energy is required to run the process) and there is no biodegradable material being sent to landfill, so in the LCI model no biogas, therefore no electricity, is generated.

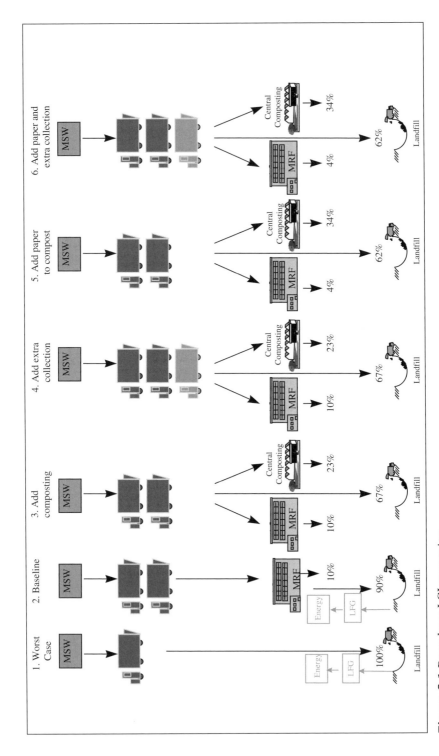

Figure 6.1 Pamplona LCI scenarios.
LFG = landfill gas

The scenario modelling the composting of organic material and paper with an extra collection logically uses the most energy, approximately 26,000 GJ.

When the benefits associated with recycling, principally the replacement of virgin materials by recycled materials, are included in the model the net energy use of each of the scenarios changes radically. The worst case scenario remains the same as there is no recycling, but the existing waste management scenario now produces almost 130,000 GJ as the energy saved by replacing virgin material with recycled material is significant. The net energy balance of the remaining scenarios all result in the production of energy, due to the energy savings associated with recycling. The energy produced by adding composting to the existing waste management system reduces the amount of energy produced (for the same reasons as in the paragraph above) to 82,000 GJ; the addition of an extra collection reduces this further to 70,000 GJ. The energy produced by adding composting of organic material and paper to the existing waste management system reduces the amount of energy produced (again for the same reasons as in the paragraph above) to 36,000 GJ; the addition of an extra collection reduces this further to 24,000 GJ.

The relative change in the Global Warming Potential (GWP) of the greenhouse gases (CO_2, CH_4 and NOx) produced by each of the waste management scenarios modelled was also calculated. Using the existing waste management scenario as zero, sending all MSW for disposal to landfill would result in an approximate 10% increase in GWP for the whole system. The scenarios that remove organic material from landfill (preventing the production of methane, a potent greenhouse gas) by composting result in a decrease in GWP of almost 38%, while the addition of paper to the composting process further decreases the GWP of the whole waste management system by approximately 45%.

Conclusions

The LCI tool was successfully used to model the addition of another treatment technology to the existing waste management system in Pamplona. It revealed that adding a composting facility would result in the net use of more energy by the whole waste management system than is used by the current waste management system, which is based on materials recycling and landfill. However, the LCI results also indicated that adding composting to the existing waste management system would result in a significant reduction in the GWP of the overall system. The LCI tool did not indicate which was the 'best' or 'correct' waste management scenario; that decision must be based on local conditions and local environmental and economic priorities.

Acknowledgement

This case study was based upon original research carried out by Elizabeth Wilson, during her time as a graduate student at Vrije Universiteit Brussel. She now works for the Atmospheric Protection Branch, MD-63, APPCD, US Environmental Protection Agency, Research Triangle Park, NC 27711, USA.

Gloucestershire county, UK – LCI scenarios for composting, recycling and incineration

In Gloucestershire in 1998 the majority of wastes were disposed of to landfill. There were 35 landfill sites throughout the county. Gloucestershire produced 1.5 million tonnes of waste material per year. Almost half of this was construction and demolition waste, while only 15% was household waste.

Development of waste management scenarios for Gloucestershire

The potential advantages and disadvantages of alternative waste management scenarios in Gloucestershire were compared by first developing an LCI model of the existing waste management operations in the county. A number of alternative waste management scenarios were then developed using the same waste management data relating to the types and quantities of wastes managed within the system area, but modifying the waste management processes required for each scenario. An outline of the baseline scenario and each of the alternative scenarios is presented below and in Figure 6.2.

Baseline and alternative scenarios

The baseline scenario was the existing waste management system in Gloucestershire:

- landfill: 92% of total waste inputs (household and commercial/industrial) sent directly to landfill and 81% of household waste stream sent directly to landfill
- recycling rates (as percentage of household waste stream) 5% through household waste recycling centres, 5% through collection banks and 9% through kerbside collection
- sorting in materials recovery facility of 4.5% of total waste input to system
- composting 18% of organic material in household waste stream.

Scenario 1: Composting of all organic waste, restwaste to landfill, utilisation of landfill gas.
Scenario 2: Recycling of all recoverable materials through kerbside sort systems, restwaste to landfill.
Scenario 3: Incineration (without energy recovery) of 50% of total solid waste, restwaste (and residues from the incinerator) to landfill.
Scenario 4: Centralised Material Recovery Facility (MRF), no separate collection of recyclables, Refuse-Derived Fuel (RDF) facility, restwaste and all residues to landfill.
Scenario 5:; Separate collection of recyclables, centralised MRF for unsorted waste, RDF facility, restwaste and all residues to landfill.

Results

The results of the LCIs for each of the five alternative waste management scenarios were compared with the results from the baseline scenario. The major differences between the scenarios were related to the production of CO_2, CH_4 and net energy (see Table 6.3).

Scenario 1 (composting) and 2 (recycling) were most similar to the baseline scenario as they did not include incineration. These two scenarios resulted in small reductions in CO_2 emissions but larger reductions in CH_4 emissions of 54% and 29%, respectively. In scenario 1 this was caused by the diversion of most of the organic material from the landfill to composting, where

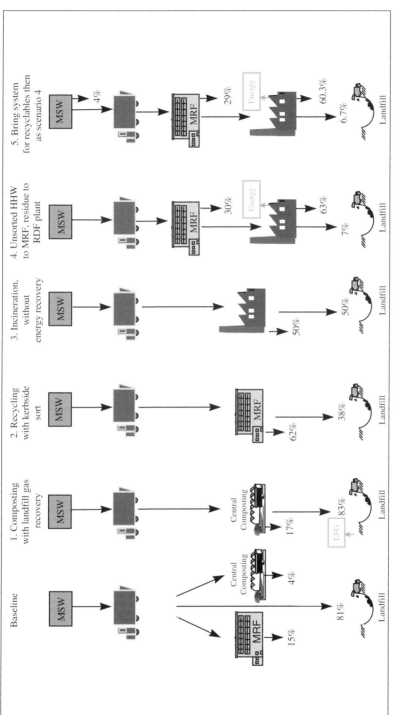

Figure 6.2 Gloucestershire LCI scenarios.

LFG = landfill gas.

	Carbon dioxide production ('000 tonnes)	Methane production ('000 tonnes)	Net energy production ('000 GJ)
Baseline	75	28.32	165
Scenario 1 (composting)	57	15.42	515
Scenario 2 (recycling)	64	19.28	1545
Scenario 3 (incineration)	396	14.36	−45
Scenario 4 (central sorting +RDF)	286	8.77	4320
Scenario 5 (separate collection + RDF)	250	9.04	3291

Table 6.3 Comparison of Gloucestershire's existing 'baseline' waste management system with the five alternative waste management scenarios

breakdown is aerobic (producing CO_2) rather than anaerobic (producing CH_4) and also the utilisation of landfill gas from that fraction of the organic material that was not collected and was disposed of to landfill. The smaller decrease in CH_4 emissions in scenario 2 was due to the recycling of paper, which was therefore not landfilled and did not get broken down anaerobically.

The differences between the energy consumption of the different waste management scenarios varied considerably. The baseline scenario provided a net saving of 165,000 GJ due to the energy credited to this scenario for materials collected and sent for recycling, i.e. the difference between the amount of energy used to produce virgin material and the amount of energy required to produce recycled material. The utilisation of landfill gas resulted in a threefold energy saving for scenario 1, whereas the large increase in recycling in scenario 2 resulted in an energy saving nine times greater than that of the baseline scenario.

In scenarios 3 (incineration without energy recovery), 4 (central sorting + RDF) and 5 (separate collection + RDF) the addition of an incineration stage resulted in very different levels of air emissions and energy consumption. Scenario 3 modelled the incineration of 50% of the waste stream without energy recovery while scenarios 4 and 5 both modelled a 30% diversion of the waste stream to a Materials Recovery Facility (MRF), the residue from which was converted to RDF, which was incinerated with energy recovery. The most significant changes relating to these scenarios were the emissions of CO_2 and CH_4. All three scenarios resulted in a large increase in CO_2, three to four times greater than the baseline scenario. CO_2 emissions were highest for scenario 3, as a higher percentage of waste was incinerated. Only a small difference in CO_2 emissions between scenarios 4 and 5 was calculated due to differences in the collection and sorting processes. In scenario 5, separate collection of recyclables using collection containers achieved an increased materials recovery rate. This resulted in a diversion of some of the organic fraction (paper) from incineration and energy recovery, lowering CO_2 emissions. This diversion of materials resulted in a lower net energy savings for scenario 5 as less material was available for energy recovery. Scenarios 4 and 5 also resulted in lower CH_4 emissions than in the other scenarios, representing a 300% decrease in comparison with the baseline scenario. This was due to energy recovered from the organic fraction of waste and results in lower emissions of CH_4 from landfill.

In terms of net energy consumption, scenario 4 (central sorting plus RDF) resulted in the saving of 4.32 million GJ per year through both recycling savings and energy generation. These energy savings were more than 20 times greater than savings achieved by the baseline scenario.

Overall, scenarios 4 (central sorting + RDF) and 5 (separate collection plus RDF) resulted in the lowest quantity of waste going to landfill, achieved a 50–51% diversion rate compared with only 6% for the baseline scenario, and a 21–22% recovery rate for materials. This recovery rate was only matched by scenario 2, which included 100% recycling and kerbside sorting of MSW.

Conclusions 1

The main differences in environmental burdens between the scenarios are due to the difference between the recovery of materials and the recovery of energy from the incineration process. Different collection and sorting of waste (the comparison between scenarios 4 and 5) has a relatively minor influence on the environmental burdens of the whole system.

The environmental burdens caused by CH_4 and CO_2 emissions change dramatically between the different scenarios. The generation of CH_4 decreases as recycling increases and decreases further when the organic fraction is diverted from landfill to either composting or incineration with energy recovery. The generation of CO_2 also decreases as recycling increases and again decreases further when the organic fraction is diverted from landfill to composting. The generation of CO_2 increases when incineration is added as the carbon content of the incinerator feedstock is converted to CO_2 during the incineration process. As CH_4 has a Global Warming Potential 21 times greater than CO_2, waste management options that would result in a lowering of CH_4 emissions may be regarded favourably. The advantage of incineration in scenarios 4 and 5 is that electricity is generated resulting in much greater net savings of energy. Energy savings are higher in scenarios 4 and 5, where incineration utilises the organic fraction of the waste directly to generate electricity, compared with scenario 1, which utilises landfill gas.

Scenarios 4 and 5 result in the lowest levels of CH_4 emissions and the highest levels of energy saving; their main drawback is the higher levels of CO_2 emissions. The least favourable scenario would appear to be scenario 3, incineration without energy recovery, as this has a net energy consumption, high levels of CO_2 emissions and significant emissions of CH_4. Scenarios 1 and 2 have low levels of CO_2 emissions, but much higher emissions of CH_4 and lower energy savings than scenarios 4 and 5. In terms of meeting UK national targets for reductions in gases with Global Warming Potential, scenarios 4 and 5 appear most suitable. This conclusion is based on the assumption that state of the art air pollution control is carried out.

Application

This LCI study was commissioned by a waste management company, which was planning to tender for a contract to provide waste management services within Gloucestershire county. Before the results of the LCI were presented to representatives from the County, the waste management company was taken over by a larger company and the change in management resulted in an end to their interest in waste management in Gloucestershire.

Further studies

A second LCI study on waste management has been carried out in Gloucestershire to develop recycling plans for each of the six Local Authorities (LAs) within Gloucestershire county, that fit within the 1997 county waste disposal strategy. The study again used the IWM-1 model to develop four waste management scenarios for the six LAs. The results of the LCI model were used to fulfil the environmental assessment requirement for the Draft Recycling Plan submitted to the Department of the Environment Transport and Region, by the LAs. One of the aims of this study was to examine how the results of the LCI were used by each of the six LAs within Gloucestershire.

Gloucestershire's county waste management strategy for the period 1997–2025 (Gloucestershire County Council, 1997) presents the aims of the County Council rather than laying down specific requirements for the future. The aim is to implement sustainable waste management within the county through applying the concept of best practicable environmental option. This includes adopting the government targets of 40% MSW recovery by 2005, reducing the proportion of controlled waste going to landfill by 60%, providing accessible recycling facilities for 80% of households by 2000, recycling or composting 25% of household waste by 2000 and home composting to be carried out in 40% of domestic properties with a garden by 2000.

The county strategy requires increased recycling and composting. These are demanding targets given the current low levels of recycling and composting in the district councils. In the county waste management strategy, the difficulties in reaching high recycling levels are recognised and so the strategy includes the possibility of constructing a new 200,000 tonne/year Energy from Waste (EfW) plant. The LAs are responsible for delivering realistic strategies and so developed three waste management scenarios, which they believed had the potential to deliver the aims listed above. Life Cycle Assessment was proposed as a suitable tool to examine the potential environmental burdens of each scenario. Three new scenarios were modelled for each of the six LAs:

Baseline scenario: the existing waste management system as of February 1999
Scenario 1. 25% Recycling
Scenario 2. 25% Recycling + 8% home composting
Scenario 3. 25% Recycling + Centralised EfW incinerator.

The model showed a similar pattern of environmental burdens for each of the six LAs under scenarios 1,2 and 3 despite the differences between districts with respect to waste composition, waste quantities, distance from landfill sites and collection methods.

Use of LCI results by local authorities

The report detailing results from the LCI model was submitted to each LA as it became available. Three months later key personnel from the waste management section of each LA (either the Waste Manager or the Recycling Officer) were interviewed to determine the extent of utilisation of the LCI information in developing actual local recycling strategies. The key problems in using the data from the LCI studies were seen to be:

1. a difficulty in understanding the model itself
2. a failure to appreciate the units in which numerical data were expressed (e.g.: kWh, MJ, GJ)

3. a lack of understanding of the effects of different emissions on the local environment
4. a lack of understanding of the difference between local and global emissions
5. a limited understanding of the significance of emissions.

Conclusions 2

This LCI study showed in all six districts, that Global Warming Potential (GWP) declines as recycling increases from the baseline to the 25% target. The majority of this reduction is due to 'recycling savings', the reduction in emissions and energy consumption by using recycled materials rather than virgin materials. The highest GWP was generated by the baseline scenario. The lowest GWP varied with centralised incineration having the lowest GWP in Gloucester, and the home composting scenario the lowest in the other five districts. These differences were due to variations in waste composition and transport distance to a centralised incinerator. In terms of SOx emissions, incineration appeared the least favourable option from a local perspective even though global savings may be achieved. The most favourable energy balance utilised incineration and the least favourable utilised home composting. As the organic fraction is removed for recycling and/or composting, methane and thus energy generation from landfills declines.

The local authority must therefore balance changes in emissions and energy generation locally, with reduced emissions and energy consumption from recycling that may occur elsewhere. They are being asked not just to consider the local environment but global environmental effects in making decisions about waste management. This is a complex task made more difficult by having to factor in the financial implications of all options at the local level.

Two main constraints on actual utilisation of the data from this LCI study were recognised. First was the application of the model itself. The results of the LCI were not published in time to play a part in development of recycling plans, and most importantly were not communicated in a way that made the results easy to understand. Second was the policy making process itself. Both political and financial aspects were felt by most to be the priority in decision making. At the local level waste management is political and limited by a lack of financial resources. There is clearly a role for LCI information in raising awareness about the environmental effects of proposed waste management options, but attention must be paid to the transparency and credibility of the LCI model if direct utilisation of the results is the aim.

Acknowledgements

This case study was based upon original research carried out by John Powell, Alex Steele, Nick Sherwood and Tony Robson of the Countryside and Community Research Unit, Cheltenham and Gloucester College of Higher Education, Francis Close Hall, Cheltenham, GL50 4AZ.

Barcelona Metropolitan Area – LCI for long-term Integrated Waste Management strategy planning

The metropolitan area of Barcelona is made up of 33 municipalities with a population of approximately 2.9 million people. The city of Barcelona accounts for 54% of the 1.29 million tonnes of MSW generated in this metropolitan area. Of the total amount of MSW collected in Barcelona, 3% is recycled, 27% is incinerated and the remainder is landfilled. The legal obligations that are established within the Catalan regulations for waste management authorities in the metropolitan area include: the collection of waste; and the establishment of recovery and recycling centres; the valorisation (incineration with energy recovery) and recycling of the organic and inorganic fractions of waste and its final disposal.

The EC Packaging and Packaging Waste Directive (EC94/62) had a significant impact on the waste management system in Barcelona, as it resulted in the separate collection of packaging material and the establishment of materials recycling, biological treatment and valorisation (incineration) targets for the year 2001. The Catalan regulations also require separation of the organic fraction of waste.

Collection and disposal

A bring system operated in Barcelona in 1996. Householders were encouraged to separate their waste into glass, paper and cardboard, and light containers (defined as mixtures of plastic, beverage cartons, aluminium foils and cans). The remaining material was collected as rest-waste, at kerbside collection points. The separated material was sent to a small recycling and recovery centre. The residue from the recovery and recycling centre, plus approximately 25% of the collected household restwaste was incinerated with energy recovery. The remaining restwaste was sent to landfill. A small composting facility (with a capacity of 30,000 tonnes per year) was operational but the actual amount of material treated at the site was very small.

Objectives of the new waste management system

The Catalan Metropolitan authority (EMA) has been appointed as the sole authority in charge of the treatment and disposal of MSW for the whole metropolitan area. The new waste management system aims to 'increase the reutilisation, recycling and valorisation of waste in order to reduce its environmental burdens'. The disposal of waste in landfills will be limited to the waste fractions that cannot be treated by the methods listed above. This approach is expected to reduce the quantities of waste requiring final disposal to landfill. It is anticipated that the new waste management system will result in the creation of new jobs in activities related directly or indirectly to the overall waste management strategy. This social benefit is of major importance to all of the municipalities involved.

Use of an LCI tool to help develop the new Integrated Waste Management system

With the above objectives in mind, EMA used IWM-1 (White et al., 1995) to model different Integrated Waste Management scenarios for the metropolitan area of Barcelona. The existing waste management system was modelled, to act as a baseline (see Figure 6.3) against which all other scenarios could be compared. The scenarios that were modelled focused mainly on the development of the collection system to increase the amount of recyclable material collected and the treatment of the organic fraction of the waste stream.

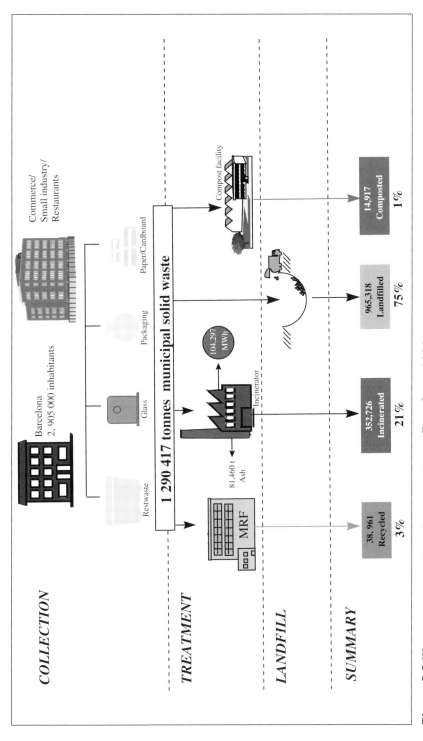

Figure 6.3 Waste management system operating in Barcelona, 1996.

By modelling many different scenarios, each of which was a realistic progression from the baseline scenario, EMA was able to develop an Integrated Waste Management strategy that would meet their objectives. As the new strategy evolved from the existing waste management operation and was designed by taking a holistic approach it is inherently compatible with the specific circumstances of the metropolitan area of Barcelona. Therefore this strategy has been developed taking environmental, economic and social considerations into account.

The new Integrated Waste Management system

The new waste management system based upon the results of the LCI tool will improve the existing infrastructure and the collection system through a series of major investments and public information strategies. A campaign amongst the people of the metropolitan area to promote waste minimisation and reduction is planned. EMA is aware that the long-term success of the programme will depend on public participation. The system was developed to be completed in four stages (see Figure 6.4).

1. **(1998–1999).** Construction of a new composting facility, two anaerobic digestion facilities, 28 new Recovery and Recycling centres and an increase in the number of kerbside collection containers. MSW treatment and disposal forecast at the end of this stage:

   - Recycling   16%
   - Composting/anaerobic digestion   6%
   - Incineration   28%
   - Landfilling   50%.

2. **(2000–2001).** Construction of the third and fourth composting facilities, the third anaerobic digestion facility and two new recovery and recycling centres. The first anaerobic digestion facility will become operational. MSW treatment and disposal forecast at the end of this stage:

   - Recycling: 20%
   - Composting/anaerobic digestion   23%
   - Incineration   31%
   - Landfilling   26%.

3. **(2002–2003).** The three anaerobic digestion facilities will be fully operational. MSW treatment and disposal forecast at the end of this stage:

   - Recycling: 24%
   - Composting/anaerobic digestion   27%
   - Incineration   31%
   - Landfilling   18%.

4. (2004–2006). Construction of the fifth composting facility. By the end of the year 2006, the cost of waste management in Barcelona will be approximately 8343 pesetas (18.8 euros)/tonne of waste. MSW treatment and disposal forecast at the end of this stage:

- Recycling: 30%
- Composting/anaerobic digestion   30%
- Incineration   33%
- Landfilling   7%.

Conclusions

The application of a life cycle approach to the development of a long-term waste strategy for the metropolitan area of Barcelona has been successful and this exercise has resulted in the adoption of a clear and achievable plan, which will dramatically alter the management of solid waste in the area by 2006. Starting with the existing waste management scenario for Barcelona, many different scenarios were modelled and the environmental and financial aspects of each scenario were compared and analysed. These results were then used to modify new scenarios to allow the eventual development of a long-term strategy based upon a four-step plan. The four final scenarios enable financially viable infrastructure development to proceed in parallel with the optimisation of environmental performance at a local level.

This is a good example of a waste management authority using LCI as a tool to provide it with both environmental and economic data. LCI data can be used to improve the decision-making process by making it more data based and therefore more reliable and enable the development of a comprehensive long-term Integrated Waste Management strategy. The data from successive iterations enabled the users to build appropriate and realistic scenarios, which could take into account the requirements of both the legislator and the decision maker. Considerable time and effort is necessary to develop the model scenarios and communicate the results of the model. This investment will result in a waste management strategy built upon the solid foundations of data-based decisions. This approach can deliver more sustainable waste management systems.

Concepts and Case Studies

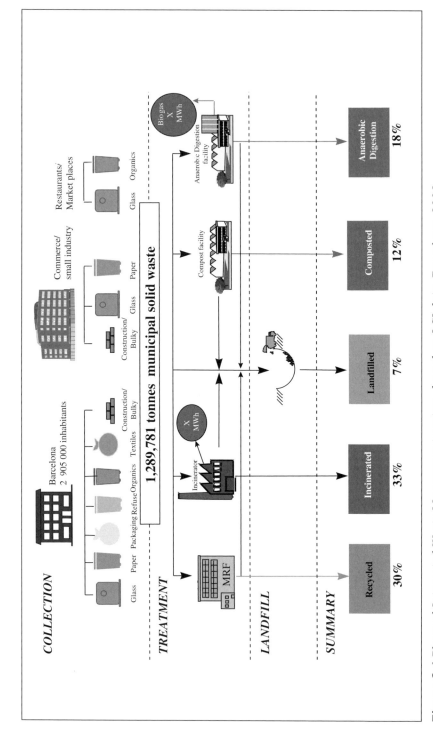

Figure 6.4 Planned Integrated Waste Management system, based on LCI data, Barcelona, 2006.

London, Ontario, Canada – LCI for assessment of different materials recycling options

Two Canadian industry groups, Corporations Supporting Recycling (CSR) and the Environment and Plastics Industry Council (EPIC) have co-sponsored the development of an LCI model that municipalities can use to evaluate the environmental and economic effects of proposed changes to their Integrated Waste Management system, strategies and practices. This holistic system approach takes into account the upstream and downstream effects of waste management decisions. It is based on the best data currently available and the consideration of selected environmental burdens, and has a municipal focus.

The model has been designed with significant municipal input from the co-participant in the project, the City of London, Ontario. London's participation has provided an excellent case study wherein data inputs, analysis, interpretation and results have been provided by municipal staff and communicated to municipal stakeholders. The 'Beta version' of the model was completed in September 1998. The Peer review was completed in April 1999. Its release began soon after and municipal training workshops have been held across Canada. The model's developers, Envirosphere and Procter & Redfern Ltd, have carried this work out under the direction of a steering committee for the project consisting of representatives from EPIC, CSR, City of London, Environment Canada, Resource Integration Systems and Procter & Gamble.

The development and testing of the model began in this southwestern Ontario municipality (population 340,000 in 1998) in late 1996. The first test of the model began with a benchmarking process, using both the environmental and economic modules of the tool. Waste management data gathered in 1995 and 1996 were analysed by the model and the environmental and economic burdens predicted by the model were compared against actual operating data. The model was shown to provide an accurate profile of the City of London's waste management system and it described in detail the environmental benefits associated with the changes made to the system between 1995 and 1996.

In December 1997, London Municipal Council approved the long-term waste strategy known as the Continuous Improvement System for waste management; a dynamic framework that recognises Integrated Waste Management as an important environmental service in the community, which contributes to the protection of human health and the environment by effectively allocating operating and capital budgets and employment resources. One key aspect of the Continuous Improvement System is measuring environmental performance of the total MSW management system. Environmental parameters are calculated by the computer model developed as described above.

The City, faced with a renewal of its kerbside recycling programme contract in 1999, which was going to have increased costs as a major decision factor, provided the second major test of the model's capability. The LCI model was used to evaluate the differences in the potential environmental burdens of collecting and recycling four different groups of materials, as shown in Table 6.4.

The LCI model was used to calculate the effect of changes to the recycling programme in terms of changes in emissions of various pollutants within the City's waste management system and net changes in life cycle emissions, which could potentially occur globally (in London and outside), if materials recovered from the City's waste stream were used to replace the production of virgin materials.

	Scenario 1 (minimum mandated requirements)	Scenario 2 (selected materials)	Scenario 3 (maximum paper recycling)	Scenario 4 (existing system, in 1998)
Newsprint, magazines and telephone books (ONP)	9780	9780	9780	9780
Old Boxboard (OBB)	0	570	570	570
Old corrugated containers	0	0	810	810
Fine and mixed paper	0	0	525	525
Steel food and drinks cans	915	915	915	915
Aluminium beverage and food cans and foil	305	305	305	315
Glass bottles and jars	2840	0	2840	2840
PET containers	285	285	285	285
HDPE bottles, PP bottles and LDPE bottles	0	0	0	370
Total tonnes recycled	14,125	11,855	16,030	16,410

Table 6.4 City of London: materials collected by the four different recycling scenarios. Source: Envirosphere (1998), after City of London

In order to calculate net changes in 'potential' life cycle emissions due to recycling, the model relies on data available in the public domain on the life cycle emissions associated with the production of various materials (paper, aluminium, glass, steel, plastic). The users of the model recognise that while a number of government and industry agencies are in the process of developing reliable life cycle data for the different materials, the data that are currently available are old and in some cases incomplete. The users then correctly warn the decision makers that 'at this stage therefore these data should be used only as a very rough indication of potential burdens of recycling programmes on emissions from virgin material production systems'.

The potential environmental burdens identified by the model are expressed in terms of everyday equivalents such as the emissions from the production of electric power for the average Canadian home, or emissions from the average passenger car. This is a helpful method to help decision makers quantify the scale of the potential burdens being described by the data from the model. The significance of these potential burdens has been further evaluated based upon:

1. How the magnitude of the change compared with the variability of the emission data used in the model. The model uses emissions data from processes including the extraction of fuels and raw materials, the production of materials and energy and from the management of solid waste. Emissions from each of these processes are highly site specific and therefore

variable, and depend upon factors such as the technology used, the source and nature of raw materials, climatic conditions, etc. The data used in the model generally represent 'industry averages'. Changes that were judged to be smaller than the estimated variability in the data were therefore reported as not significant.

2. How the magnitude of the change compared with emissions from other sources within the community. According to Statistics Canada census data, the City of London had a population of 325,650 in 1996. Between 1991 and 1996, London's population grew at an average rate of 0.9% per year, or 2900 per year. Based on an average household of 2.5 persons, this converts to an increase of over 1100 homes per year. Similarly, based on the number of passenger car registrations per capita reported for Canada by Statistics Canada, there are approximately 150,000 cars in the City of London and the annual population growth represents an increase of approximately 1300 passenger cars per year. Changes that are small relative to these natural 'growth' numbers were therefore considered to be insignificant.

3. How the magnitude of the change compared with annual emissions of those pollutants from all sources in Ontario and/or Canada.

Results from the LCI model

Under scenario 1, the effect of limiting the grades of paper collected to old newspapers, magazines and telephone directories, and limiting the plastics collected to PET was examined. This has the effect of reducing the quantity of recyclables collected from 16,410 tonnes to 14,125 tonnes, a reduction of approximately 2285 tonnes a year or 14%. This option meets the minimum recycling requirements mandated by the province.

In scenario 2, the effect of collecting only selected materials was examined. In addition to limiting the grades of paper collected, glass is also removed from the recycling programme. This has the effect of reducing the quantity of recyclables collected further to 11,850 tonnes a reduction of approximately 28%.

In scenario 3 the amount of paper collected was maximised. In this scenario, all grades of paper are collected. In addition, all other materials remain unchanged, except for the removal of mixed plastics from the recycling programme. The total quantity of recyclables collected under this scenario would be only marginally (2.5%) lower than the existing system.

In evaluating the effects of reducing the quantity of materials collected for recycling, it was assumed that the materials excluded from the recycling programme were sent to an Energy from Waste (EfW) facility. The changes in energy and emissions that occur as a result of each of these changes within the waste management system and over the entire Life Cycle of the materials are described.

Energy

Because the City of London sends a portion of its MSW to an EfW facility, the waste management system is a net producer of energy. The energy that is produced and captured as usable steam from the combustion of waste is more than is consumed by all waste management processes (collection, sorting, processing and landfilling).

In scenario 1 (Minimum Requirements), the amount of energy (diesel, natural gas and electricity) consumed by the recycling programme (for collecting and sorting recyclables) is reduced by approximately 19%. In addition to this, the incineration of paper and plastics sent to the EfW facility generates energy. These changes improve the energy balance of the system by the equivalent of the amount of electric power required by approximately 800 homes for a year. Reducing the quantity of recyclables collected, however, increases the net energy consumption (over the whole Life Cycle) because the production of virgin paper and plastics requires more energy than the production of recycled paper and plastics. Global consumption increases by the equivalent of approximately 1200 homes. Thus, the overall net effect is an increase in energy consumption equivalent to power for 400 homes.

In scenario 2 (Selected Materials), the energy used for recycling is reduced by a further 6.5%. However, the addition of glass to the MSW stream going to the EfW facility (as glass is not recycled in this scenario) reduces the amount of energy generated by that facility (relative to scenario 1) by almost 30%. As a result, the energy balance improves by the equivalent of only 600 homes (less than the improvement in scenario 1). Net energy consumption (over the whole Life Cycle) increases by the equivalent of power required for 600 homes.

Scenario 3 (Maximum Paper Recycling) represents a very small reduction in the quantity of material recycled. Recycling energy savings are therefore small and only 350 additional tonnes of material are sent for energy recovery. Energy savings within the waste management system are therefore correspondingly small (equivalent to the power required for 200 homes). The net increase in energy consumption (over the whole Life Cycle) is also relatively small, the equivalent of power for 300 homes. By comparison, if there was no recycling programme in place, and all the materials that are currently recycled were landfilled, then energy consumption within the system would fall by the equivalent of the power required for 300 homes. The net increase in energy consumption (over the whole Life Cycle) would be much larger, at the equivalent of power for 7900 homes.

Global Warming Potential (GWP)

Reducing the amount of paper and plastics collected by the recycling programme in Scenarios 1, 2 and 3 reduces the GWP associated with the collection and sorting of these materials and increases emissions associated with combusting these materials in the EfW facility. Reducing the quantity of material collected for recycling therefore has little effect (ranging from the equivalent of 0 to 100 cars) on GWP within the waste management system.

The IWM model indicates that net emissions of GWP gases (over the whole Life Cycle) could potentially increase by the equivalent of emissions of between 200 and 1000 cars/year as a result of reduced replacement of virgin materials by recycled materials.

The users noted that 'in addition to the general data quality concerns associated with the estimation of burdens of replaced virgin material mentioned above, the data used for paper in the IWM model have two shortcomings':

1. The credit for replacing virgin paper is calculated on the basis of data for only two grades of paper, old newsprint (ONP) and old corrugated containers (OCC).
2. The greenhouse gases that are taken up by trees grown for the production of virgin paper are not taken into account by the data. It is likely therefore that it overstates the GWP associated with the production of virgin paper and consequently the potential savings that may result from replacing virgin paper with recycled paper.

As a consequence of these shortcomings, the results of the model with respect to the Life Cycle burden of recycling paper on GWP are considered unsuitable for use in decision making at the present time. As better data becomes available, it will be incorporated into the model, thereby increasing the level of confidence with which such analysis can be undertaken.

This detailed level of explanation of the results from an LCI model for waste management is essential for the purpose of transparency for other expert users and to ensure that decision makers understand the limitations of the results from such a model. Clear and comprehensive reporting of the results from an LCI model allows decision makers to have confidence in the data upon which they are basing their decisions.

The reported results of this application of the LCI model also included data on:

1. acid gases (nitrogen oxides, sulphur dioxide and hydrogen chloride)
2. smog precursors (nitrogen oxides, volatile organic compounds and particulate matter)
3. heavy metals (lead, cadmium and mercury) and trace organic (dioxins) emission to both air and water
4. total restwaste.

Conclusions

The results from the IWM model showed that in general, there was little difference in the emissions from the overall waste management system, between the four different recycling scenarios. The results indicated that within the waste management system, emissions of approximately half the pollutants decrease slightly, while emissions of the other half show a small increase, relative to the existing system.

Larger differences were found when net emissions (over the whole Life Cycle) were considered. This takes into account potential savings in emissions that may occur outside the waste management system as a result of recycling. However, these differences are generally small when considered in relation to the variability of the data available for this type of analysis and the magnitude of emissions of these pollutants from other sources.

However, the comparison of the existing recycling system with the hypothetical 'no recycling' scenario demonstrates that the environmental benefits associated with recycling are significant. Increases in net emissions of all pollutants (over the whole Life Cycle) are indicated where there is no recycling.

Based on these results from the LCI model, The City of London Standing Committees and Council have approved an expanded recycling system based on public acceptability and environmental benefits despite the *higher cost* to the municipality. The use of the LCI model (within a decision making framework) ensured that elected officials made their decisions based upon both environmental and economic data.

Acknowledgements

This case study was based on research carried out by Jay Stanford of the City of London, CSR/ Corporations Supporting Recycling and Ruksana Mirza of Envirosphere EMC Inc.

United States Environmental Protection Agency case studies

This case study describes the preliminary results of a research project that is ongoing at the time of writing.

Background

The US Environmental Protection Agency's (US EPA's) Office of Research and Development, is working with the Research Triangle Institute (RTI) and its partners to develop a computer-based Decision Support Tool (DST) designed to evaluate the cost and environmental perform-ance of integrated MSW management systems. By adopting MSW management strategies that improve the integration and efficiency of MSW management operations, local governments can help reduce the release of greenhouse gases, conserve energy and other natural resources and reduce burdens to air and water quality or ecosystems.

In addition to the DST, this research is producing a stand-alone database, which enables users to search for data specific to a waste management unit operation, structure, piece of equipment, or Life Cycle Inventory (LCI) parameter (air emission, waterborne effluent, solid waste). The information and tools developed through this effort will enable the evaluation of the trade-offs among environmental burdens, energy, and costs for different integrated waste strategies for MSW management including collection, separation, transportation, material recovery facilities, remanufacturing, composting, incineration, and landfilling.

Decision Support Tool

The Decision Support Tool (DST) is a tool designed to aid in evaluating the cost and environ-mental burdens of Integrated Waste Management strategies. It enables users to simulate exist-ing MSW management strategies and conduct scenario analyses of new strategies to optimise the cost or environmental performance of the system. The tool is designed to be used in con-junction with community-specific data on waste generation and composition, recycling or diversion programmes and facility (e.g. landfill) design and operation.

The processes that can be modelled using the tool include multiple alternatives for waste collection, transfer stations, Materials Recovery Facilities, mixed municipal and garden waste composting, incineration, Refuse-Derived Fuel incineration, and disposal in a MSW landfill or a hazardous waste landfill. Existing facilities and equipment can be incorporated as constraints to ensure that previous capital expenditures are not negated. A screen shot of summary-level results from the tool is shown in Figure 6.5.

In addition to viewing summary-level results, users can click down to obtain more detail about each waste management operation selected. Data on all environmental burdens are available on a total or process level basis. This information can be used to help evaluate the trade-offs between different strategies and evaluate environmental performance. In addition, the full costs associated with the Integrated Waste Management system are also provided for the whole strategy or on a process level basis.

Local governments and solid waste planners can use the tool, for example, to evaluate the effects of changes in the existing MSW management on cost; to identify least-cost ways to man-age recycling and waste diversion; and to evaluate options for reducing GWP, priority pollu-tants, and environmental burdens to water. The tool will also be valuable to other user groups, including environmental and solid waste consultants, industry, Life Cycle practitioners, and

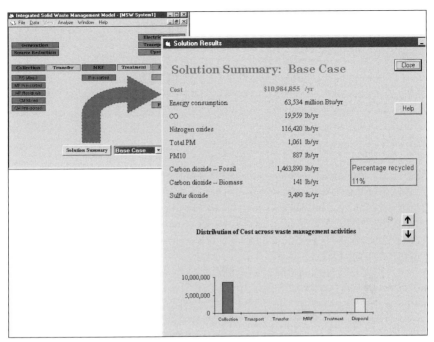

Figure 6.5 The Solution Summary, which shows cost, energy consumption, and environmental emissions data for all waste operations included in the solution. The user can view the cost and environmental data for the suggested waste management strategy. In addition, the user can view process-level data on the full costs and environmental burdens over the whole life cycle.

environmental advocacy organisations. The tool may be used in evaluating policies and programmes for reducing total costs and environmental burdens in responding to the following example issues:

1. changes in waste diversion or recycling goals
2. changes in market value for recovered materials
3. quantifying potential benefits associated with recycling and
4. identifying strategies for optimising energy recovery from MSW.

Through ongoing case studies, the potential applications of the tool will be evaluated to help clearly understand the potential uses and limitations of the tool and information. The Decision Support Tool (DST) contains general cost parameters, and therefore is not a community-specific model and is not intended for setting prices for any specific waste management service. The cost results provided by the tool represent indicative engineering costs, which accrue to the public entity (i.e. local government). A more detailed cash flow analysis substituting local parameters would be needed to determine the appropriate prices for services and materials. The tool is also not designed to conduct Life Cycle comparisons of any specific products or materials. It is considered to be a comprehensive tool of significant value in finding improved solutions for Sustainable Waste Management.

Testing the Decision Support Tool in local communities

A prototype of the DST is being tested in a number of case studies with state and local governments. At the time of writing, case studies are ongoing with Lucas County, Ohio; the Great River Regional Waste Authority, Iowa; Anderson County, South Carolina; the State of Wisconsin; the Integrated Waste Services Association, and the US Navy. Issues being analysed with the DST for these different groups and studies are as follows:

- Lucas County, Ohio is developing a 15-year plan for their solid waste management system. They believe that their waste operations are not cost effective and ignore pollution and Life Cycle implications. The analyses and results of this case study are helping in the development of integrated, cost-effective, and environmentally preferable plans and targeting opportunities for increasing recycling rates, reducing costs, and improving environmental performance.
- The Great River Regional Waste Authority in Iowa is exploring the efficiency of integrated collection systems versus multiple collection options. Their goals are to evaluate effects of reconfiguring service areas and applying existing systems to them, and in addition to develop a waste management plan for a 50% recycling scenario, which is to be presented to the State authority.
- Anderson County, South Carolina is evaluating the cost and environmental implications of implementing a residential kerbside recycling programme for the more densely populated areas of the county as well as setting up a garden waste composting programme. The results of this study will assist the County in determining the most cost-effective strategies for implementing the programmes while simultaneously considering environmental performance.
- The State of Georgia used the tool to analyse the effects of banning the collection of garden waste on air emissions within Gwinnett County. Current NOx emissions attributable to garden waste collection were estimated to be 105 tons per year and the elimination of a garden waste ban would result in an 11% decrease in NOx. The number of collection trucks needed for collecting garden waste with MSW increases from 171 trucks (with no garden waste collected with MSW) to 201 trucks. Discussions are underway to conduct additional case studies in Georgia to assist with evaluating regional solutions to Integrated Waste Management.
- The State of Wisconsin is investigating the environmental benefits of State-wide recycling programmes. The DST is being used to analyse how changes in levels of State-mandated recycling goals can potentially affect environmental burdens. The results of this study will assist the State in deciding which solid waste strategies should be used in the future to meet environmental improvement goals.
- The Integrated Waste Services Association is interested in analysing the effects that advances in MSW management technologies have on GWP gas emissions. The DST is being used to investigate GWP gas emissions from various technologies including landfill gas recovery, Energy from Waste incineration and recycling.
- The US Navy has requested a case study for the Pacific Northwest. There is major interest in reducing costs, increasing recycling rates, and ensuring that environmental goals are being met. In addition, due to the closure of small local landfills and the need to transport waste by

rail to a larger regional landfill, the Navy is interested in the consequent change in environmental burdens, energy and economics. The Navy is also investigating options that would include waste from nearby communities to develop more cost-effective and environmentally preferable solutions leading to a more regional approach for integrated waste management. The Navy is also considering additional case studies in San Diego and the Pacific Rim.

These case studies are providing cost and environmental information about alternative waste management strategies, which will assist in the development of management plans and policies. The case studies are also enabling the research team to refine the methods and data used in the DST as well as the user interface tool. The details of the State of Wisconsin case study are described below. However, the findings are preliminary and are not considered final. Once comments are received from the State of Wisconsin, and the tool with its associated data have received final clearance, the full findings from these US case studies be released.

Additional case studies are planned and these will reflect the issues of urban and rural settings throughout the United States to ensure that the DST is flexible enough to address the wide range of variation among local communities.

Wisconsin case study methodology and results

The purpose of the State of Wisconsin case study was to estimate the environmental effects of recycling and waste management in Wisconsin. In this study, the full Life Cycle benefits (or burdens) of additional recycling being carried out in Wisconsin in 2000 as compared to 1995 were quantified. A Life Cycle approach was taken to estimate the emissions to air and water, solid waste arisings, and the energy consumed for managing Wisconsin's waste in 1995 and 2000. This approach includes waste collection, processing, treatment, disposal and remanufacturing of recyclables.

Waste composition, generation, and recycling data

A case study methodology document was prepared that described the data sources and assumptions made in entering the waste composition, generation, and recycling data for the state of Wisconsin for the years 1995 and 2000. Waste generation data estimated for year 2000 was used for both 1995 and 2000, but the recycling rates are different for the 2 years. Thus, year 1995 and 2000 levels of recycling were applied to waste generation data for model year 2000.

Collection, recycling, and disposal options for residential, multi-family, and commercial waste

In establishing a baseline model for the entire State of Wisconsin, a general waste management strategy was defined and used as the basis for calculating results. This included:

- collection of presorted recyclables and remaining (residual) mixed waste
- processing of recyclables in a presorted Materials Recovery Facility
- disposal of restwaste in a landfill
- composting of garden waste in a garden waste composting facility. Note that in addition to garden waste sent to the compost facility, garden waste is also composted in residential backyards.

Key assumptions employed

When applying the DST to the real-world waste management practices of Wisconsin, some assumptions were required to 'fit' the real-world practices into the modelling environment of the tool. For example, a community may manage a waste material that is not included in the model (e.g. pallets, household hazardous waste) and thus assumptions must be made on how best to handle such materials outside the model. The key assumptions employed in the Wisconsin case study are as follows:

- Some materials that were actually recycled are assumed to be non-recyclable because the DST does not contain remanufacturing data for those items, e.g. batteries, tyres, etc. These items were excluded for purposes of the case study and represent approximately 5% of the waste generated.
- The LCI for composting 290,000 tonnes of garden waste composted in residential backyards was added to the LCI of garden waste, which was collected and treated at a composting facility for each year modelled.

Waste generation by category (data used to generate results)	1995 (tonnes)	2000 (projected) (tonnes)
Waste generated		
Residential	1,858,000	1,883,000
Multi-family apartments	213,000	220,000
Commercial	1,443,000	1,599,000
Materials recycled that are		
not accounted for in the model	201,000	221,000
Materials recycled		
Residential	349,000	353,000
Multi-family apartments	36,000	37,000
Commercial	557,000	617,000
Materials recycled that are		
not accounted for in the model	201,000	221,000
Garden waste diverted from landfill		
Backyard composting	290,000	290,000
Garden waste composting at facility	199,000	200,000
Waste landfilled		
Residential	1,310,000	1,329,000
Multi-family apartments	177,000	183,000
Commercial	886,000	982,000

Table 6.5 Summary of waste flows modelled for the State of Wisconsin. Notes: this output data from the DST was used to generate the Life Cycle profile for waste management in Wisconsin. The quantities of material recycled are slightly different from actual numbers expected using Wisconsin data due to rounding and the requirements of the model

Parameter	Units	Savings in 2000 due to recycling over 1995 levels
Energy consumption	MBTU/year	2,065,000
Air emissions		
Total particulate matter	lbs Total PM/year	438,000
Nitrogen oxides	lbs NOx/year	1,588,000
Sulfur oxides	lbs SOx/year	2,129,000
Carbon monoxide	lbs CO/year	5,558,000
Carbon dioxide biomass	lbs CO_2 Bio/year	563,309,000
Carbon dioxide fossil	lbs CO_2 Fossil/year	−97,408,000
Carbon equivalents	tonnes /year	−16,000
Hydrocarbons (non-CH_4)	lbs HC/year	101,000
Lead (air)	lbs Pb /year	16,000
Ammonia (air)	lbs NH_4 /year	323
Methane (CH_4)	lbs CH_4/year	−926,000
Hydrochloric acid	lbs HCl/year	−7000
Solid waste		
Total Solid Waste	lbs SWTotal/year	1,885,000
Water releases		
Dissolved solids	lbs DS/year	−183,000
Suspended solids	lbs SS/year	−226,000
BOD	lbs BOD/year	−451,000
COD	lbs COD/year	1,536,000
Oil	lbs Oil/year	−69,000
Sulphuric acid	lbs H_2SO_4/year	46,000
Iron	lbs Fe/year	−70,000
Ammonia	lbs NH_4 /year	7000
Phosphate	lbs P/year	−1000
Zinc	lbs Zn/year	−1000

Table 6.6 Summary of preliminary results predicting the Life Cycle benefits resulting from increased recycling in the State of Wisconsin from 1995 to 2000. Notes: negative figures represent Net increases in LCI parameters for year 2000 over 1995. These data are totals for the entire waste management system including collection, recycling, treatment, disposal, and remanufacturing

Discussion of results

The results from the preliminary analysis of recycling in Wisconsin in 1995 in comparison with 2000 are presented in Tables 6.5 and 6.6. Based on these results, the following observations can be made:

1. The recycling levels for 2000 were higher than those estimated for 1995 for all waste components. The net emissions include environmental burdens associated with collection, processing, treatment, disposal, and remanufacturing of waste and recyclables in Wisconsin.
2. For several air emission parameters, the net numbers are negative. The negative number indicates that there was a net increase in emissions for those LCI parameters.
3. The results show that recycling at year 2000 levels results in lower LCI parameter values for some parameters, and higher values for others.
4. The higher net values of LCI parameters in 2000 can be explained by the increased quantities of waste generated, collected, recycled, and disposed of in 2000 over 1995 levels. For example, in year 2000 there was a 4% increase in the quantity of waste landfilled over 1995 levels. There was a corresponding 11% increase in net methane releases for year 2000. This increase in methane emissions can be partly explained by the higher quantity of waste landfilled (that generates methane during decomposition) in year 2000.
5. Emissions and energy use in the remanufacturing of recyclables recovered from waste dominate emissions and energy use in the waste collection, processing, treatment, and disposal stages of waste management.

Acknowledgements

This text is based upon a paper entitled 'Using A Life Cycle Approach to Achieve Sustainable Municipal Solid Waste Management Strategies at Local, State, and National Levels in the United States' by Keith A. Weitz and Subba Nishtala of the Research Triangle Institute, 3040 Cornwallis Road, Research Triangle Park, NC 27709 and Susan A. Thorneloe of the US Environmental Protection Agency, Office of Research and Development, Air Pollution Prevention and Control Division (MD-63), Research Triangle Park, NC 27711. The research is being conducted by RTI, North Carolina State University, University of Wisconsin, Franklin Associates, Ltd, and Roy F. Weston, Inc., through a co-operative agreement with US EPA's Office of Research and Development. Support was also provided from other groups, including the Environmental Research and Education Foundation, which was responsible for funding the research for developing the Life Cycle data needed for modelling modern sanitary landfills. EPA and the research team are also very appreciative of the more than 70 stakeholders who have provided enthusiasm, expertise and insight in ensuring that the outputs from this research will address their needs. Further information can be found at www.rti.org/units/ese/p2/lca.cfm#life, including a PowerPoint presentation of the decision support tool.

United Kingdom Environment Agency case studies

The Environment Agency of England and Wales used case studies to help develop its LCA software, WISARD, which it produced jointly with the Scottish Environmental Protection Agency and the Northern Ireland Environment and Heritage Service. The main objectives of the programme (which started in October 1994) were:

1. To get decisions on waste management put on a sound, scientific basis
2. To provide waste managers with the basis to arrive at sound decisions using the same means.

The purpose of the case studies were threefold:

1. To prove whether the tool could be used by local authorities
2. To improve further the usability of the tool
3. To provide a source of data and experience that others could call upon when carrying out their own studies.

Introduction

All local authorities in England and Wales (some 400 counties, districts and unitaries) were asked whether they were willing to carry out a case study in their area. Authorities were asked to commit a suitable member of staff for a 3-week period during the 6 weeks allowed for the case study. They were also told they would have to supply data on waste vehicle movements and weights of different types of MSW produced. One hundred local authorities volunteered to take part in the case studies. Fifteen authorities were selected, taking into account the need to include different types of authority (waste collection, waste disposal and waste planning authorities), different areas of the country and different demographic situations. One member of staff from each authority was given 2 days' standard training on the software and asked to prepare a report covering:

- the area studied and its setting
- details of population and its distribution
- the present system
- any constraints on the waste management system
- the purpose of use
- details of data sources
- changes made and new systems
- results and assumptions.

Summary data from 11 of the studies are presented in Tables 6.7 and 6.8. All the case studies are available as a joint research report published under the Environment Agency's research programme.

From Tables 6.7 and 6.8, it can be seen that a wide range of different waste management scenarios were modelled as part of the overall case study procedure. A key element of this work was to determine how useful the model was to local authority staff with respect to inves-

Area/Authority	Total population (study area if different)	Total MSW arisings (tpa) (study area if different)
Brighton & Hove Council	41,000	128,000 Household waste = 82,000
Carmarthenshire County Council	170,000 (40,000)	80,000 (16,800)
Nottingham City Council	287,000	124,000
Dorset County Council	686,000	400,000
Gateshead Metropolitan Borough Council	211,800	63,100
Pendle Borough Council	83,250 (2266)	40,000 (1100)
Broadland District Council	118,000	44,675
Nottinghamshire County Council	1,031,600 (110,300)	– (57,960)
Powys County Council	125,000	47,513
Shropshire County Council	412,300	223,054
Surrey County Council	1,000,000	1,800,000 Household waste = 505,000

Table 6.7 Local authorities participating in the WISARD case study programme.
tpa = tonnes per annum

tigating different waste management treatment and disposal options at a local level. Similarly, another important point was whether the users could interpret the (very comprehensive) results of the model and whether the users could practically apply these results to the development of a future local waste management strategy.

Interpretation of the data from WISARD

In the reports each local authority modelled their existing waste management scenario and compared it with one or more alternative waste management scenarios, some of them requiring relatively small operational or infrastructure changes, others requiring more significant changes. Each report described the output from WISARD and documented the interpretation of this data by the local authorities. It was clear from the reports that the users had not only understood the principles of LCA for waste management but they had also understood that the model does not decide which is the best or most suitable waste management system. That decision must be made by the user based on their interpretation of the results from the model within their local or regional context.

Area/authority	Existing waste management scenario	Proposed scenario
Brighton & Hove Council	Landfill 100%	Landfill 75% Recycling 25%
Carmarthenshire County Council	Landfill Recycling	Landfill Anaerobic digestion
Nottingham City Council	Landfill 49% Incineration 100%	Incineration 51%
Dorset County Council	Landfill 75% Recycling 15% Composting 10%	Incineration 70% Recycling 15% Composting 15%
Gateshead Metropolitan Borough Council	Landfill 100%	Landfill 30% Recycling 70%
Pendle Borough Council	Landfill 91% Recycling 6% Composting 3%	Landfill 70% Recycling 25% Composting 5%
Broadland District Council	Landfill 96% Recycling 4%	Landfill 20% Incineration76% Recycling 4%
Nottinghamshire County Council	Landfill 35% Incineration 54% Recycling 6% Composting 5%	Landfill 36% Incineration 36% Recycling 17.5% Composting 10.5%

Powys County Council	Landfill 91%
	Recycling 9%
		1. Landfill 87%
		   Recycling 9%
		   Composting 4%
		2. Landfill 59%
		   Incineration 32%
		   Recycling 9%

Shropshire County Council	Landfill 92% Recycling 6% Composting 2%	Incineration 80% Recycling 6% Composting 2% Anaerobic digestion 12%
Surrey County Council	Landfill 90% Recycling 10%	Incineration 90% Recycling 10%

Table 6.8 The existing and proposed waste management scenarios modelled by WISARD

The local authorities' reaction to the case study exercise was documented in the conclusions of their reports to the Environment Agency. Some of the comments that highlight the depth of understanding reached in a relatively short period of time by the local authority staff working on this project are summarised below.

Brighton & Hove Council

'Before drawing firm conclusions about the environmental performance of the different scenarios, the meaning of these apparent effects, their magnitude and underlying reasons, would need further investigation.'

Carmarthenshire County Council

'In conclusion the overall assessment of the results favour Dinefwr 2000 (the proposed scenario) as the preferred waste management option. The scenario for Dinefwr 2000 could be further explored to reduce the burdens attributable to logistics by changing the frequency of the organic waste collection from a weekly collection to a fortnightly collection or by investigating the potential to change the traditional collection vehicle to one that has the capability of collecting the paper and organic waste fractions in a separate compartment.'

Nottingham City Council

'The study undertaken was to compare the authority's current waste disposal system, which includes incineration, landfill and recycling, with another system encompassing only incineration and recycling. The second model (system 1), was chosen because the authority recognises that with the ever increasing landfill tax, the present system, sending 49% of the authority's waste to landfill, will be an unattractive future option. There is an EfW incinerator in the authority, which currently operates with two lines, but has the capacity for a third line if there is a long-term future. Therefore, it may be a viable option for the authority to send the majority of its waste to the incinerator and reduce the dependence on the landfill option from the authority's waste management strategy.'

Dorset County Council

'From the analysis of four major environmental burdens, it can be concluded that the proposed scenario appears to be a preferable waste management system to the current system in environmental terms.'

Gateshead Metropolitan Borough Council

'On examination of all the data, it is clear that Gateshead 2000 (the proposed scenario) presents savings in environmental burdens when comparing it to our current practice. Whether these savings are statistically significant is unclear at present. As economic implications have not yet been considered, we are unable to state whether Gateshead 2000 would be the best practicable environmental option. This tool, however, has gone some way towards providing very useful information that should be taken into consideration when developing our waste management strategy.'

Pendle Borough Council (Lancashire)

'This Life Cycle Assessment indicates that there are environmental advantages in the development of PostREAP (the proposed scenario) as a collection and treatment system for the area of the study. However it will be necessary to combine this assessment with other methods of appraisal to help select the most appropriate system.'

Powys County Council

'Composting a limited quantity of only 2000 tonnes has beneficial effects over landfill as methane emissions are reduced. The remaining flows tend to be marginally affected, but the increased use of transport has a tendency to increase pollutant burdens for those linked to transport, such as acid gas emissions and consumption of energy. This has implications for the future design of the municipal composting schemes in Powys beyond the initial 2000 tonnes scale-up from the pilot project. In future, the collection and handling systems must be designed to minimise the transport distance and maximise loads to reduce consumption of fossil fuels and their associated adverse emissions.

Incineration appears to have a favourable impact, principally in the pollutant flows linked to avoided burdens from energy generation. Powys' waste generation is too small and spread out to justify the capital investment to build a facility in the County. However, should a facility be built in the neighbouring South Wales Valleys, diversion from landfill to such a plant would have a beneficial impact on some of the pollutant flows.

The magnitude of the differences between the three scenarios for most flows is not large. In addition, there is no single option that has clear benefits over the others. Future judgements about what options are best for Powys would have to be made with a set of clear priorities identifying what flows are most important locally, regionally, nationally and internationally. For all three scenarios, the current recycling route has a substantially positive effect against most flows, particularly in avoided burdens linked to primary materials and avoided energy use. If recycling were to be stopped and materials diverted to landfill, there would be a far greater adverse effect upon the environment. Although it is planned to significantly boost recycling to 25% by 2003/04, this scenario has not been tested in the model but will be examined in follow-up work. This is expected to show a significant beneficial effect.'

Shropshire County Council

'The clearest indication given by the output from this case study exercise is of a slight improvement in environmental performance between the baseline and scenario 1, with a more dramatic improvement between scenarios 1 and 2. This is most clearly demonstrated for total primary energy consumption, coal use and fossil CO_2 emissions. Clearly, the changes in these flows derive from the avoided use of resources and energy associated with greater recycling and the replacement of landfill with an Energy from Waste facility as the main disposal route. There is a minor reduction in methane emissions between the baseline and scenario 1, which derives from increased recycling and a resultant decrease in landfilling. The scale of reduction is much greater between scenario 1 and scenario 2, due to the replacement of landfill by an Energy from Waste facility as the main disposal route.'

Surrey County Council

'In comparing the results from the scenarios, scenario 2 (the proposed scenario) is thought to be the more favoured waste disposal option for Surrey County Council on an environmental basis. Scenario 2 reduces the amount of emissions to air and water, saves on natural resources and saves on most forms of energy. From the impact assessment scenario 2 was shown to have less of an impact on the environment than scenario 1. The main difference between the two scenarios was the introduction of an EfW incinerator to take approximately 60% of the county's waste. This seems to result in a system which is far more beneficial to the environment.'

Concepts and Case Studies

Conclusions

The case study exercise was successful in demonstrating that WISARD can help waste managers to make data-based decisions with respect to planning and optimising waste management systems. The 2-day training course in conjunction with support from Environment Agency personnel enabled local authority staff to model their existing waste management systems and compare these systems with a number of alternative strategies. The problems of interpretation of the data and results that were seen in the Gloucestershire case study were avoided because of this good working relationship between the Environment Agency and the local authority. As local authority staff become more familiar with the model, detailed interpretation of output data will play a more significant role in the development of future waste management strategies.

Acknowledgements

This text is based upon a compilation of local authority reports prepared for Terry Coleman, Waste Strategy Manager, UK Environment Agency. The reports were prepared by: Mary Van Beinum of Brighton & Hove Council; Sian Wise of Carmarthenshire County Council; Steve Palfrey of Dorset County Council; Alison Makepeace of Gateshead Borough Council; Sue Procter of Lancashire County Council (Pendle); Theresa Barnes of Nottingham County Council; Adrian Tyas of Norfolk County Council (Broadland); Simon Clunie of Nottingham City Council; Steve Simmons of Powys County Council; Stuart Rackham of Surrey County Council and Adrian Cooper of Shropshire County Council.

Where to from here?

The earliest LCI models for IWM were no more than a first attempt to apply the technique to this field. If LCI results are going to be used as the basis for discussion between the many and varied stakeholders in waste management decisions, the tool needs to be credible. The methodology and assumptions must be transparent, and the basic data relevant and reliable. Achieving endorsement from the UK Environment Agency and the US Environmental Protection Agency may help in some way to establish the credibility of models. The real proof of the pudding is in the eating. It is through the experiences of waste planners and managers with the tool of LCI that its full value will be understood, and the best ways to include it in the decision-making process determined. Case studies have started to be published – some are referenced above – but clearly more are needed. The tool is clearly appropriate for use in countries with developed or developing economies, at a local level. The case studies described above use a number of different LCI models, and once again each LCI model demonstrates that the sustainable management of MSW depends on both local circumstances and priorities. To facilitate the acceptance and application of such LCI models more case studies and practical experience is required. Together with user-friendly, credible, reliable and flexible models, this will help fully explore the potential of this environmental management tool in conjunction with the concept of Integrated Waste Management to develop more sustainable waste management systems.

CHAPTER 7

The Overall Picture

Introduction

This book began by defining sustainable development as 'meeting the needs of the present without compromising the ability of future generations to meet their own needs' (WCED, 1987). The depletion of the planet's non-renewable resources, although still an issue, is no longer an immediate concern (Beckerman, 1995; UNDP, 1998; UNDESA, 1999). The two most critical issues now facing some regions on the planet are the ability of sinks to handle emissions and waste and the degradation of renewables (such as soil, groundwater, fish stocks, forestry and biodiversity) (UNDP, 1998).

Waste management is closely linked to the whole debate and can significantly reduce or increase the overall burdens placed on the environment (both in terms of emissions and resource use) depending on how the system has been designed and operated. Sustainable development requires that we produce **more** value from goods and services, with **less** environmental burdens and depletion of resources. When applied to waste management, sustainability requires the production of **more** value from recovered materials and energy, with the consumption of **less** energy, **less** resources and the production of **less** emissions to air, water and land (Box 7.1).

The Life Cycle Inventory (LCI) technique gives us a way to quantify the 'more' and the 'less' – to predict the amounts of materials that will be recovered, the amount of energy consumed and the likely emissions that will be released. This book has constructed a Life Cycle Inventory (LCI) for Municipal Solid Waste, starting with a definition of the objectives of the LCI and the system boundaries (Chapter 4), then defining the quantity and composition of the waste that is being managed (Chapter 8) and continuing until each stage in the Life Cycle of waste has been described and discussed. For each of the processes from wastebin to 'grave', i.e. from waste pre-sorting in the home, through collection and waste treatment, to final disposal, the environmental inputs and outputs have been quantified, and generic values have been suggested where appropriate. When all of these individual modules are assembled together, it is possible to calculate the overall environmental burden of the whole waste management system.

Environmental sustainability is not the only issue, however. Waste management systems also need to be affordable if they are going to be economically sustainable. Consequently the facility to carry out a simple economic assessment has been built into the model; this can be carried out in parallel with the LCI to calculate the overall economic costs of waste management systems.

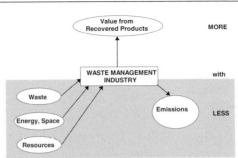

1. The basis for Sustainable Waste Management, **More** value from Recovered Products with **Less** consumption of resources and production of emissions (Chapter 1).

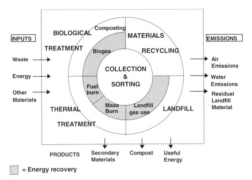

2. Identifying the inputs and outputs of an Integrated Waste Management System (Chapter 4).

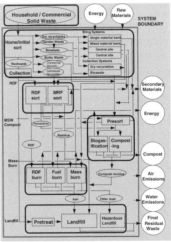

3. Quantifying the '**More**' and the '**Less**'. Using a Life Cycle Inventory to predict the quantity of useful materials recovered and the amounts of energy consumed and emissions released to air, water and land (Chapter 5).

Box 7.1 From Sustainable Development to Life Cycle Inventory results.

From Life Cycle Inventory results to sustainability

The results produced by the LCI of Municipal Solid Waste represent a considerable, and perhaps daunting, amount of information, but then the waste management systems that these results describe are themselves complex. The results quantify the input and outputs of an IWM system (see Box 7.1), namely the amounts of waste, energy and money entering the system, together with the amounts of useful products (recovered materials, compost and energy), emissions to air and water, and final residual waste to be landfilled. Environmental sustainability involves producing more useful products, but less emissions and final residual waste to be landfilled, with less energy and less resources consumed in the process. Economic sustainability is achieved by keeping the amount of money needed to operate the system to a level that is acceptable to all stakeholders.

The progress so far

Since the first edition of this book was published in 1995, the methodology of LCI has been accepted by the scientific community and guidelines for its use have been agreed by the International Standards Organisation and published as the ISO series 14040-14043. The interest in, and development of, Life Cycle tools for waste management by Government Environmental Agencies and the addition of requirements to carry out LCI modelling of future waste management strategies in a growing number of strategic waste management legislation documents indicates that the concept has come of age. This is despite the fact that the details of how best to model some of the individual waste management processes are still being refined.

Life Cycle models of waste management systems are now being used to predict the environmental performance of IWM systems, specifically with respect to emissions (on a local and a global level) and final waste arisings. As more specific data for each part of the Life Cycle are becoming available, the amount of generic data needed is decreasing, therefore the results of inventories are becoming more accurate. This improvement in data quality is allowing LCI models to help optimise existing waste management systems and carry out comparisons between different waste management options. LCI models are also being used to demonstrate the interactions that occur within waste management systems. As they represent the whole system, a Life Cycle model shows how different parts of each system are inter-connected and therefore improves our understanding of the behaviour of the system.

Using a computer model for the LCI allows the user to compare a number of hypothetical waste management systems, their environmental burdens and economic costs. These 'What if ... ?' calculations are now providing revealing information about interactions within Integrated Waste Management systems.

The Life Cycle approach has encouraged wider debate with respect to the appropriateness of a hierarchy of waste treatment options. It has also helped move the discussion on from the relative merits of individual technologies such as recycling and incineration where it has been concentrated for some time. The benefits of an integrated approach to waste management have been recognised and the tool of LCI/LCA has been identified as the most appropriate method of appraisal.

The results from LCI tools have been used as part of a communication programme at local and regional government levels. LCI results have also been used to educate the public with respect to the environmental burdens associated with different waste management options and the differences between the effects of waste management at local and global level. Clear communication of results is an important aspect of Life Cycle work, as the large amount of data that are generated can easily lead to confusion and the loss of audience interest. The authors advise that only summaries of results (the key data) should be presented to non-expert audiences but that full details of the results should be available in hard copy format for reference where necessary.

Future directions

The application of Life Cycle Assessment in general, and Life Cycle Inventory in particular, is still a relatively new approach to waste management. It offers the possibility of taking an overall view of the waste management system, and allows measurement of progress towards the goal of sustainable development.

The overall value of the LCI model developed in this book will become clear as it is applied to different solid waste management systems. Whilst improvements to this and the other available LCI models can and will be made, they represent workable tools, which can be used both to compare future Integrated Waste Management options and optimise existing systems. As with all Life Cycle studies, it is likely that when the models are applied, they will come up with a few surprises and offer some interesting new insights.

There are undoubtedly improvements to be made. The authors anticipate that this will not be the final form of this model. It is hoped that experts in individual waste treatment processes will continue to contribute their knowledge to improve each of the modules of this and the other models. An inevitable, and valuable, consequence of developing this second version of the model has been the reduction in the number of gaps in the data. As more appropriate data are forthcoming, they can be incorporated in the model. There is also a need for continued input from LCA experts. Currently there is still much debate about the methodology for Impact Analysis, and it is to be hoped that scientifically based, transparent and generally accepted methods will be available sooner rather than later. When this happens, a more complete Impact Assessment can be added on to the Life Cycle Inventory described here.

The best way to develop the field of LCI for waste management will be by using the existing models. We hope that waste managers, regulators, legislators and waste producers will find the approach both stimulating and helpful. Their experience and feedback will further improve this and the other Life Cycle Inventory tools, thus providing a powerful methodology to assess the environmental and economic sustainability of Integrated Solid Waste Management systems.

CHAPTER 8

Solid Waste Generation and Composition

Summary

This chapter provides information on the quantity and composition of solid waste likely to be generated in a given area. The lack of comprehensive and standardised data collection is one of the limiting factors in this process, and in the development of effective solid waste management in general. This chapter presents the data currently available on the generation and composition of solid waste in general, and of Municipal Solid Waste (MSW) in particular, for Europe, North America, South America and Asia. These data are limited; they are incomplete and are based on different definitions of waste categories. Definitions are given for the type of waste that will be dealt with in this book, namely MSW consisting of household (collected and delivered), commercial and institutional waste. The limitations of present classification schemes are discussed and new developments in waste analysis outlined.

> **Definition: Waste.** The Organisation for Economic Cooperation and Development (OECD) defines waste in general terms as: 'unavoidable materials for which there is currently or no near future economic demand and for which treatment and/or disposal may be required'.
>
> The United Nations Environment Programme (UNEP) defines waste as: 'objects which the owner does not want, need or use any longer, which require treatment and/or disposal'.
>
> The European Community broadly defines waste (Directive 75/442/EEC on Waste) as: 'any substance or object which the holder disposes of or is required to dispose of pursuant to the provisions of the national law in force'. This was later amended (Directive 91/156/EEC) to: 'any substance or object in the categories set out in Annex I which the holder discards or intends or is required to discard'. Annex I, entitled 'categories of waste', lists a series of different types of waste. The broad definition of waste is reinforced by the final category: 'any materials, substances or products which are not contained in the above categories'.

Introduction

The definitions of waste presented above show that due to the very heterogeneous nature of waste it is difficult to define other than in general terms. The simple definition of Municipal Solid Waste is both accurate and vague. It is accurate because each municipal waste stream is made up of the waste arisings collected by that municipality. It is vague as the waste arisings collected by municipalities vary within countries and vary considerably between countries. This is described in Chapter 3, The Development of Integrated Waste Management Systems: Case Studies and their Analysis.

The functional unit of the LCI study of solid waste was defined in Chapter 5 as the management of the amount of household and similar solid waste generated in the specified geographical area. The first requirement, therefore, is to determine both the amount and composition of the solid waste generated in the region being studied. Ideally this should be data specific to this region, collected from local solid waste analyses, but in most cases despite increasing interest in this area, such information is not available. Assessment of solid waste generation rates and composition must consequently be based on generic data, presented in the form of country specific averages. This information suffers from being at best an overall average, at worst an estimate, but poor data is still better than no data at all.

Accurate definition of the major components of the solid waste stream is essential in an LCI study of solid waste, as the solid waste itself is the majority of material inputs into the system. Whatever enters the system in or with the solid waste, whether contaminants or energy, has to leave the system somewhere. Thus most of the total emissions from the system reflect the composition of the incoming solid waste. This underlines the need to reduce the amount of solid waste produced in the first place, and the need to eliminate any potentially harmful materials from the waste.

Alterations to the waste management system can change how the waste materials leave the system (e.g. in compost or as air emissions), or whether energy is harnessed for use or dissipated, but will not change the total amount of waste materials or energy that arise from the solid waste, since this is fixed. The LCI for solid waste described in this book considers the major components of solid waste to be: paper, plastic (both rigid and film), glass, metal (ferrous and non-ferrous), organic, textiles and the term 'other', to describe the normally small amount of remaining material not covered by the previously named categories.

Solid waste generation

A complete overview of data on the amounts of solid waste generated globally is difficult to obtain. This is because much of the data is simply not available and also due to the variety and quality of the available data sources. Different sources in each country use different definitions of total solid waste, Municipal Solid Waste (MSW), commercial waste (sometimes included in MSW, sometimes not) and industrial waste. Solid waste data are often presented in different formats depending on the objectives of the reporter, further obscuring the true figures. Data are often reported as one or even several of the following: Total waste arisings, Total household waste, Total municipal waste, Total residual waste, Total collected waste, Total landfilled waste and other permutations, which depend on local circumstances. The main data sources available are the Organisation for Economic Co-operation and Development (OECD), the European Environment Agency (EEA), countries' own Environmental Agencies and statistical organisations, e.g. Eurostat.

The total amount of waste (excluding agricultural waste) generated in European Union (EU) countries in 1995 was estimated to be 1.3 billion tonnes, while the total amount of MSW generated in Europe was estimated to be 0.21 billion tonnes tonnes (EEA, 1998). Figures 8.1 and 8.2 present data on the waste arisings (including agricultural waste) from a range of OECD (Organisation for Economic Co-operation and Development) countries.

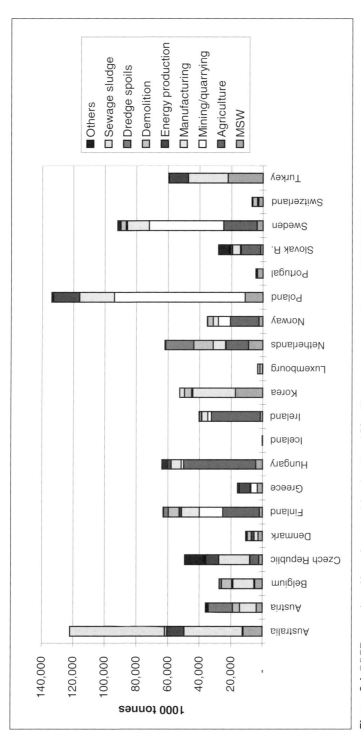

Figure 8.1 OECD countries with total waste arisings less than 150 million tonnes per year. Source: OECD (1997).

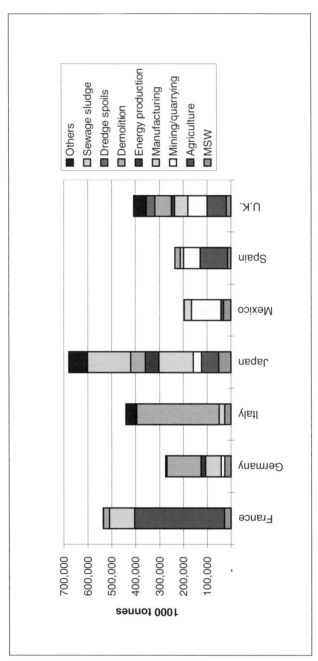

Figure 8.2 OECD countries with total waste arisings less than 1 billion tonnes per year. Source: OECD (1997).

Canada and the USA are not included in Figure 8.2 as their total solid waste arisings are far greater than 1 billion tonnes per year. Canada generates over 1 billion tonnes of mining and quarrying waste per year and the USA generates over 7 billion tonnes of manufacturing waste per year (OECD, 1997).

Although Figures 8.1 and 8.2 represent a recent data set, it remains incomplete. Furthermore, the figures in many cases represent only estimates of the amounts of waste generated. National Governments have openly stated that many figures quoted for waste generation are based on estimates (e.g. UK DoE, 1990).

Historically, solid waste arisings have been measured in tonnages, on disposal, rather than when and where they are generated. Thus where wastes are produced and dealt with at source,

Solid waste category	Description
Agricultural	Waste arising from agricultural practices, especially livestock production. Often either used (applied to land) or treated *in situ*.
Mining and quarrying	Mainly inert mineral wastes, from coal mining and mineral extraction industries.
Dredging spoils	Organic and mineral wastes from dredging operations.
Construction and demolition	Building waste, mainly inert mineral or wood wastes.
Industrial	Solid waste from industrial processes. Sometimes will include energy production industries.
Energy production	Solid waste from the energy production industries, including fly ash from coal burning.
Sewage sludge	Organic solid waste, disposed of by burning, dumping at sea (soon to cease in the EU), application to land or composting. May result from industrial or domestic waste water treatment.
Hazardous/ Special waste	Solid waste, which can contain substances that are dangerous to life, is termed 'Special waste' in UK, or 'Hazardous waste' in EU directives.
Commercial	Solid waste from offices, shops, restaurants, etc. often included in MSW.
Municipal Solid Waste (MSW)	Defined as the solid waste collected and controlled by the local authority or municipality and typically consists of household waste, commercial waste and institutional waste.

Table 8.1 Categories of solid waste

Elements of IWM

as with many agricultural wastes, they are not measured or included in statistics. Also, since waste disposal has not been high on the political agenda in many countries in the past, efforts to maintain up-to-date national statistics on waste disposal, let alone waste generation, have been limited.

Waste classification has traditionally been by source rather than by composition. Because of the different administrative methods used in the countries of Europe and other countries around the world, no universal classification has been adopted. The category 'Municipal Solid Waste' typifies the confusion (Carra and Cossu, 1990; ERRA, 1998). In some countries figures are collected for household waste only (the MSW figure for the UK in Figure 8.2 refers only to household waste), whereas other countries include waste derived from commercial and even light industry. Similarly, figures for solid waste generated during energy production may be quoted separately, or alternatively included under the umbrella of industrial wastes. Clearly 'Solid Waste' is a very diverse category; it has been said that every consignment of waste is unique. Although classification of waste types is difficult, the lack of consistent categories renders comparisons across countries problematic. The most commonly used categories are listed in Table 8.1.

As previously discussed, the lack of reliable statistics for solid waste generation has resulted in wide variability in reported figures. Examples of this variability are shown in Table 8.2 below, which presents different data reported for solid waste generation over a similar time period. Most of this variability can be accounted for by differences in the estimates for agricultural and mining wastes generated. Both may be treated at source, so their generation is hard to assess. Table 8.2 shows how different sources in different countries can arrive at very different estimates of total MSW arisings.

Two key factors contribute to this lack of reliable data: the absence of systematic data collection and the lack of a standard classification for waste.

Country	Source	MSW (in 1000 tonnes)	Year
Austria	ARA	2500	1995
	Dobris II (EEA)	4472	1995
	OECD	3841	1995
Denmark	Ministry of Environment	1200	1993
	Dobris II (EEA)	2708	1994
	OECD	2788	1995
Germany	Dobris II (EEA)	25,777	1993
	Ringel/DSD	34,820	1993
	OECD	25,777	1995
United Kingdom	Dept of Environment	26,000	1996/1997
	Dobris II (EEA)	35,000	1990
	OECD	20,000	1996

Table 8.2 Examples of the differences in data provided by different sources

Discussion about waste management is also hampered by uncertainty over data sources. Lack of reliable data has resulted in the proliferation of reports which, while quoting data, do not reveal their sources. In this book, data that can be traced back to their original source will be used wherever possible.

The lack of consistent definitions and reliable statistics on a national scale does not prevent the development of local or regional Integrated Waste Management schemes, since these require accurate local rather than national data. Most regional authorities will have weighbridge data on the amount of selected wastes that are produced. They will often have to rely on national averages for the composition of the waste as waste analyses are expensive and labour-intensive to conduct.

Solid wastes dealt with in this study

Although the term is widely used in waste management, MSW is not a naturally defined category. It has been defined above simply as the waste controlled and collected by the local authority or municipality. Consequently, there is no uniformity in material composition, merely in how it has been collected, or more precisely, by whom. MSW is potentially the most diverse category of waste, as it consists of waste from different sources, each of which is heterogeneous. For the purposes of this book, MSW sources are defined as:

- **Household waste** – generated by individual households. This includes all solid wastes originating on the property, including garden waste. It may also be termed domestic waste. This is further subdivided into:
  - **Collected household waste** – household waste collected from the property by the waste collection service. Essentially this is the combined contents of the wastebin/bin bag/Blue box.
  - **Delivered household waste** – household waste delivered to a collection point by the householder. This may include bulky waste items (e.g. cookers, refrigerators) and garden waste (in some areas), plus recyclable materials deposited in bring systems (e.g. bottle banks, waste paper collections). This category of waste would not be included in analyses of wastebin contents, and may be overlooked in some statistics.
- **Commercial waste** – waste generated by commercial properties, such as shops, restaurants and offices, which is collected together with household waste.
- **Institutional waste** – waste generated in schools, leisure facilities, hospitals (excluding clinical waste), etc. This is often included within the commercial waste category.

Quantities of MSW generated

MSW represents a small but significant proportion of total solid waste arisings, accounting for about 7% of total solid waste production in Europe (Figures 8.1 and 8.2) and 3% in North America. When broken down by country, it can be seen that there are regional differences in the amounts generated per person (Figures 8.3–8.6). Making allowances for the possible differences between countries in what is included within MSW, Norway and The Netherlands appear to

Elements of IWM

Elements of IWM

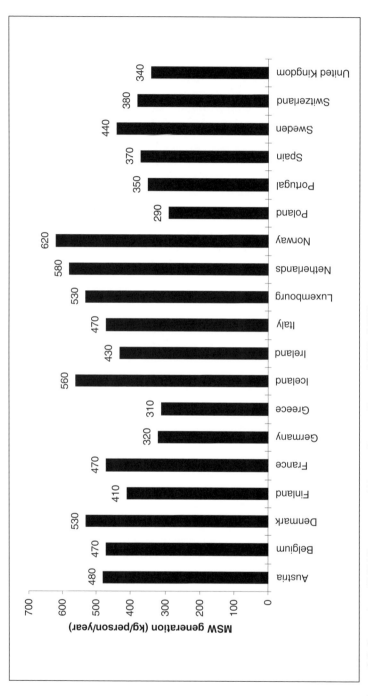

Figure 8.3 Generation of MSW by country – Western Europe. Source: OECD (1997).

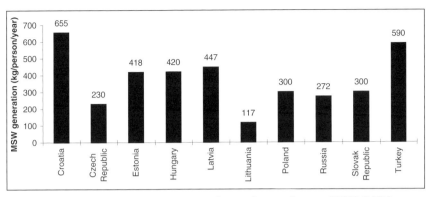

Figure 8.4 Generation of MSW by country – Eastern Europe. Source: OECD (1997).

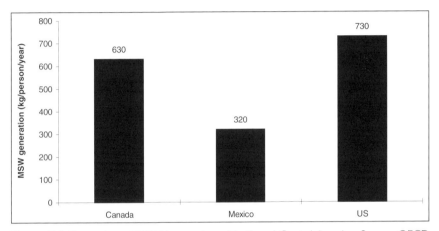

Figure 8.5 Generation of MSW by country – North and Central America. Source: OECD (1997).

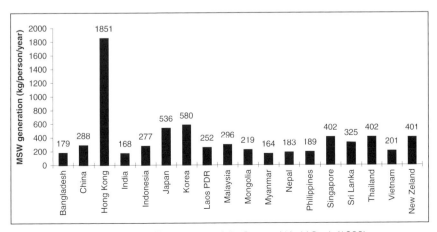

Figure 8.6 Generation of MSW by country – Asia. Source: World Bank (1999).

have the highest *per capita* generation of MSW in Europe. All Western European countries have lower generation rates than both Canada and the USA. Eastern European countries have on average lower generation rates than Western European countries. Developing Asian countries have lower generation rates than Eastern European countries, while the developed Asian countries have generation rates similar to that of Western European countries, except for Hong Kong, which has the highest *per capita* waste generation rate in the world. This is likely to be related to its rapidly growing economy and high population density.

Composition of MSW

Data on the generation of MSW are difficult to interpret because of the different definitions used for MSW. National data that breaks down MSW according to the source of the material, that is into collected household waste, delivered household waste, commercial waste and institutional waste are equally hard to obtain. Collected household wastes are normally the best docu-mented, with poorer data available for delivered household and commercial wastes (see Table 8.3).

Complete and up-to-date figures on this split into MSW sources are essential when planning an Integrated Waste Management system, since the source of the MSW will determine the collection strategy necessary. Such data may be available in a geographical area, though not consolidated into a national figure.

Composition of MSW – by materials

Knowledge of the material composition of MSW is essential for effective management and disposal. The composition of MSW by weight, compiled from various sources for a range of countries is given in Tables 8.4–8.7.

Country	Collected household	Delivered household	Commercial	Total	Reference
Denmark	32%	22%*	46%†	100%	Christensen, 1990
France	30%	6%‡	64%§	100%	Barres *et al.*, 1990
Germany (West)	62.5%	8.5%¶	29%§	100%	Stegmann, 1990
The Netherlands	66%	8.5%¶	25.5%**	100%	Beker, 1990
UK	65%	20%	15%	100%	DoE/ERL, 1992

Table 8.3 Composition of MSW by source for selected countries
*Bulky waste plus garden/park waste; †Commercial plus industrial waste; ‡Bulky household waste, plus car wrecks and tyres; §Industrial waste similar to household refuse; ¶Bulky waste; **Shop/office/service waste

Europe	Waste type	Year	Paper/board (%)	Plastics (%)	Glass (%)	Metals (%)	Food/garden (%)	Other (%)
Austria	Household	1995	27	14	11	7	30	11
Belgium	MSW	1995	16	7	7	4	37	29
Denmark	Household	1994	20	5	4	2	36	33
Finland	MSW	1992	26		6	3	32	33
France	Household	1993	25	11	13	4	29	18
Germany	Household	1995	18	5	9	3	44	21
Greece	MSW	1995	20	9	5	5	49	12
Iceland	Household	1995	34	10	9	13	24	10
Ireland	MSW	1995	33	9	6	3	29	20
Italy	MSW	1995	22	7	6	3	43	19
Luxembourg	Household	1994	19	8	7	3	44	19
The Netherlands	Household	1994	26	6	6	3	38	21
Norway	MSW	1995	31	6	4	5	18	36
Poland	MSW	1995	10	10	12	8	38	22
Portugal	MSW	1994	23	12	5	3	35	22
Spain	MSW	1994	21	11	7	4	44	13
Sweden	MSW	1994	44	7	8	2	30	9
Switzerland	MSW	1995	29	15	3	3	38	12
United Kingdom	Household	1995	37	10	9	7	19	18

Table 8.4 Composition of MSW (by weight) in Western Europe. Source: OECD (1997)

Eastern Europe	Waste type	Year	Paper/board (%)	Plastics (%)	Glass (%)	Metals (%)	Food/garden (%)	Textiles	Other (%)
Croatia	MSW	1995	19.6	7.3	3	2.3	31.2		36.6
Czech Republic	MSW	1994	8	4	4	2	18		64
Estonia	MSW	1994	8.1	3	7.4	4.3	53	5.1	19.1
Hungary	MSW	1997	19	4.4	3	3.8	32.3	3.6	33.9
Latvia	MSW	1998	5	12	7	8	50		18
Lithuania	MSW	1998	16.9	8	4	2.5	50		18.6
Russia	MSW	1997	31.9	4.3	5.5	3.6	33.8	3.9	17
Slovak Republic	MSW	1985	14	7	9	7	16		47
Turkey	Household	1993	6	3	2	1	64		24

Table 8.5 Composition of MSW (by weight) in Eastern Europe. Source: Enustun (1999)

North and Central America	Waste type	Year	Paper/board (%)	Plastics (%)	Glass (%)	Metals (%)	Food/garden (%)	Other (%)
Canada	MSW	1995	28	11	7	8	34	12
Mexico	MSW	1995	14	4	6	3	52	21
USA	MSW	1995	39	9	6	8	21	17

Table 8.6 Composition of MSW (by weight) in North and Central America. Source: OECD (1997)

Asia	Waste type	Year	Paper/board (%)	Plastics (%)	Glass (%)	Metals (%)	Food/garden (%)	Textiles	Other (%)
Bangladesh	Dom,* Comm†	1992	6	2	3	3	84		2
China	Dom, Comm	1998	4	4	2	0.5	35.5		54
Hong Kong	MSW	1994	26	14	2	2	29	4	23
India	MSW	1995	6	4	2	2	42		44
Indonesia	MSW	1993	11	9	2	2	70		6
Japan	MSW	1990	38	11	7	6	32	6	
Korea	MSW	1995	24	5	5	8	32	26	
Laos PDR	Dom, Comm	1998	3	8	9	4	54		22
Malaysia	Not reported	1990	24	11	3	4	43		15
Myanmar (Burma)	Dom, Comm	1993	4	2	0	0	80		14
Nepal	MSW	1994	7	2	3	1	80		7
Philippines	Not reported	1995	19	14	2	5	42		18
Singapore	MSW	1990	28	12	4	5	45		7
Sri Lanka	Dom, Comm	1994	11	6	1	1	76		5
Thailand	Not reported	1996	15	14	5	4	48		14
Vietnam	Domestic	1995	3	0	0.5	1	52	1	42.5
Australasia									
Australia	MSW	1990	22	7	9	5	50		7
New Zealand	Household	1995	26	11	5	4	47		7

Table 8.7 Composition of MSW (by weight) in Asia and Australasia (*domestic; †commercial). Source: OECD (1997) and World Bank (1999)

The data are not directly comparable, since most of the figures relate to MSW, while some relate specifically to household waste, which is only one part of MSW in most countries. Some of the data also refer to a combination of domestic and commercial waste arisings. There are clear differences in the composition of collected household waste, delivered household waste and commercial waste (see Figures 8.7 and 8.8) so it is important to specify precisely the type of waste analysed for any data quoted. As with the above data for amounts generated, household waste (and in particular collected household waste) is the best documented part of MSW in terms of materials composition. Data for the commercial fraction of MSW are harder to obtain and vary from area to area. Commercial waste often forms a significant part of MSW, however (see Table 8.3) and can have an important role to play in improving the economics of recovery and recycling operations (IGD, 1992). There is therefore also a need to understand the amount and material composition of this portion of MSW within the context of an Integrated Waste Management system.

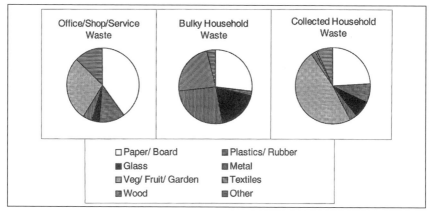

Figure 8.7 Composition of Different Parts of MSW in The Netherlands. Source: Beker (1990).

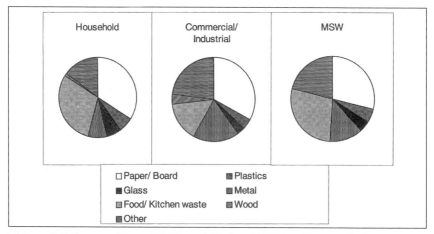

Figure 8.8 Composition of Different Parts of MSW in Denmark. Sources: Carra and Cossu (1990); Christensen (1990).

Despite these differences in the way data have been compiled, some trends in MSW composition can be seen from Tables 8.4–8.7. The two major fractions in all countries are paper/board and food/garden waste. Plastics, glass and metals occur at much lower levels. There is, however, evidence of geographical variability. Southern European countries (e.g. Spain, Portugal, Italy) and the developing countries generally have a higher level of food/garden waste than northern countries (e.g. Finland, Denmark, France, UK, USA and Canada) where consumption of pre-processed and packaged food is high, whereas paper and board show the opposite trend (Figure 8.9). The food/garden waste fraction of the waste stream in Asian countries tends to be even higher than that of the Southern European countries as virtually all of the food consumed in this region is fresh and unprocessed. The preparation of meals therefore generates significant quantities of organic waste. The 'Other' fraction of the waste stream can be seen to be relatively high in Eastern Europe and Asia. In these cases a significant amount of this fraction is often ash/dust from solid fuel cooking or heating.

Patterns are less easy to detect in the proportion of plastics, glass and metals in MSW, although there are interesting individual points, such as the high proportion of plastic waste in

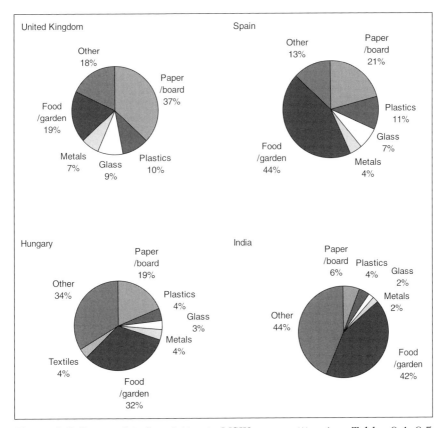

Figure 8.9 Geographical variation in MSW composition (see Tables 8.4, 8.5 and 8.7). Sources: OECD (1997); World Bank (1999).

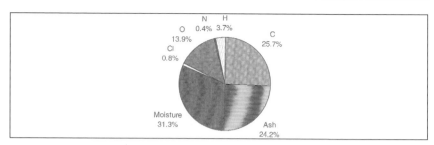

Figure 8.10 Ultimate analysis of Canadian MSW. Source: Environment Canada (1988).

Switzerland, high levels of glass in France and Poland, and high levels of metals in Iceland. By contrast, in the developing countries low levels of plastic waste occur in Bangladesh, Mayanmar (Burma) and Vietnam; low levels of glass in Sri Lanka, Myanmar and Vietnam, and low levels of metals in China, Myanmar, Nepal, Sri Lanka and Vietnam.

Composition of MSW – by chemical analysis

A third valid way to classify MSW is into its constituent chemical elements. Since the solid waste represents the largest input into the overall solid waste system, the composition of the waste will determine the majority of the emissions from the overall system. If the chemical composition of the incoming waste, and of individual fractions, is known, some of the emissions from waste treatment processes can be predicted (see Figure 8.10).

This is particularly true for inorganic trace contaminants such as heavy metals (see Table 8.8), which generally pass unchanged through the waste management process. Consequently, the total amount released will reflect the total input in the waste. Knowing how the heavy metals enter the waste stream allows efforts to be made either to reduce the levels of these contaminants, or to ensure that they are effectively handled.

Two other useful characteristics of waste fractions are their carbon content (which allows calculation of emissions of carbon dioxide, methane, etc.), and their water content (which varies markedly between different waste fractions and will affect their calorific value).

Variability in MSW generation

When planning an MSW management system for any given region, it is important to have relevant and recent data for that region, since waste generation will vary geographically, both between and within countries. The composition of commercial waste will clearly vary according to the nature of local commerce. Also, both the amount and composition of household waste will vary with population density and housing standards. Rural areas, for example, are likely to have a greater amount of vegetable, fruit and garden waste than inner city areas (see Table 8.9).

Differences have also been measured between areas depending on the type of domestic heating installed. In former East Germany, for example, areas with domestic open fires burning brown coal produced up to 190 kg of waste per person per year, whereas areas with central

	Cadmium	Nickel	Zinc	Copper	Lead	Mercury	Chromium
Total load (mg/kg dry)	3–15	80	1000–2000	200–600	400–1200	4–5	250
			% of total load				
Fines < 10 mm	1–2	**12–13**	5	7	5–7		2–3
Fines 10–20 mm	1–2	**16–19**	5	**33–39**	**13–16**	1	
Organics	2–3	9–11	5	4–6	5–13	2	4–6
Paper/carton	1–2		8–9	7–8	**18–19**	2–4	8–12
Textiles	2	3–4	1	1–2	1	1	3–4
Leather	4	3	1–2	3–8	1		**39–50**
Rubber			**11–13**		2		
PVC	**36–40**						
Other plastic	**13–14**	**24–25**	3–4	4–7	8–9	1	8–9
Glass		6–10	1	1	2		9–12
Ferrous metal			1	**27–31**	**35–41**	1	**3–26**
Non-ferrous	6–7		**12–13**	2–31	1		1–2
Batteries	**39–48**	**20–22**	**44–47**	12	1	**93**	

Table 8.8 Presence of heavy metals in household waste fractions as a percentage of total load (figures in bold type indicate significant load). Source: Rousseaux (1988)

Population density (persons/km^2)		Waste fractions		
		Paper	Glass	Fruit/vegetable/garden
Inner city	2000	24%	20%	28%
Suburban	1000	20%	14%	34%
Urban	500	16%	11%	39%
Rural	150	12%	8%	52%

Table 8.9 Effect of population density on waste composition. Source: Rheinland-Pfalz, Ministry of Environment (1989)

Waste fraction	Regions with central heating	Regions with open coal fires
Metal	3.2%	3.2%
Glass	12.3%	10.2%
Plastics	5.5%	3.6%
Textiles	3.5%	2.8%
Paper/board	24.5%	10.9%
Wood	0.5%	0.2%
Bread	6.4%	3.4%
Fine material (<16 mm)	11.7%	41.0%
Other	32.4%	24.7%
Total	100%	100%

Table 8.10 Variation of household waste composition in former East Germany with domestic heating type. Source: von Schoenberg (1990); data from Dresden (1988)

heating produced up to 260 kg/person/year (Bund, 1992). Differences in waste composition also occurred, as might be expected, with less paper waste in houses with open fires, but more fine material, i.e. ash (Table 8.10).

Even within a given area, there will also be seasonal effects. The composition and amount of household waste generated will vary especially over holiday periods, and the amount of garden waste included will clearly vary with the seasons. Thus, while the figures quoted here will give a guide to both amounts and composition, they will not always reflect local conditions, nor a particular time of the year.

Effects of source reduction

The amount and composition of solid waste generated by a region will also be affected by any attempts to promote source reduction. A form of source reduction that is on the increase is the promotion of home composting. A variety of home composting units are commercially available that are capable of dealing with both garden and kitchen wastes, and some municipalities have offered these to residents in an effort to reduce the amount of such wastes that need collection and treatment. The Adur Home Composting Scheme (Adur District, West Sussex, UK) for example, has a district-wide participation rate of 22% (ERRA, 1994). An initial survey suggested that around 13% of household waste (by volume) could be diverted from the normal collection system in this way. A similar home composting trial in Luton, UK where 40% of residents accepted a free composter, showed that the trial area actually produced 11% more waste during the trial period of 8 months than before the trial started (Wright, 1998). An analysis of the trial areas waste during this period may have enabled the identification of the extra waste being generated. It is likely that the organic fraction being diverted into the composter was partially replaced by non-biodegradable garden waste (soil and rubble) and the tendency for people to fill whatever size of bins they are given.

Since the LCI boundary used in this study is the waste leaving the householder's property, home composting is not considered within the waste management system modelled. However, it is an effective means of source reduction, and can reduce the amount of waste requiring treatment.

MSW classification – the need for standardisation

It is clear from the previous pages that to approach waste management even on just a European scale would require considerable standardisation of terminology. Whilst waste management practices will continue to vary from country to country, it is important that lessons from one area can be disseminated widely and implemented elsewhere. A clear and common understanding of what is included under the categories household waste (collected and delivered), commercial waste and MSW is required.

Standardisation of waste material categories is similarly essential, but more detailed classifications are also required. The most effective method of dealing with any particular waste item will depend on what it is made of, so detailed knowledge of the composition of waste is a prerequisite for effective waste management. Whilst the classification scheme used in Tables 8.4–8.7 gives useful general trends, there is insufficient detail for waste management purposes. Plastics, for example, may exist as thin films, rigid bottles or as a multitude of other objects. Knowing that they are made of plastic may give a guide to their suitability for Energy from Waste schemes, but further knowledge of their form is necessary to determine their suitability for material recycling. To fill this need, more detailed waste classification schemes have been devised. In the UK, Warren Spring Laboratory have analysed household waste using a 33-category classification (see Table 8.11).

Analysis of this kind gives a much more detailed picture of what items are in waste, but it still does not define the composition of all items (for example, plastic resin type is not specified). To meet this requirement the European Recovery and Recycling Association (ERRA) has proposed

Elements of IWM

Category	% by weight
Newspapers	12.29
Magazines	4.99
Other paper	10.08
Liquid containers	0.64
Card packaging	3.81
Other card	2.98
Refuse sacks	1.15
Other plastic film	4.27
Clear plastic beverage bottles	0.65
Coloured plastic beverage bottles	0.12
Other plastic bottles	1.15
Food packaging	1.96
Other dense plastic	2.04
Textiles	2.17
Disposable nappies	3.87
Other miscellaneous combustibles	3.56
Miscellaneous non-combustibles	1.66
Brown glass bottles	0.36
Green glass bottles	1.08
Clear glass bottles	1.32
Clear glass jars	1.55
Other glass	4.83
Garden waste	3.15
Other putrescible material	16.62
Steel beverage cans	0.50
Steel food cans	3.86
Batteries	0.05
Other steel cans	0.39
Other ferrous metal	0.99
Aluminium beverage cans	0.40
Foil	0.46
Other non-ferrous metal	0.64
−10 mm fines	6.39
Total	100.00

Table 8.11 Household waste classification scheme developed by Warren Spring Laboratory. Figures give an average of UK waste analyses. Source: Warren Spring Laboratory (UK) (1994)

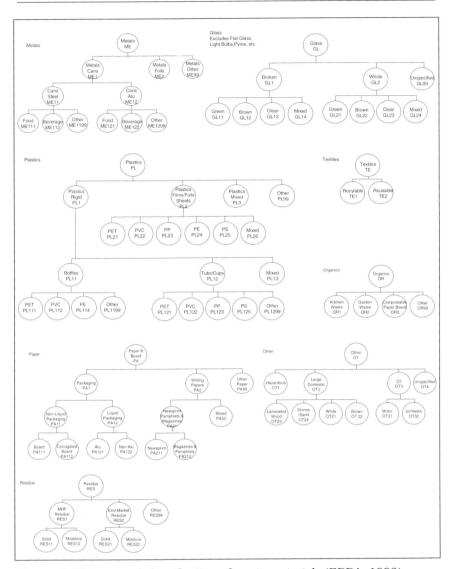

Figure 8.11 Proposed classification of waste materials (ERRA, 1993).

an hierarchical classification system, which specifies not only the form of waste items (films, bottles, cans, etc.) but also the material (see Figure 8.11). If this proposal is adopted as a standard, information flow on waste composition will be significantly improved. A simplified version of this ERRA classification is used as the basis for the LCI model in this book (see Table 8.12).

Elements of IWM

Category	Description
MSW fractions	
Paper (PA)	Paper, board and corrugated board, paper products.
Glass (GL)	Glass bottles (all colours), sheet glass.
Metal (ME)	All metals including cans
	Further subdivided into: ferrous (ME-Fe) and non-ferrous (ME-nFe).
Plastic (PL)	All plastic resin types, including bottles, films, laminates
	Further subdivided into rigid plastics (PL-R) and plastic film (PL-F).
Textiles (TE)	All cloth, rag, etc. whether of synthetic or natural fibres.
Organic (OR)*	Putrescible kitchen and garden waste, food processing waste.
Other (OT)	All other materials, including fines material, leather, rubber, wood.
Waste treatment residues	
Compost (CO)	Residues from biological treatment (composting or anaerobic digestion, that cannot be marketed as products due to contaminant levels or lack of suitable markets).
Bottom ash (AS) (calculated)	Bottom ash, clinker or slag from incinerators RDF or alternate fuel boilers (non-hazardous material).
Filter dust (FD) (calculated)	Fly ash and residues from gas cleaning systems from incinerators, RDF or alternate fuel boilers (hazardous material).

Table 8.12 Classification of solid waste used in the IWM-2 LCI model. *Paper and plastic fractions are also strictly of organic origin, but to maintain alignment with the ERRA classification system, the term 'organic' is used here to describe putrescible kitchen and garden waste only

Elements of IWM

MSW analysis methods

An integrated approach to waste management clearly requires better waste statistics in the future. Use of a more detailed and standardised classification scheme is only part of the solution. Standardised and appropriate sampling and analysis techniques are also required.

Methods for household waste analysis have been developed from techniques originally used to sample mineral cores (e.g. Poll, 1991). Such methods suffer from two major limitations. Firstly, MSW is much more heterogeneous in particle size than mineral samples, so different techniques need to be applied in the taking of samples. Secondly, many sampling procedures require the use of specialised equipment. The need for widespread sampling at many waste collection and treatment facilities precludes the routine use of such equipment; simple but robust sampling techniques are required. In line with this, ERRA proposed a simplified waste analysis procedure (ERRA, 1993), which gives guidelines on both the techniques to be used, and the number of samples required for statistical analysis.

Sampling methods have been applied mainly to collected household waste rather than the more bulky and heterogeneous delivered household waste. There is little evidence of systematic sampling and analysis of commercial waste, although this tends to be more homogeneous than household waste.

Information on waste composition also comes from various collection and recovery schemes (bottle banks, kerbside collection schemes, Material Recovery Facilities). These are valuable sources of data, but here again there are problems in the interpretation of data due to lack of standardisation in the terminology. The terms capture rate, recovery rate and recycling rate can vary widely in their intended meaning. To overcome this, ERRA has also published proposed definitions of such 'programme ratios' (ERRA, 1992b) to clarify the confusion existing in many reports.

Elements of IWM

CHAPTER 9

Waste Collection

Summary

This chapter emphasises the importance of the collection operation in Integrated Waste Management. It looks at the processes of home sorting and waste collection, from the creation of waste up to its delivery at a central sorting or treatment site. The characteristics and effectiveness of different collection methods are discussed, including both collection of separated fractions and the collection of co-mingled materials. The limitations of the common division into 'bring' and 'kerbside' schemes when comparing systems are emphasised, as is the need for effective communication between waste collectors and waste generators.

Introduction

There are good reasons why collection lies at the very centre of an Integrated Waste Management system (see Figure 2.4). The way that waste materials are collected and sorted determines which waste management options can subsequently be used, and in particular whether methods such as materials recycling, biological treatment or thermal treatment are feasible with respect to economic and environmental sustainability. The collection method significantly influences the quality of recovered materials, compost or energy that can be produced; this in turn determines whether markets can be found. The importance of markets for recovered materials cannot be over-emphasised: in the absence of suitable markets, useful products cannot be produced. Therefore, either the collection method defines the subsequent treatment options, or taking the reverse case, the existing or potential markets will define how materials should be collected and sorted if they are to be recovered. In any event, there must be a match between market need and the materials collected and sorted.

Waste collection is also the contact point between the waste generators (in this case households and commercial establishments) and the waste management system, and this relationship needs to be carefully managed to ensure an effective system. The householder–waste collector link needs to be a customer–supplier relationship (in the Total Quality sense (Oakland, 1989)). The householder needs to have his/her solid waste collected with a minimum of inconvenience, whilst the collector needs to receive the waste in a form compatible with planned treatment methods. There is clearly a balance to be struck between these competing needs; waste management systems that fail to achieve balance in this relationship are unlikely to succeed.

193

Collection operations are rarely independent of sorting operations, since the type of collection will determine the amount of subsequent sorting needed, and some collection methods themselves involve a level of sorting (e.g. 'Blue Box' recovery schemes). This chapter focuses on the collection process, including any household, kerbside or bank sorting i.e. pre-sorting, but leaves centralised sorting until Chapter 10.

Home sorting

From the householder's viewpoint, co-mingled collection of all solid waste together probably represents the most convenient method, in terms of both time and space requirements. This collection method will limit, however, the subsequent options for treatment. Most treatment methods will require some form of separation of the waste into different fractions at source, i.e. in the home, prior to collection. At its simplest this might involve removing recyclable materials, e.g. glass bottles for delivery to a bottle bank; more extensive sorting involves the separation of household waste into several different material streams. The degree of home sorting achieved in any scheme will be a function of both the ability and, especially, the motivation of householders.

Sorting ability

Several schemes and pilot tests have demonstrated that, given clear guidance, householders are able to accurately sort their solid waste into different categories. A study carried out in Leeds, UK, for example, showed that householders could sort their waste into six different categories with a 96.5% success rate (Forrest et al., 1990). A US study showed a similar result (Beyea et al., 1992). Clear instructions to the householder are essential for success, to which end many schemes run extensive communications programmes and publish frequent newsletters.

Sorting motivation

The above sorting experiment in Leeds showed that accurate sorting of household waste is possible, but participants in the trial were volunteers, and therefore likely to be highly motivated. Is it realistic to expect that the majority of householders will be similarly motivated? Whilst the degree of sorting in the Leeds study may be excessive, participation rates in other schemes with more limited sorting demands suggests that householders can be motivated.

Participation rates are very difficult to measure, since what people claim that they do, and what they actually do, are not the same. ERRA define the participation rate by the number of waste generators putting out their recyclable materials at least once in a 4-week period (ERRA, 1992b). A range of ERRA schemes report participation rates in voluntary home sorting and collection schemes between 60 and 90% (ERRA, 1993c); in one scheme (Adur, UK) the rate was measured and found to be 75% (Papworth, 1993). Similar high levels of voluntary participation are reported from North America; Blue Box schemes in Ontario had measured participation rates of 85–91% in 1992 (Quinte, 1993) and 80–90% in 1995 (RIS, 1996).

Where consumer research has been carried out to seek the views of householders the most frequent comments voiced were that recycling was seen as a good idea, and that it 'helps the environment' (IGD, 1992; RIS, 1996).

Participation rates for voluntary schemes will depend also on the economics. If householders have to pay for an additional recylables container participation rates will be lower; if households are offered a cost reduction for having less non-recoverable material in their restwaste bin, participation rates are likely to be higher. In some schemes such as Lemsterland in The Netherlands, participation is not voluntary, no alternative waste collection is provided. Separation of certain fractions of the waste at source is required by law in some countries (e.g. separation of organic material in The Netherlands). In such cases participation rates are likely to be higher still.

Overall recovery rates for waste materials depend not only on the number of households participating, however, but also on the householder's sorting efficiency. The actual amount of any material recovered from household waste by home sorting can be calculated by:

Amount of material in household waste stream × participation rate × separation efficiency

(ERRA, 1993b).

Even if participation is compulsory, motivation is still required to ensure a high level of sorting efficiency, see Table 9.1.

Amount of material recovered = Amount of material in waste stream
× Participation rate × Separation efficiency

Amount of material in waste stream: See Chapter 8.

Participation rate: % of householders providing sorted material at least once per month

Separation efficiency: % of material correctly sorted and separated

Both participation rate and separation efficiency will be influenced by:

Level of convenience:
Amount of sorting
Difficulty of sorting
Frequency and reliability of collection
Extra storage space required
Distance of collection point
Hygiene problems

Level of motivation:
Quality and frequency of communications
General environmental awareness/concern
Peer pressure
Legal requirements
Availability of alternative disposal routes
Cost reduction/rebate for producing less waste

Table 9.1 Influences on material recovery

Elements of IWM

Suburban housing		High-rise apartment buildings	
Location	Participation rate (%)	Location	Participation rate (%)
Amsterdam	89	Amsterdam	65
Apeldoorn	91		
Purmerend	90	Purmerend	62
Medemblik	97		
Ede	87	Ede	72
Nuenen	90		

Table 9.2 Participation in source-separation programmes in The Netherlands. Source: Kreuzberg and Reijenga (1989)

Motivation, and hence both participation rates and separation efficiency, will be influenced by factors such as the level of convenience or inconvenience to the householder. Schemes with extensive home sorting may require too much time or too much space to store the separate waste streams before collection. Research in North America has shown that co-mingled collection schemes actually increase final diversion rates (Skumatz et al., 1998). The frequency of collection may influence participation rates, but less frequent collection of recyclables has been shown not to significantly lower the final diversion rate of kerbside recycling schemes (Skumatz et al., 1998). Reliability of collection has been seen to be closely linked to participation rate, a reduction in participation rate from 68% to 37% in one area of a kerbside collection programme in Sheffield, UK was related to an unreliable collection service (ERRA, 1996). Any loss in comfort level, for example if odour becomes a problem when organic material is not collected regularly, will lower motivation levels. Housing type also has an effect: data from The Netherlands suggest that occupants of high-rise buildings are less likely to participate in source separation programmes than those in suburban areas (see Table 9.2). This may reflect a lack of storage space, but is also likely to be due to lack of social peer pressure in such buildings, as it is not possible for neighbours to see who is participating i.e. who is environmentally responsible. Data from Canada (Table 9.8) shows that even after intensive communication campaigns the amounts of recyclables collected from multi-family high-rise buildings remain significantly lower than that collected from single-family housing in urban, suburban or rural areas.

Levels of environmental awareness vary geographically across Europe and North America, but overall, there appears to be a willingness on the part of householders to participate in some type of home sorting. Assessing the prevailing level of motivation in any particular area and matching the collection scheme to this will achieve the best level of home sorting obtainable from a given area.

Bring versus kerbside collection systems

Collection methods are often divided into 'bring' and 'kerbside' collection schemes. ERRA (1993b) defined bring collection systems as those where 'householders are required to take recyclable materials to one of a number of (communal) collection points'. In kerbside collection schemes, the 'householder places recoverables in a container/bag which they position, on a specific day, outside their property for collection'. Note that collection need not be, literally, from the kerbside, the key distinguishing point being that in bring systems, the householders transport the materials from their home, whereas in kerbside collection they are collected from the home. In reality however, bring and kerbside are just the two ends of a spectrum of collection methods (Figure 9.1). The extreme form of bring system is the central collection site, variously called a Civic Amenity site (UK), Déchétterie (France), or Recyclinghof (Germany), to which householders transport materials such as bulky items and garden waste. Such sites often also have collection containers for recyclable materials such as glass bottles and cans. Next in the spectrum of bring systems come materials banks at low density (i.e. a high number of connected inhabitants), often situated locally at supermarkets. As the density of bring material containers increases, they become 'close-to-home' drop off containers (ERRA, 1993b), to which householders can walk rather than drive. This applies particularly to high-rise housing, where residents of apartment blocks usually take their waste (and recyclables) to large communal containers positioned outside the apartment blocks or at the side of the street. This is essentially a waste container in the street, outside, rather than inside, the property. The only difference between this 'bring' system and a kerbside collection from individual properties is that the containers are communal, rather than for individual households.

The term 'bring system' clearly includes a range of different schemes. Kerbside collection is more narrowly defined, but collection can also be of separated fractions or of co-mingled waste. As a result, blanket comparisons of 'bring' versus 'kerbside' approaches must be made with caution. Table 9.3 lists some of the common attributes of these categories. It can be seen that some attributes, particularly contamination, depend more on whether the material is collected

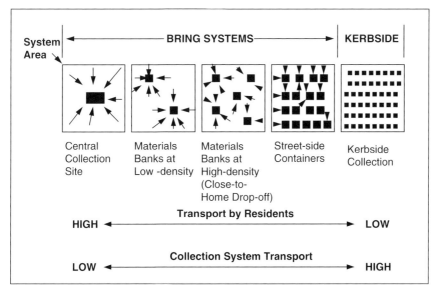

Figure 9.1 The spectrum of collection methods from 'bring' to 'kerbside' systems. (Arrow lengths indicate distances travelled by residents to collection points.)

	Bring	Kerbside
Definition:	Materials taken from property to collection point by householder	Materials collected from property/home
Sorting:	Sorted by householder. May or may not be centrally sorted	Sorted by householder. May also be sorted at kerbside and/or centrally
Materials collected:	Separated materials or mixed materials	Separated materials or mixed materials
Containers:	Communal	Individual (may be communal for apartments)
Consumer Transport needed:	High ◄————► Low	None
Collection Transport needed:	Low ◄————► High	High
Amount collected:	Low ◄————► High (depends on bank density)	High (assuming effective motivation)
Contamination level:	Low: (separate collection) to High: (mixed collection)	Low: (kerbside sorted, e.g. Blue Box) to high: (mixed collection)

Table 9.3 Attributes of 'bring' and 'kerbside' collection systems

separated or mixed, than on whether a 'bring' or 'kerbside' approach is used. Collection systems will therefore be discussed in the following section according to the materials collected, rather than whether bring or kerbside systems are involved.

Household waste has traditionally been collected mixed, but where household sorting has occurred, the different waste streams are collected separately, whether by the same or different collection vehicles. The categories collected separately vary by geography: in Germany, for example, the Duales System Deutchland (DSD) collects packaging material as a separate stream, whereas in Japan householders separate out combustible material for separate collection. In Europe and North America, separate collections are most commonly used for dry recyclables (paper, metal, glass, plastic), biowaste (kitchen and garden waste, with or without paper) and in some countries, household hazardous waste (batteries, paint, etc.). A collection for remaining residual waste (known as restwaste) is also needed. Garden waste and bulky waste may be handled as separate streams, or alternatively included within the biowaste and restwaste streams, respectively.

Collection systems

Dry recyclable materials

This category employs the greatest range of collection methods, from central or low density materials banks, to kerbside collection of recyclable materials in specially designed trucks.

Single (mono) material banks

Materials banks ('drop-off') that collect a single material per container, represent one of the best known forms of materials collection, mainly due to the success of 'bottle banks' for glass. High levels of glass collection have been achieved using this method, though there is considerable variation between countries (Table 9.4). A similar pattern of increasing collection rates can be seen for paper and cardboard, which are often considered as a single material type (Table 9.5). Other industries (steel and aluminium) have tried to match these successes.

The success of materials banks at low density (high number of connected inhabitants) for materials recovery is hard to assess. At the country level, the recovery rate can be calculated by dividing the total amount of material recovered by the national consumption of that material. At the local level, in contrast, it is not clear what area a bring container covers, and, therefore, the base amount of material from which the collected material has been recovered. Committed recyclers, for example, may travel some distance to a bring container, so importing waste from outside the area considered to be covered by the bring scheme. The best estimates of success rates will come from relatively isolated communities, which have a saturation density of bring containers, and the data is best expressed in terms of amounts collected per person, or per household, rather than percentage recovery rates, since the latter depends on what figure is used for the base amount of waste.

In a bring system, the amount collected will depend on the density of banks or containers, since this will determine how far individuals will have to transport their recyclables to the bank, and thus their motivation. Differences in bank density probably account for a large amount of the geographical variation in glass recovery rates in Table 9.4. Lemsterland has one bank of three containers for every 500 inhabitants; the density for The Netherlands as a whole is around one bank per 750 inhabitants (Cooper, 1998). Increasing bank density will increase the amount collected, but with diminishing returns, i.e. the extra amount collected will decrease with every extra bank collected. At a certain point, the additional economic cost and environmental burdens of emptying and servicing banks will outweigh the environmental gains from the collection of material, though at present there is insufficient evidence to identify this optimal bank density.

A more immediate problem with increasing bank density is finding suitable sites. At high densities, small containers on street corners may be suitable, and inhabitants should be able to walk to these with their recyclable materials. At lower densities with necessarily longer transport distances, car transport is likely to be used. These banks need to be placed in strategic sites that are already regularly visited (petrol stations, supermarkets, etc.) so that specific car journeys to bank sites are not necessary. Unfortunately, consumer behaviour in using bring systems is another area in which reliable data are lacking, though of prime importance. Given that the energy saving possible from the process of recycling glass (transport not included) is approximately 3.5 MJ/kg (BUWAL, 1998, see Chapter 14), and the fuel consumption of a car is approximately 2.7 MJ per kilometre (ETSU, 1996), the need to minimise specific car journeys to materials collection banks is clear.

Year	80	81	82	83	84	85	86	87	88	89	90	91	92	93	94	95	Tonnes collected in 1995*
Austria		20	20	20	30	38	39	44				60	64	68	76	76	199,000
Belgium		33	32	32	36	42	44	39				55	54	55	67	67	225,000
Canada	12					12							17				
Denmark	8	8	10	10	20	19	32	32				35	48	64	67	63	104,000
Finland	10					21		25			36	31	44	46	50	50	30,000
France		20	20	24	25	26	28	26		29	29	41	44	46	48	50	1,400,000
Germany	23	28	32	36	38	43	45	49	49	53	54	61	60	65	75	75	2,784,000
Greece	15					15					15	17	20	20			38,000
Iceland											70	75	75				
Ireland	8	8	8	8	7	7	8	8			23	23	27	29	31	39	38,000
Italy	20	20	21	22	24	25	26	38			48	53	53	52	54	53	869,000
Japan	35		42	41	42	47	55	54	49	48	48	52	56				
Korea											46	45	43	44	46	57	
Mexico												4	4	4	4	4	
Netherlands	17	27	35	42	47	49	49	50	52	55	67	70	73	76	77	80	372,000
Norway												22	44	67	72	75	39,000
Portugal			12	12	10	10	13	14	14	24	27	30	30	29	32	42	91,000
Spain			12	12	13	13	20	22			27	27	27	29	31	32	402,000
Sweden						20			22	22		44	58	59	56	61	96,000
Switzerland		36	42	42	45	46	46	47		55	65	71	72	78	84	85	263,000
Turkey						33	25	27	27	33	31	28	25	23	22	12	36,000
UK	5	5	6	8	9	12		14	14	17	21	21	26	29	28	27	501,000
USA	5				8						20			22	23	25	2,750,000

Table 9.4 National glass recycling rates (% of apparent consumption). Source: OECD (1997); *FEVE (Fédérations Européene du Verre d'Emballage, Brüssel (1999)

Year	80	81	82	83	84	85	86	87	88	89	90	91	92	93	94	95
Austria	30	30				37					37				66	65
Belgium		15		14	14	14	14	14							14	12
Canada	20					23					28		33			
Denmark	26					31	30	29	30	30	35	35	36	46	43	44
Finland	35					39				40	41		48	46	43	57
France	30	32	33	33	34	35	34	35	34	34	34	34	34	36	36	38
Germany	34	36	36	37	38	43	42	42	43	44	44	47	50	55	59	67
Greece	22					25					28	29	30	30	20	19
Iceland											10	30				
Ireland				9	10	10	6	11							13	12
Italy						25	25	27	27	26	27	28	28	30	28	29
Japan	48	48	49	49	51	50	50	49	48	49	50	51	51			
Korea											44	43	44	46	51	53
Mexico												2	2	2	2	2
Netherlands	46					50	53	57	59	48	50	53	58	53	67	77
Norway	22			19	19	21	21	23	21	24	25	29	34	34	39	41
Portugal	38	38				37			42	39	41	41	39	38	39	37
Spain	47	47	54	52	53	57	55	54	54	51	51	46	47	49	48	52
Sweden	34										43	51		50	57	54
Switzerland	35					38					49		54	54	59	61
Turkey									30	31	27	28	26	25	36	34
UK	32	28	28	27	27	28	27	27	26	28	35	36	35	33	36	35
USA	22				21						29			34	35	

Table 9.5 National paper and cardboard recycling rates (% of apparent consumption). Source: OECD (1997)

Mixed recyclables banks

Communal 'bring' systems have also been used for collection of mixed (co-mingled) recyclables from high-rise housing areas, which present special collection problems. This represents the highest density of bring containers, with the density equalling that of the regular refuse containers. The dry recyclables collection schemes set up in some such areas have tried to match the normal refuse collection, with large wheeled bins (up to 1100 litre) located next to the refuse containers.

Kerbside collection

A range of collection methods have been used to collect recyclable material, varying in the degree of sorting involved, and including boxes, bags and wheeled bins. In its simplest form, recyclables are separated by the household and stored together in a bag, box or wheeled bin ready for collection. As with the collection of mixed recyclables from streetside containers, collection can use existing collection vehicles, in some cases even with compaction. Co-mingled collection of recyclables, whether from communal kerbside containers or household bags or bins requires extensive subsequent sorting at a Materials Recovery Facility (MRF).

The collection method involving the highest level of sorting is probably the 'Blue Box' system, that has been imported into Europe from North America. Blue Box schemes supported by the European Recovery and Recycling Association (ERRA) have been operating in the UK since 1989 (for a review see Birtley, 1996). Householders sort out the targeted materials and store them in the box, which is put out at the kerbside for collection in a specially adapted vehicle. At the kerbside the box contents can be sorted by the vehicle operator into several different compartments on the vehicle, and the empty box left at the kerbside. Since this is a positive sort, any unwanted materials can be left in the box and subsequently returned to the restwaste collection by the householder. This practice further educates the householder as to what materials are accepted by the recycling programme and should encourage good sorting efficiencies. The material collected in this way has already been sorted, and therefore limited further sorting is required at a central MRF.

Communication campaigns (using mass-media approaches) have been shown to be important in maintaining high levels of participation and low levels of contamination (the set-out of non-requested material) in kerbside recycling schemes. Such campaigns are necessary to enable people to participate correctly in all recycling schemes, as illustrated in Table 9.6. Unfortunately, it is common for only a fraction (1–3%) of the total MSW management budget to be directed towards public communication programmes (Homes, 1996).

Material	Recovery rate before campaign	Recovery rate after campaign
Newspaper	74%	84%
PET Bottles	63%	81%
Ferrous metal	39%	68%

Table 9.6 Blue box scheme in Ontario, Canada – material recovery rates before and after mass media communication campaign. Source: RIS (1996)

Amount of material collected

Reliable data on the performance of single material bank systems across Europe is not widely available. Lemsterland (NL), using igloo containers, collected 54 kg of glass per household per year, whilst in Germany, banks at a density of one per 800–1000 individuals reported collecting 18–25 kg of glass and 50–60 kg of paper per person (ORCA, 1992). Glass collection in Bapaume, France, (a rural area) has been reported as 42 kg per person per year; this was measured as 80% of all glass in the municipal waste stream (Schauner, 1997). Plastics banks in Hamburg were reported to collect 0.5–1.5 kg per person per year in 1986 (i.e. prior to the establishment of the DSD system (Härdtle et al., 1986).

Area and type of scheme (kerbside unless otherwise stated)	Programme Recovery* Rates (%)					Overall diversion rate‡
	Paper	Glass	Plastic†	Cans	Textiles	
Separate wheeled bins						
Leeds (biweekly)	70–80	n/c	70–80	¶	40–50	50
Blue box (weekly)						
Stocksbridge,						
Sheffield	28	45	21	17	n/a	6.6
SE Sheffield	52	66	28	14	32	15.3
Milton Keynes	57	44	57	24	n/c	18.7
Adur	67	71	60	54	n/c	27
No container (biweekly)						
Chudleigh, Devon	36	55	n/c	21	3	6.9
Green bag (biweekly)						
Cardiff	52	52	13	6	21	17.7
Bring systems						
Ryedale	13	40	n/c	n/c	3	4.1
Richmond-upon-Thames	18	61	n/c	3	8	8.2

Table 9.7 Comparison of UK bring and kerbside collection systems. Source: Atkinson and New (1993a, b). *Recovery rate gives % of each material recovered after both collection and sorting, compared to amount of that material in the household waste stream; †Programme recovery rates for plastics are for beverage bottles and food containers only; ‡Diversion rates are calculated differently for bring and kerbside collection schemes and this will tend to flatter kerbside collection schemes. Bring recovery rates are calculated as a proportion of total household waste (collected and delivered). Kerbside recovery rates are usually calculated as a proportion of collected household waste only. §As reported by the scheme operator; ¶30–40% for aluminium cans, 50–60% for steel cans. n/c = not collected; n/a = not available.

Area	Housing type	Average Blue Box collection (kg/household/year)*
Barrie	Suburban (single-family dwelling)	289
Sudbury	Urban (single-family dwelling)	209
North Simcoe	Rural (single-family dwelling)	200
Etobicoke	Urban (multi-family apartments)	186

Table 9.8 Blue Box material (mixed dry recyclables) collection rates in different areas of Ontario, Canada
Source: RIS (1996) (*Canada Municipal Waste Generation (1995) = 630 kg/person/year).

Amounts of recyclables collected in material banks and kerbside collection schemes in the UK, Canada and Germany are given in Tables 9.7–9.9.

The total amounts of material collected by a Blue Box scheme in four different areas of Ontario, Canada are presented in Table 9.8. The figures show significant variation in the amounts of material set out for collection, depending on housing type.

Contamination levels

The contamination level can be defined as the percentage of non-targeted material that is collected by a given method. This non-targeted material may be:

1. the wrong material type for that part of the system, e.g. paper in a glass bank
2. the right material but in the wrong form, e.g. plastic film in a plastic bottle bank
3. dirty material, e.g. containers with contents still inside
4. non-recyclable material.

The contents of a single-material bank may be bulked and sorted, but normally the collected material is shipped direct to the processors, and thus effectively leaves the waste management system as defined in this book. Any contaminants in the bring containers will also, therefore, leave the waste management system, though they may re-enter it when they are screened out at a material reprocessing facility. Levels of contamination will vary with the material collected. In the case of glass for example, it is necessary in many cases to collect this colour-separated to achieve the highest market prices (particularly in the case of clear glass), so any failure to separate clear, brown and green glass where this is requested will constitute contamination. Additional contamination is likely to come in the form of organics (original container contents), ceramics and plastics (labels, closures) and metals (caps). Typical levels of contamination in recovered glass cullet that reaches the reprocessors (mainly via glass bank collection schemes) is around 5–6% (Ogilvie, 1992). In the case of paper and plastic collection, where it is likely that only certain types of the material (e.g. newspapers only) are required, contamination may arise from the public depositing materials not requested. Since the containers are left unattended in the open, there is also the possibility of contamination by people using the containers to dispose of litter or other waste. As with other collection schemes, clear instructions from the collectors, and a reasonable level of motivation from the public are paramount.

	Paper		Glass		Metals	Plastics	Mixed dry recyclables	Biowaste
	kg/person/ year	Recovery rate (%)	kg/person/ year	Recovery rate %	kg/person/ year	kg/person/ year	kg/person/ year	kg/person/ year
Bring systems:								
2000 pers/bank	5–15	8–25	5–15	13–38	0.5–2.5	1–2	15–50	5–30
1000 pers/bank	10–25	17–42	10–20	26–51				
500 pers/bank	15–50	25–50	15–25	38–64				
Kerbside collection:								
Paper (collected in bundles)								
– every week	20–35	33–58						
– every 2 weeks	15–25	25–42						
– every 4 weeks	10–20	17–33						
Paper (in containers)	35–55	58–92						
Glass (in containers)			15–30	38–77				
Multi-material containers (glass, metal, plastics)	30–50	50–83	12–30	31–77	5–10	5–10		
Bag collection	5–25	8–42	5–20	13–51	1–2	5	30–60	
Biobin							50–140	

Table 9.9 Comparison of German bring and kerbside collection schemes. Source: Schweiger (1992)

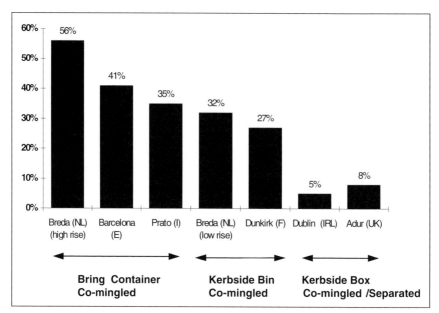

Figure 9.2 Contamination levels for dry recyclables collection schemes (figures are for packaging fraction only). These figures are for the residue left after collection and sorting, so include both non-targeted contaminants and some targeted materials that are not selected. See Chapter 10 for further discussion. Source: ERRA (1993c).

Levels of contamination in mixed material banks and kerbside collection schemes for dry recyclables show a clear pattern (Figure 9.2). Kerbside box schemes generally have the lowest contamination levels (5–8%). The open nature of the box allows inspection of the contents and in some cases a kerbside sort. Any unwanted materials can thus be left in the box, so do not enter the recyclables stream. Kerbside inspection and sorting is not possible where co-mingled recyclables are collected in a bin or bag. Such schemes generally have a higher contamination level (27–32%), due to the inclusion of non-targeted materials. The highest contamination levels (35–56%) have been recorded from co-mingled material banks (i.e. bring system). These are communal bins; lack of 'bin ownership' and perhaps some contamination from litter probably explain why contamination is higher than in collection of mixed recyclables from kerbside bins serving individual households.

The composition of the contaminants in the packaging collection bins in the Barcelona scheme is shown in Table 9.10. The wide range of possible contaminating materials is clearly shown, underlining the need for effective communication with householders as to what materials are required in which container.

Biowaste and garden waste

Garden waste, if not dealt with at source (e.g. home composting), can be managed by a bring system at a central collection site. If kept separate, this material can be used as the feedstock for

Contaminant	% of bin contents
Clothes	5.66%
Wood	1.47%
Stones/sand	9.92%
Miscellaneous	5.09%
Plastic film	3.83%
Tubs/cups	0.78%
Other plastics	1.65%
Other metals	3.26%
Total contaminants	31.66%

Table 9.10 Analysis of contaminants in Barcelona mixed recyclables (packaging) bins. Source: ERRA (1993c)

composting plants, to produce so called 'greenwaste' or 'yardwaste compost'. Alternatively, it may be collected along with other biowaste, or restwaste, via a kerbside collection.

For biowaste, there are two main collection methods available: kerbside or close-to-home collection. There has been a strong trend recently in Europe towards the separate collection of the organic fraction of waste for biological treatment. This trend has been particularly strong in German-speaking countries, The Netherlands and in some parts of Scandinavia, where schemes have been in operation since the late 1970s and early 1980s (ORCA, 1991b). In The Netherlands, for example, legislation has already been passed that requires municipalities to introduce source-separated collection of biowaste. In the Flemish region of Belgium, compostable material is banned from entering landfills or incinerators, and municipalities implemented separate collection of this material by 1996 (OVAM, 1991). In Germany, by 1996, some 104 communities, servicing 10.9 million people, used 'Biotonne' containers for the collection of source-separated biowaste (Würz, 1999), and by 1997 Germany had 520 plants processing 6 million tonnes of source-separated biowaste (ENDS, 1997). This large increase in composting of source-separated biowaste was due to a combination of heavy investment in composting plants and the effect of the introduction of separate collection systems in 1988 (see Figure 9.3). The pattern of a large increase in compost plant capacity developing after the introduction of separate collection systems for biowaste was also seen to occur in The Netherlands, between the years 1993 and 1995 (Koopmans, 1997).

Biowaste definition

Whilst there has been a clear trend towards the separate collection of biowaste, there has been less agreement as to what should be included in the biowaste definition. Narrow definitions include only vegetable, fruit and garden waste (VFG). Collection of garden waste only can occur, but this is usually done through a central bring system (e.g. at a civic amenity site in the UK); collection schemes usually involve at least some household organic waste. At the broader end of the spectrum the biowaste definition can encompass the VFG material, plus part or all of the non-recyclable paper fraction. A range of biowaste definitions used in different schemes is presented in Table 9.11.

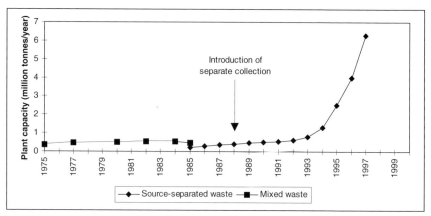

Figure 9.3 Increase in the composting capacity in Germany after the introduction of source separation. Source: Grüneklee (1997).

A survey of collection schemes in Germany showed that 40% involved kitchen and garden waste only, 55% collected kitchen and garden waste, plus soiled paper (e.g. tissues, etc.); the remaining 5% utilised kitchen and garden waste plus the entire paper fraction (Fricke and Vogtmann, 1992). The collection of non-recyclable paper along with the organic fraction as biowaste for biological treatment has many advantages, including providing structure to the biowaste, without which the wet biowaste tends to become anaerobic and produce offensive odours. Including paper in the biowaste also reduces seepage water from the bin, seasonal variability in the amounts of biowaste collected, production of leachate during composting and improves the final compost quality (see below). It will also ensure that a larger proportion of the waste is diverted from final disposal (assuming that this dirty paper fraction would not otherwise be treated by materials recycling or burning as fuel), provided that the inclusion of paper does not adversely affect the quality of the final compost produced. During 1998, almost 200,000 tonnes of non-recyclable paper were composted in the USA. This included 50,000 tonnes from industrial, commercial and institutional sources, 75,000 tonnes from municipal waste and 73,000 tonnes from garden waste bags (Warmer Bulletin, 1999a).

The advantages of including non-recyclable paper in the definition of biowaste
Advantages for collection

1. Reduction of seepage water during storage and transport. Biowaste without paper has a high moisture content, especially in inner city areas where garden waste is sparse. In the German inner city of Soln, for example, the biowaste had a total solids content of only 23% (Doh, 1990). The high water content leads to leakage during storage, collection and transport.

2. Reduction of malodours. Odours are linked to the high moisture content of biowaste. The biowaste is highly putrescible, and with the high water content will rapidly become anaerobic especially in summer months, producing offensive odours. Addition of the paper fraction will absorb this moisture and so reduce odour generation.

Biowaste elements	The Netherlands	Copenhagen (DK)	Frederikssund (DK)	Mainz-Bingen (D)	Witzenhausen (D)	Diepenbeek (B)
Food waste (meat, fish, bones, cheese, fruit, vegetables, etc.)	×	×	×	×	×	×
Egg-shells	×	×	×	×	×	×
Coffee-filters and grounds, tea-bags	×	×	×	×	×	×
Nutshells	×	×	×	×	×	×
Flowers and house-plants	×	×	×	×	×	×
Pet litter	×	×	×	×	×	×
Grass, straw and leaves	×		×	×	×	×
Hedge cuttings, garden plants, small branches	×			×	×	
Paper diapers			×			
Kitchen paper			×	×		
Sanitary towels			×			
Newspaper				×		
Newspaper with potato peelings	×	×		×	×	×
Wet newspaper			×			
Cardboard				×		

Table 9.11 Definitions of biowaste. Items marked with × are included in the definition of biowaste. Source: ORCA (1991b)

3. Reduction in seasonal variability in amounts collected. Where only kitchen and garden waste are collected, large seasonal variation will occur in the amount collected. In Germany, about three times more biowaste is collected in the spring and autumn compared to the winter (Selle *et al.*, 1988). The quality also varies, being limited to very moist kitchen waste in winter, but including drier garden waste at other times. Inclusion of paper reduces the variability in both quantity and quality of biowaste collected (De Wilde *et al.*, 1996).

Advantages for biological treatment (composting)
1. Reduced production of leachate. Biowaste including 20% or more paper can be composted in windrows without production of leachate (Fricke, 1990; see Chapter 12).
2. Reduced requirement for bulking agents. Biowaste without paper, i.e. with a high moisture content, requires bulking agents to absorb water and ensure free circulation of air. Otherwise, anaerobic conditions will occur. Some composting processes require up to 250 kg of wood chips to be added to every tonne of biowaste (Haskoning, 1991).
3. Improved carbon/nitrogen (C/N) ratio. Addition of non-recyclable paper corrects the C/N ratio from about 15–20 for biowaste, to 25 or more, which is optimal for biodegradation. Less ammonia is produced (De Wilde *et al.*, 1996). At the lower C/N ratio the composting process is slowed down (see Chapter 12), and more odours such as ammonia are released (Jespersen, 1991).
4. Increased organic content of final compost. Biowaste compost in Germany has an organic content of around 26% of total solids (Selle *et al.*, 1988), whereas in some countries a minimum of 30–40% may be required (ORCA, 1991b). Adding paper to the biowaste can increase the organic content to this level (De Wilde *et al.*, 1996).
5. Reduced salt level of final compost. High salt levels (above 2 g of NaCl/litre) found in biowaste composts can limit their potential usage. Due to a dilution effect, adding paper will reduce the salt concentration below this critical level (Fricke, 1990).

Overall advantage
Increased diversion rate. Cities focusing on food waste alone in their biowaste will collect only 15–25% of their waste in this fraction. A broader definition of biowaste to include non-recyclable paper, paper products and some garden waste can result in the biological treatment of 40–50% of household solid waste (ORCA, 1991b). A 13-month study of collection and composting wastepaper with the organic fraction of household waste in a semi-urban area North of Antwerp, Belgium resulted in a diversion from landfill of 46% (De Wilde *et al.*, 1996). The system operating in Bapaume, France diverts 40% of total household solid waste from final disposal to landfill (Schauner, 1997).

Possible disadvantages of including non-recyclable paper in the definition of biowaste
Disadvantages for collection
1. Increased contamination level. A wider biowaste definition could cause confusion in households as to what materials are required. This should be overcome by a well-organised and frequent education and communications programme.
2. Subsequent compost quality: heavy metal levels from inks and other contaminants. The general heavy metal content of non-recyclable paper and paper products is low, but inks used in magazines and wrapping paper often use metallic pigments (Rousseaux, 1988), which will contribute to the heavy metal content of the finished compost.

Town		Area	% of total household waste	
Amsterdam	(NL)	Inner-city	23.5	(1)
Amsterdam	(NL)	High-rise buildings	16	(1)
Apeldoorn	(NL)	Suburban	43	(1)
Apeldoorn	(NL)	High-rise buildings	27	(1)
Nuenen	(NL)	Suburban	40	(1)
Purmerend	(NL)	High-rise buildings	20	(1)
Frederikssund	(DK)	Suburban	40	(2)
Copenhagen	(DK)	Inner-city	20	(2)
Mainz-Bingen	(D)	Suburban	50	(3)
Witzenhausen	(D)	Suburban	29	(3)
Diepenbeek	(B)	Suburban	60	(4)

Table 9.12 Quantity of biowaste. References: (1) Kreuzberg and Reijenga, 1989; (2) Jespersen, 1989; (3) Selle *et al.*, 1988; (4) Rutten, 1991

Amounts of biowaste collected

The actual amount of biowaste collected in any separate collection scheme will clearly depend on the amount of organic waste generated (i.e. the potential maximum amount) and the definition of biowaste applied. Organic waste generation rates vary both between urban and rural areas (Table 9.12) and seasonally. Given this variability, however, a German Government Report (1993) suggests that on average, separate collection of biowaste is likely to recover 90 kg of organic material, per person, per year; an average figure of 80 kg per person per year has been reported for The Netherlands (Koopmans, 1997).

The wider the definition of biowaste used, however, the greater the amount of biowaste likely to be collected. The range of amounts collected for different biowaste definitions and collection areas in Germany are presented in Table 9.13. Similar overall recovery rates have been achieved in source-separated collection schemes for organics in the USA. In a scheme run by the Audubon Society in Fairfield and Greenwich, Connecticut, an average of 6.4 kg of organics and soiled paper waste was collected per household per week (equivalent to 333 kg/household/year) (Beyea et al., 1992). Given the higher household waste generation rate for the USA (in this project 1110 kg/household/year), this meant that biowaste collection was able to divert 30% of the total household waste from landfill.

Contamination levels

As with other separate collection systems, contamination of the biowaste with unrequested materials will occur. The evidence suggests, however, that the contamination level is low (Table 9.13). In Germany and France (biowaste and waste paper), contamination levels are around 5% (Selle et al., 1988; Fricke, 1990; Schauner 1997), consisting mainly of plastic. Less has been reported elsewhere. Results from The Netherlands show that the sum of glass, metal and plastic contaminants account for less than 1% of the biowaste (Kreuzberg and Reijenga, 1989), and a similar level has also been reported for the Diepenbeek scheme in Belgium (Rutten, 1991).

	Quantity (kg/person/year)	Rate of recovery (%)*	Level of contamination (%)
Urban districts	73	69	2.24
Inner city areas	46	49	4.02
Rural districts‡	102	73	1.77
National average	92	70	2.02
Biowaste including paper	184	85	7.50

Table 9.13 Comparison of collected biowaste in Germany. *Recovery rate = amount collected/amount available in waste; ‡Excluding projects which include the entire waste paper fraction in the biowaste definition. Source: Fricke and Vogtmann (1992)

Much of the contamination will consist of refuse bags, used to collect the waste and transport it to the refuse container.

Like the overall amounts collected, the levels of contamination also vary between city and rural areas and with biowaste definition. Reports from rural areas of Germany suggest contamination levels of between 2 and 3% while contamination in inner city areas can rise to 7–10% (ANS, 1999), apparently due to lack of household motivation and effective peer pressure in high-rise accommodation. Similarly, with a broader definition for the biowaste, there is scope

Parameter	Range*	Average*	MSW compost†	Source separated compost‡	German (UBA) recommended limits
Zinc	29.8–178.0	117.8	1570	222	300
Lead	3.5–94.4	37.4	513	68	100
Copper	13.5–44.6	29.0	274	50	75
Chromium	13.0–20.8	17.1	71	71	100
Nickel	6.8–16.0	11.3	45	21	50
Cadmium	0.14–0.25	0.19	5.5	0.7	1
Mercury	0.07–0.18	0.11	2.4	0.2	11
Moisture content (% of wet weight)	66.6–72.7	69.7			
Organic content (% of dry weight)	68.5–82.7	78.1			

Table 9.14 Heavy metal levels in German biowaste and German compost (in mg/kg dry weight). Sources: *Tidden and Oetjen-Dehne (1992); †Fricke (1992); ‡Bundesgutegemeinschaft Kompost (1997)

for more confusion as to what should be included, leading to an increased level of nuisance materials (Table 9.13). It should be possible to counter this trend, however, by clear instructions and an active communications programme to householders.

Along with such 'nuisance materials', biowaste will also be contaminated with heavy metals. This is of particular importance if the biowaste is to be processed into marketable compost, since heavy metal levels may determine whether the resulting compost is of acceptable quality (Chapter 12). Typical heavy metal levels for German biowaste and recommended maximum levels are given in Table 9.14.

Collection methods

Biowaste is generally collected in bins or bags. Bins, either split into compartments or not, have the advantage that they do not add to the level of plastic contamination that must be removed at the biological treatment plant (unlike bags), but they may need washing out, especially in hot weather. A possible disadvantage of using bins, especially large wheeled bins, is that householders are tempted to add their garden waste too, rather than composting it at home. This can result in an increase in the waste entering the system (Selle *et al.*, 1988; Wright, 1998). Another method has been to use paper bags, with a moisture barrier, which are biodegradable in the subsequent composting process. The Audubon scheme in Connecticut cited above successfully used such 'wet bags' (Beyea *et al.*, 1992), and trials have also been run using bags made from biodegradable polymers in other parts of the USA (Goldstein, 1993). The main reasons given in Denmark and Sweden for the adoption of paper bags for household waste collection are weight and odour reduction (Rand, 1997). The bag enables the contents to breath, allowing evaporation as the composting process begins. The likelihood of the contents of a paper bag becoming anaerobic are much less than that if plastic bins or bags are used. The onset of anaerobic digestion, which is common in wheeled bins and plastic bags, leads to strong odours, condensation and the production of a noxious black liquor.

A survey of biowaste collection schemes in Germany reports that most schemes (81%) use 120- or 240-litre wheeled bins, with 2% using 35-litre bins and 9% multi-chambered bins. Only 7% used bags, evenly split between paper and plastic (Fricke and Vogtmann, 1992).

Collection of biowaste can also require specially modified vehicles, involving rotating drums, pneumatic presses or multi-chambered bodies. One of the major problems is the leachate leaking from the trucks, so in many cases they need to be specially sealed to prevent this. Alternatively, collection of some paper with the biowaste can reduce leachate production, as previously discussed.

Packaging waste

Packaging waste has become the focus of strategic legislation, i.e. rules or guidelines on the waste management options that can be used to treat at least part of the municipal waste stream. This is diverting attention from treating the whole municipal waste stream in the most economically and environmentally sustainable manner. The European Union (EU) Directive on Packaging and Packaging Waste (94/62/EC) adopted in December 1994 requires that Member States (MS) recover between 50% and 65% of packaging waste arisings. Recovery shall mean recycling, energy recovery and biological treatment. A minimum of 25% and a maximum of 45% of total packaging waste must be recycled. Within each material type (paper, plastic, metal, and glass) a minimum of 15% recycling is required (see Figure 9.4). At the time of writ-

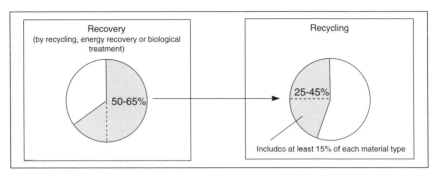

Figure 9.4 EU Packaging and Packaging Waste Directive – 2001 recovery/recycling targets. Source: *Warmer Bulletin* (1997d).

ing, the Packaging Directive is under revision.

The twin objectives of the Directive were '*to harmonise national measures concerning the management of packaging and packaging waste in order, on the one hand, **to prevent any impact thereof on the environment of all Member States** as well as of third countries or to reduce such impact, thus providing a high level of environmental protection, and on the other hand, **to ensure the functioning of the internal market and to avoid obstacles to trade and distortion and restriction of competition within the Community**'. In brief, to minimise the environmental impact of packaging waste on EU countries and to avoid the creation of trade barriers between EU countries.

Unfortunately this source-orientated (packaging) end-of-life (waste) legislation has resulted in segregated waste management systems (one system for MSW and another for packaging) within countries and restrictive fee structures (different packaging materials within segregated waste management systems are charged at different rates) between countries.

An analysis of European packaging recovery systems (PriceWaterhouseCoopers, 1999) concluded that:

1. exact data on the amount of waste, packaging waste and recycling are hard to get and ambiguous
2. data on the amount of waste, packaging waste and recycling are not comparable
3. costs of packaging waste management systems vary from country to country and have a different scope
4. the benefits of the Directive on packaging waste management are not clear.

Practical experience is now showing that mandated packaging recovery systems lead to segregated waste management systems with their associated high costs and environmental burdens.

Status of implementation

Theoretically the Member States (MS) of the EU and the European Economic Area (EEA) were obliged to implement this Directive nationally by June 1996. In Summer 1999 this national implementation can be considered as having been completed in all countries except for

Greece. However, the nationally adopted regulations and the resulting recovery schemes vary widely with regard to timings and approaches taken. As a consequence there is no coherent approach towards managing packaging waste throughout the EU. However, the directive has added momentum to recovery efforts. The total amount of recovered packaging waste in 1997 is estimated to be at minimum 15.5 million tonnes.

The recovery schemes for used packaging differ with regard to funding and coverage of ongoing costs. They reflect the national legislation, which ranges from applying a policy of Shared Responsibility (of all members of the distribution chain, including consumers and Local Authorities alike) to the application of an Extended Producer Responsibility policy.

An example of Extended Producer Responsibility is the German approach that in practice requires the marketer (packer/filler) to organise and fully finance the recovery of all used packaging arising at the household level. Shared Responsibility can be seen as represented by the approach taken in The Netherlands and also by the approach chosen in the UK. In The Netherlands, industry is committed to recover/recycle those materials that either have been separately collected or that have been separated from the waste stream by the local authorities. In The Netherlands the system works essentially without extra charges or fees. It is anchored in a set of negotiated covenants concluded between industry and Government. The UK is so far the only Member State of the European Union applying an essentially market-driven approach (a market for emission rights) with tradable so-called Packaging Recovery Notes (PRN).

The recovery schemes for used packaging differ in maturity of their operation. The national schemes in Austria (ARA), Germany (DSD and several smaller ones) and the covenant negotiated approach in The Netherlands have long since reached national coverage. Hence it would make sense to research cost aspects and environmental benefits of these schemes in order to close data gaps and to start building a data base for future improvement measures. The schemes in Belgium (Fost Plus), France (Eco-Emballages and several smaller ones) and Sweden (REPA) have been operational for several years and are continuously increasing coverage. The schemes in the remaining countries have become operational more recently or are being set up during 1999. It is because of this staggered maturity (Figure 9.5) that only limited data on recovery achievements and operating costs are available.

From an analysis of publicly available information on the national implementation of the

Elements of IWM

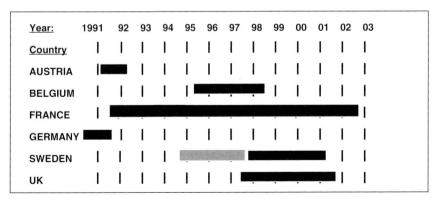

Figure 9.5 EU Packaging and Packaging Waste Directive: mandated implementation time.

Directive it appears that short implementation times for setting up recovery schemes for used packaging with national coverage must be considered a cost driver (Draeger, 1997).

Inconsistencies between packaging recovery schemes

Recovery fees (per kg of material) charged by different schemes cover a very broad range and do not even follow an identical pattern (Figure 9.6).

Different executions for a 1.5 kg detergent pack further illustrate this inconsistency (Figure 9.7). The laminate bag is cheapest in Sweden and most expensive in Austria, whereas the most expensive execution in Germany is a plastic bag.

The absolute recovery fees for plastic bottles demonstrate the wide range that can occur for the same package in different countries (see Figure 9.8). The French scheme operates with fees that are lower by a factor of 30–40 than those of the German and the Austrian schemes, whereas the Danish scheme and the one in The Netherlands operate without any additional fee.

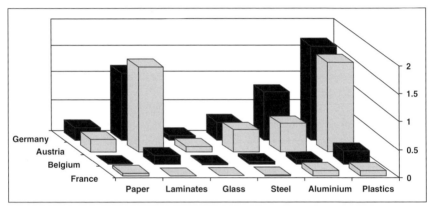

Figure 9.6 Differences in fees (per kg of material) charged by different packaging recovery systems. Source: Draeger (1997).

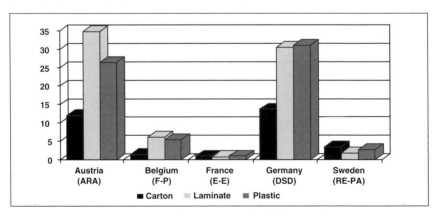

Figure 9.7 Relative recovery scheme costs for different detergent packagings. Source: Draeger (1997).

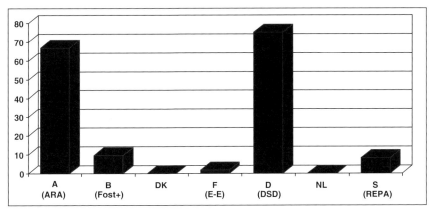

Figure 9.8 Recovery fees (S/1000 pieces) for a 300 ml plastic bottle. Source: Draeger (1997).

Costs of different recovery schemes

Although cost data are not directly comparable, they can serve as indicators. If a recovery scheme in one country requires three or four times as much in percent of the GNP than a recovery scheme in another country (Figure 9.9), this may be interpreted as a competitive disadvantage and suggests that such a scheme is not economically sustainable. It does at least indicate that more research on cost aspects of packaging and packaging waste recovery systems is needed.

A PriceWaterhouseCoopers study (1999) states the compliance costs for the EU in 1997 to be at least 5.3 billion euros with a tendency to increase. This appears to be a significant cost for questionable environmental benefits, especially in the light of traditional recycling activities (e.g. OECD, 1997), which had been implemented already prior to any legislation in the field and which typically generated net revenue. The Danish approach and the one taken in The Netherlands are both essentially Integrated Waste Management systems where little or no specialised packaging recovery occurs.

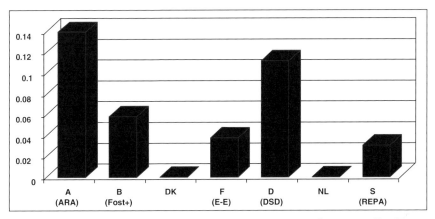

Figure 9.9 Estimated cost of EU Packaging Recovery Schemes as % of per capita GNP. Source: Draeger (1997).

Elements of IWM

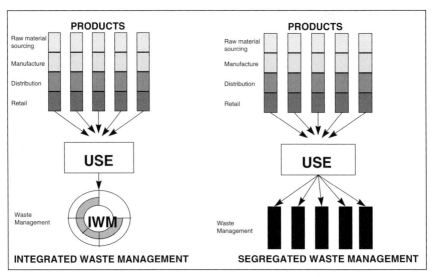

Figure 9.10 Schematic representation of how segregated systems lead to inefficiencies.

Results of used packaging recovery schemes

The wide variety of approaches chosen in the EU Member States and the widely varying cost indicators suggest that further research is needed to identify best practice in this field. However, legislation that requires source-orientated collection, sorting and recovery or disposal of products, results in a tendency to develop separate, parallel or segregated waste management systems. Evidence coming from countries implementing mandated take-back systems for used packaging have shown this to be the case. Segregated waste systems lose the benefits of economies of scale, and synergies between different treatment options enjoyed by integrated systems, so tend to be less efficient, both economically and environmentally (Figure 9.10). A number of independent reviews of the DSD system in Germany are available (e.g. Staudt, 1997; Axt, 1998; Raymond Communications Inc., 1998; Schmitt-Tegge, 1998; Schroll, 1998; Scarlett, 1999; Staudt and Schroll, 1999) and they all support these conclusions. Moreover, the multiplication of parallel systems can confuse the householder and result in reduced sorting efficiency of certain materials (see Chapter 12, Contamination identified by wastestream analysis in Lahn-Dill Kreis).

Hazardous materials in household waste – the exception that proves the rule

Household waste contains hazardous materials such as used motor oil, pesticides and solvent and paint residues in used cans and bottles. Contaminants such as heavy metals also occur; they are found in small quantities in a range of household waste items but are mainly concentrated into a few items such as used batteries, discarded light bulbs and tubes and mercury thermometers (see Chapter 8). Normally such materials are included in the residual or rest-waste collection, but their presence can limit the options available for treating this waste stream.

For example, if mixed MSW composting is to be carried out, the operator must be aware that high heavy metals levels (from batteries, etc.) may prevent the resulting compost being sold as it would be likely to exceed the legally permissible heavy metal limits. Similarly, the presence of organic contaminants such as persistent pesticides could result in ground water contamination if the mixed waste were landfilled.

One solution is to separate out the hazardous materials at source, and deal with them separately. In the Flanders region of Belgium, 2 million households have been supplied with a 'KVA box' (Klein Gevaarlyk Afval – small, dangerous waste) for small hazardous waste items (ORCA, 1992). As the amount of hazardous waste generated will be small (around 5–10 kg/household/year), this material can be collected separately on an infrequent basis. Alternatively, they can be taken to a central collection site (i.e. a bring system). In Germany, for example, there is typically a small container for such waste at each multi-material collection point.

Clearly there can be advantages to separate collection, so long as the extra collection/bring system can be integrated into the normal collection system, and provided that there are effective ways of dealing with the hazardous waste once collected. Kerbside collection in various US cities, for example, has collected 7–18% of the available batteries (Warmer Bulletin, 1993a). Many current schemes collect small batteries, which have been sorted out of the waste by householders, but do not have access to appropriate recycling or disposal technology to deal with them, so are forced to stockpile them. Under existing legislation, at least in the UK, once these elements of waste are concentrated in this way, they are classified as a special waste, which limits the ways they can be stored, handled and treated.

Bulky waste

This solid waste can make a significant contribution, but is generally not included in estimates of household waste generation (see Chapter 8). In the UK, for example, bulky waste plus garden waste probably represents around 30% (by weight) of household waste generation (Atkinson and New, 1993a). Bulky wastes can either be delivered to a central collection site (i.e. a bring system) or picked up from households using a separate and infrequent collection. Once delivered to a central site, some bulky objects such as furniture or appliances can be recovered intact for re-use. Other bulky wastes can be recovered for metal recycling (appliances) and the residue, if sorted appropriately, can be either incinerated or landfilled. There is also the possibility of recovering other materials (e.g. CFCs from refrigerators).

Restwaste

In most traditional systems, this category would contain all of the household waste, collected in a completely mixed state. Some restwaste can be handled in a bring system in city centre or high-rise areas, such as in the Sagrada Familia area of Barcelona, where large (2200 litre) streetside restwaste containers are used (ERRA, 1994b). This system is common in southern Europe, since it allows daily removal of restwaste. In most areas of northern Europe, however, collection from each property or kerbside is the norm. Where bring or kerbside collection schemes have been introduced for dry recyclables, biowaste and/or hazardous household waste, both the amount and composition of restwaste will be altered. Whatever methods are used for separate collection of parts of the waste stream, however, there will always be residual waste for collection.

Although the amount collected may be reduced in weight there may be little saving in either environmental or economic terms compared to the traditional collection of mixed waste. This is because the same number of properties has to be visited, and so the same distances need to be driven. The decreased amount collected per household can lead to longer collection rounds before the collection vehicle needs to be emptied. Such efficiency improvements may lead to some cost saving: a report in the UK suggested marginal savings of around £9 (5.4 Euro) per tonne in restwaste collection (about 25% of current costs) when recyclables are collected in a separate round (DOE/DTI, 1992). An alternative is to reduce the frequency of collection, for example to every other week where current collections are weekly. This may not be possible, however, in regions where the frequency of waste collection is fixed by legislation, or where odour problems from the restwaste make regular collections necessary.

Variable rate pricing systems (pay-as-you-throw)

In communities with variable-rate programmes (also known as pay-as-you-throw or unit pricing), residents are charged for the collection of MSW based on the amount they throw away. This creates a direct economic incentive to recycle more and to generate less waste. Traditionally, residents pay for solid waste collection through property taxes or a fixed fee, regardless of how much or how little waste they generate. Pay-as-you-throw (PAYT) breaks with tradition by treating waste services just like electricity, gas, and other utilities. Households pay a variable rate depending on the amount of service they use. Most communities with PAYT charge residents a fee for each bag or bin of waste they generate. In a small number of communities, residents are charged based on the weight of their waste. Either way, these programmes are simple and fair. The less individuals throw away, the less they pay.

The US Environmental Protection Agency supports this approach to solid waste management because it encompasses the three interrelated components that are key to sustainable solid waste management systems:

1. Environmental sustainability. Communities with these systems in place have reported significant increases in recycling and reductions in waste generation, primarily due to the waste reduction incentive created by PAYT. Less waste and more recycling mean that fewer natural resources need to be extracted.
2. Economic sustainability. PAYT is an effective tool for communities struggling to cope with increasing MSW management costs. Well-designed systems generate the revenues communities need to meet their solid waste costs, including the costs of recycling and composting. Residents also benefit because within certain limits, they have the opportunity to take control of their waste bills.
3. Social equity. One of the most important advantages of a variable-rate programme may be its inherent fairness. When the cost of waste management is hidden in taxes or charged at a flat rate, people who recycle and prevent waste subsidise their neighbours' wastefulness. Under PAYT, residents pay only for what they throw away.

The following case studies from America describe some of the advantages and disadvantages of operating a pay-as-you-throw system. For a short summary of Canadian pay-as-you-throw case studies see Warmer Bulletin (1997a) (see also Chapter 3, Zurich).

Elements of IWM

Case study: San Jose, California, USA

Population: 850,000. Community: urban. Program type: four-sort. Program start: July 1993.

Why pay-as-you-throw? San Jose is the eleventh largest city in the USA; its residents are among the most educated and affluent in the country and represent a diverse community, with the two largest minority groups being Latino (27%) and Asian (14%). Before July 1993, San Jose provided an unlimited weekly waste collection service at a flat monthly rate of $12.50 per household. Residents set out an average of three 120-litre waste bins per week. The city fully implemented its Recycle Plus (RP) residential Integrated Waste Management programme for 186,000 single-family dwellings on 1 July, 1993. This programme was designed to permit the city to reach its California Integrated Waste Management Act goal of 50% waste reduction by 2000. The new RP programme resulted from over 3 years of planning, which included extensive research on all major policy changes. This programme included a fully automated waste collection system, an aggressive PAYT rate structure, a four-sort recycling system, and a contractor payment mechanism, which provided financial incentives that encouraged contractors to promote recycling.

Educating the public. The public was involved in the design of the RP programme through a questionnaire mailed to all 186,000 households, community meetings throughout the city, pilot projects in 17 neighbourhoods for collection of garden waste and mixed papers, and the use of a public review committee to select the firms that would be given 6-year collection contracts for the collection of waste and recyclables and for recyclables processing. A comprehensive public outreach campaign aimed at single-family households explained the new variable rates being introduced, the new categories of recyclables being added to the services provided, and the benefits of participating. All materials were produced in three languages (English, Spanish, and Vietnamese). The campaign was guided by the information received during a series of focus groups in the three languages, baseline and follow-up telephone surveys, and shopping mall intercept surveys. More than 250 community meetings were held in 1993, and a block leader programme and school education programme were organised.

Pricing strategy. Residents were offered 120-, 240-, 360- or 480-litre bins with an 'aggressive' unit-pricing structure. This structure provided a slight price break for each additional 120 litres of capacity at the 240- and 360-litre level, which the council considered important to help residents make the transition from flat rate to unit pricing. The council had to ensure that they had sufficient quantities of wheeled-waste bins in the sizes the residents would request. A return-reply card was sent to all single-family households in January 1993 with the estimated charges; this informed residents that no reply would result in delivery of the default 120-litre bin. Staff were able to work out a compromise with the city council, which included offering one of the most comprehensive low-income rate assistance programmes for waste service in the state. Criteria were based solely on household size and income and permitted eligible residents to receive a 30% discount on their bill. About 3400 households currently participate in this programme.

Managing the programme costs. The challenge faced by the programme is to both continue and expand its multiple recycling efforts to meet diversion goals, while reducing costs to close the projected $5 million cost-to-revenue gap in 5 years. The city already has reduced costs by over $4 million annually through contract renegotiations that resulted in extending the term of the RP and garden waste collection contracts from June 1999 to June 2002.

Success: Waste reduction and increased recycling. Staff did not anticipate how quickly residents would change their recycling participation to accommodate the 120-litre size bin, especially since prior to RP the average set-out was three waste bins. Since RP implementation, an average of 87% of residents have requested the 120-litre bin size. The difference between the 'before and after' waste set-out volume could readily be found in the quantity of recyclables collected in the new RP programme. The volume of recyclables and garden waste being collected more than doubled the levels recorded prior to RP. Most importantly, residents reported wide satisfaction with the programme and its results (80% in 1993 to 90% in 1996; figures are based on a random sample telephone survey).

Source: http://www.epa.gov/payt/tools/ssanjose.htm

Case study : Fort Collins, Colorado, USA

Population: 100,000. Community: urban. Program type: variable. Program start: January 1996.

Why pay-as-you-throw? Fort Collins is located on the Front Range of the Rocky Mountains in Colorado. The population has passed the 100,000 mark, but the community still takes pride in a small-town image, and residents are determined to manage growth well.

The city sponsored a recycling drop-off site for nearly 10 years, but without a municipal waste collection service, increased participation depended on hauliers efforts. A 1991 ordinance required hauliers to provide curbside recycling, but because they included this service at an additional cost, most customers were unwilling to pay for the service. Construction of a county recycling centre in 1992 also had little effect on residents' recycling levels. The city council adopted goals in 1994 to reduce the total waste stream by 20% by the year 2000, despite the city's growth, and to reduce landfilled waste by 20%. A specific target was set for increasing participation in curbside recycling by 80–90%. Reaching these goals was challenging, as six private hauliers work in Fort Collins, ranging from corporate operators to locally run family operations that have been in business for 40 years.

Disappointed in a slow rate of progress for recycling, the city council adopted two ordinances in May 1995 that apply to single-family and duplex residences. The first ordinance called for hauliers to 'bundle' costs for recycling and provide curbside recycling to customers upon request at no extra charge. It became effective on 12 March, 1995. The second ordinance called for volume-based rates to be charged for solid waste starting in January 1996.

Lessons learned

1. Start planning for implementation of the rate structure change at least 6 months in advance. The local waste authority did not start working with the hauliers (waste transport) until September to implement the system in January. Then, after meeting together several times, the city agreed to amend the ordinance to respond to hauliers' concerns about charging strictly by volume, but this process was time-consuming and difficult.
2. Make sure to publicise the changes to remind the public and their elected officials about what will occur in the next 2–3 months. Use news articles, advertisements, and hauliers' billings.
3. Do not underestimate the difficulty people will have understanding how new waste collection rates work, and plan for the extra work it creates for staff. Be prepared for it to take 3, 6, or even 9 months for people to realise that they can save money by generating less waste with a PAYT system.

4. Expect private waste hauliers to take the opportunity to increase collection rates at the same time the volume-based rates take effect. The public assumed the increase in collection rates was a result of the ordinance.

5. Ensure the transition between billing systems is smooth, as programme overlap results in both hauliers and city staff getting phone calls from angry, confused people who receive two different bills. However, the city was careful to reimburse customers for bins/bags of waste that they did not generate, as the most important feature of the system is to reward people with cost savings.

Success: increased recycling participation. As of July 1996, recycling increased to 79% participation in single-family and duplex households, up from 53.5% the previous year. Now the residents of Fort Collins are more conscious of reducing their waste stream, they have demanded opportunities to recycle new materials, including cardboard, office paper, and compostable items. The bundling ordinance and PAYT system have significantly increased households' recycling efforts. Now that the system has been operating for 6 months the city council is already looking ahead to the feasibility of dividing up Fort Collins into waste collection zones.

The local waste management authority is anticipating that autumn's leaf-raking and bagging will add to peoples' waste bills and that they are going to demand that the city do something about it. Fort Collins remains confident that it made the right choice by adopting their pay-as-you-throw ordinance.

Source: http://www.epa.gov/payt/tools/ssfortco.htm

Integrated collection schemes

Collection is at the centre of an Integrated Waste Management system. An integrated approach is the key to an effective collection system. The pre-sorting and collection stage needs to collect all of the waste, separated into suitable streams for subsequent treatment methods. To be efficient in both economic and environmental terms, it must do this with the minimal use of transport, including both collection trucks and householders' private cars.

An integrated collection system could include any combination of bring systems (materials banks, close-to-home drop-off centres, central collection sites for garden/bulky waste) and/or kerbside collections (for recyclables, biowaste and/or restwaste). The key is that all methods form part of one system with the objective of collecting all the waste materials in suitable streams for subsequent treatment or disposal with minimum environmental and economic burdens.

Key lessons on the importance of an integrated collection have come from a variety of collection schemes. Collection of dry recyclables in Blue Box schemes (e.g. Adur, Sheffield, UK; Ontario, Canada) in many ways represents an additional, rather than integrated collection system. A second truck often travels the same route as the residual waste collection vehicle, collecting a dry recyclables fraction. This additional truck is likely to result in increased collection costs as well as increased environmental impacts due to the vehicle's emissions. In most cases, an improvement would be to collect both the recyclables and the restwaste on the same visit. This has been introduced in Worthing (a neighbouring district to Adur, UK) using a specially designed truck that has two compartments for recyclables, plus a normal compaction compartment for the restwaste. This allows both recyclables and restwaste to be collected on a single visit to each household. An alternative, which has been developed in the USA (Omaha, Nebraska) is to collect co-mingled recyclables in blue bags, which are loaded along with the bagged rest waste into the same vehicle compartment, for separation at a sorting facility (Biocycle, 1992). This form of 'co-collection' has not been developed to any extent in Europe, however.

With more than three waste fractions, it is difficult to collect all fractions efficiently on a single visit, so an alternating collection schedule is often employed. Lemsterland in The Netherlands for example, collects four different waste streams – mixed paper, mixed dry recyclables (excluding glass), organics and restwaste. The materials are stored in two separate wheeled bins, each with interior partitions (giving four compartments) and then collected by split-compartment vehicles on an alternating basis. Leeds (UK) similarly collects three different waste fractions, with one split-compartment truck and one normal compactor truck visiting properties in alternate weeks. A range of collection methods and schedules are used for waste collection in Germany (Table 9.15).

Alternating collections can be effective in collecting several separate fractions of household waste, without increasing the overall number of visits to each property, so long as it does not affect the comfort level of the participants. Biowaste collection is one area where this comfort level is likely to be compromised, since it is necessary to retain frequent collection to prevent severe odour nuisance during summer or year round in warmer climates. This factor needs to be considered in the design of effective collection systems. In The Netherlands, for example, source separation of biowaste is mandatory and biowaste collection is often in alternate weeks as a result. During hot weather, households are instructed to put their organic waste into the residual waste, rather than into the biobin, if this is the next bin due for collection. This results in a loss of organic material from the biowaste stream (hence less compost produced), and a corresponding increase in the organic material going to landfill or incineration (hence a decreased diversion rate). Alternation of biowaste and restwaste collection is also the most frequently used method in Germany (Table 9.15). There are ways round this problem, however. Including paper and paper products in the biowaste definition can result in less odour nuisance from the bin as described above, or alternatively the schedule can be devised so that biowaste collection is kept at the previous frequency (e.g. weekly) whilst the dry recyclable and the restwaste collections are alternated.

There is a general conflict between the needs of sorting and the ease of collection. Treatment methods generally, and materials recycling in particular, require effective separation of the

Schedule	Proportion (%)
Weekly	18.8
Weekly with multiple-chamber bins	9.4
Every 2 weeks in addition to the weekly collection of non-recyclables	20.0
Every 2 weeks, alternating with the collection of non-recyclables	51.8

Table 9.15 Survey of biowaste collection methods in Germany. Source: Fricke and Vogtmann (1992)

Area (population)	Frequency of kerbside collection and number of materials collected	Organic waste collection	Diversion rate	Comments
Ann Arbor, Michigan (112,000)	n/a 28 materials	Garden waste	52% 23% from composting	Mandatory recycling ordinance and State ban on landfilling organic material
Dover, New Hampshire (26,000)	Weekly 22 materials	Autumn leaves Garden waste drop-off	52%	Pay-as-you-throw system (charged per bag)
Loveland, Colorado (44,300)	Weekly (recyclables and restwaste in one vehicle) 10 materials	Garden waste (weekly) and Garden waste drop-off	56%	Pay as you throw system (all residents pay a flat monthly fee for recyclables and garden waste, plus separate charge per bag of restwaste)
Madison, Wisconsin (200,900)	Weekly 13 materials	Seasonal Garden waste and Garden waste drop-off	49% 33% from composting	Mandated source separation of organic material
San Jose, California (850,000)	n/a 20 materials	Garden waste (weekly)	43%	Pay-as-you-throw system (volume based, three different bin sizes available)
Visalia, California (91,300)	Weekly (recyclables and restwaste in one vehicle) 13 materials	Green waste (weekly)	48% 34% from composting	
Worcester, Massachusetts (169,800)	Weekly 20 materials	Autumn leaves once per year, Garden waste drop-off	52% 28% from composting	Pay-as-you-throw system (charged per bag)

Table 9.16 Summary of different MSW collection systems in the USA. After: Platt (1998)

waste into several streams. To reduce any cross contamination, especially by organic material, it is best to separate materials as early as possible in the waste management system, i.e. at source. The number of categories for home sorting is limited, however, by householder ability and motivation, available storage space and the ability to collect many different waste streams without increasing the number of collection visits to the property. To keep the total number of fractions to a manageable level, fractions that can easily be separated by subsequent sorting (e.g. steel cans, aluminium cans and plastics) should be collected mixed.

The net result of the above conflicts is that there is no one best collection system for all areas. The best and most integrated system for any area will depend on the way the waste needs to be collected for the local treatment and disposal methods, the composition of the waste, the type of housing and population density, and the motivation of the residents. Bring and kerbside collection systems can both be appropriate, for different materials or fractions, within an integrated collection system. This is clearly demonstrated in Table 9.16, which summarises a number of different collection systems in operation in the USA. These collection systems have been carefully designed to provide an efficient and acceptable collection service at the local level. Multi-material kerbside collection and biowaste collection either as part of a kerbside system, a bring system (drop-off) or a combination of both can be seen to result in significant diversion from final disposal. With respect to the economics of such successful recycling schemes, a study by the North Carolina Division of Pollution Prevention and Environmental Assistance in 1997 concluded that 'the cost-effectiveness of a recycling programme (compared to solid waste collection and disposal) correlates with local recycling rates; i.e. local governments that achieve high recycling rates are more likely to operate recycling programs that are less expensive per ton than solid waste collection and disposal' (DPPEA, 1997).

CHAPTER 10

Central Sorting

Summary

This chapter deals with two distinct types of central sorting: sorting of mixed recyclables at a Material Recovery Facility (MRF) and the sorting of mixed waste to produce Refuse-Derived Fuel (RDF). The stages of each sorting process are described, as are the waste inputs and the outputs in terms of both products and residues. Available data are presented on the typical energy consumption of the two sorting operations. Economic data, both processing costs and revenues from the sale of recovered materials, are also included where possible.

Introduction

Sorting is an important part of any waste's Life Cycle. Solid waste is almost always mixed, and household wastes are amongst the most heterogeneous in terms of material composition. Separation of the different materials in waste, to a greater or lesser extent, is an essential part of almost all methods of treatment. This sorting can, and does, occur at any point in the Life Cycle of waste; similarly it can occur any number of times. The earliest sorting will occur in the home, when, for example, materials are separated from the residual waste stream (Chapter 9), but the same materials can be sorted further during and/or after collection. Sorting of the input also represents the first stage of many waste treatment processes, such as composting, biogasification, and in some cases sorting of the outputs also occurs (e.g. removal of ferrous metal from incinerator ash residues). Sorting is thus ubiquitous in the Life Cycle of waste, and is covered in each chapter of this book that deals with a particular waste management process. This chapter focuses on two particular central sorting operations that are not covered elsewhere – sorting at a Materials Recovery Facility (MRF), and the sorting of mixed waste to produce Refuse-Derived Fuel (RDF).

These two processes are distinct, with different inputs and outputs, so will be described and discussed separately.

General sorting techniques

This section provides a general review of the most common sorting techniques used in MRFs and in RDF production.

Elements of IWM

Manual sorting

The simplest and most widespread separation technique is hand sorting from a raised picking belt. Manual sorting can occur at the beginning of the process to remove hazardous material such as gas cylinders and any material that may damage the subsequent mechanical operations such as heavy wire rope or industrial chain, large blocks of plastic or rubber, large truck tyres and long or dense rolls of woven material. Manual sorting also takes place after several mechanical processes have pre-sorted the waste stream to recover items that are best identified using the human eye. Normally operators remove the materials required from the picking belt and the remaining unselected materials are discarded as residue.

Average manual sorting rates for different materials are shown in Table 10.1. The productivity can be seen to vary widely between different materials handled and highlights the fact that hand sorting the desired recyclable material is not always the best option. Sometimes it is more practical to pick the reject material until only the recyclable material remains. From the table it can be seen that picking paper from a mixture of paper and cardboard results in a sorting rate of only 12 kg/h. If cardboard were to be picked instead, it could be sorted at 100 kg/h. If the cardboard represented 10% by weight of office waste, then removing it would yield paper at 1000 kg/h. This is a much simplified example but highlights the importance of designing the manual sorting system carefully, with respect to the waste input.

Manual sorting rates are understandably affected by operator tiredness; this in turn depends on whether the operator is standing or sitting, how far it is to reach the items to be sorted and how much movement is necessary to deposit recovered items. Hand-sorting operations should be carried out in daylight. Artificial lights (especially fluorescent tubes) give off a narrow spectrum of light and this makes identification of materials more difficult. These parameters are difficult to quantify but must be considered when designing the manual sorting stages of MRFs.

Manual sorting is clearly labour intensive, but some schemes (e.g. Milton Keynes, UK; Omaha, Nebraska, USA) use this as an opportunity for job creation, or for training disadvantaged societal groups. There have been some concerns over health and safety issues surrounding such waste

Material	Apparent density (kg/m^3)	Sorting rate per person (kg/h)
PET containers	23	160
Paper	80	12
Cardboard	90	100
PVC/UPVC	25	240
Glass	350	500
Film plastic	20	20
Textiles	60	180
Ferrous metals	45	–
Non-ferrous metals	25	–

Table 10.1 MRF employee productivity (kg per hour per employee). Source: Manser and Keeling (1996)

sorting, and because of this (plus the need to increase efficiency), there is a trend towards increasing the mechanisation of the sorting process, to increase the possible throughput and sorting efficiency.

Mechanical sorting

Bag openers, although not a sorting technology, are often the first step in the process where material (either unsorted waste or mixed recyclables) is delivered to the sorting facility. These simply tear or cut the collection bags open releasing the contents and allowing them to begin their journey through the MRF.

There are three broad categories of mechanical operation:

1. mechanical disassembly, which separates components physically
2. separation by particle properties such as size, shape and mass (this also depends on the previous unit operations)
3. separation by material property such as magnetism or colour.

Many of the technologies used in materials recovery are borrowed or adapted from mining and chemical processing industries. Some of the common unit processes are described below.

Screening

Screening is a process of separation by particle size. The most common screen for processing MSW is the revolving or trommel screen, which is an inclined cylinder with holes in its sides that is mounted on rollers. The drum rotates at slow speeds (10–15 rpm) using little power. The material inside the drum tumbles around until it falls through the holes. Reject material does not pass through the holes while extract material does pass through the holes. This type of screen is very resistant to clogging, which is why it is more common than inclined or horizontal shaking screens, which are readily blocked by rags and paper. The application of inclined or horizontal shaking screens is normally limited to cleaner feed materials (Manser and Keeling, 1996).

Air classification

This process is used to separate the light fraction (plastics, paper and aluminium cans) from the heavy fraction (mostly inorganic material). Light materials are caught in an upward air flow and carried with the air, while the heavier fraction drops. The light fraction must be separated from the air stream and this can be achieved by a cyclone or more simply a box or bag that the particles drop into, while the air is filtered and vented. In a cyclone the air and solid particles enter the conical cyclone chamber at a tangent; this sets up a high-speed rotational air movement within the chamber. The solid particles (having a greater mass) move out to the wall of the chamber, slow down on contact and drop to the bottom under the influence of gravity. The clean air exits through a central exhaust pipe (Corbitt, 1989).

Air knife

An air knife is very similar to an air classifier, except the air is blown horizontally through a vertically dropping feed. The light fraction is carried with the air stream while the heavy fraction drops through the air stream. This technique often allows for separation into three categories:

light, medium and heavy, depending on the distance the material travels in the air stream. A system using a number of air knives linked to optical sensors is now being used to sort plastic bottles into different material types. The air knife is used to blast the bottles off the conveyor belt into the appropriate collection bunker (Corbitt, 1989).

Sink/float separation

Water (and various other liquids) can be used to remove a heavy fraction from a light fraction as the heavy material sinks and the light material floats. The partial sorting of plastics can be achieved by sink/flotation, as HDPE and PP float in water while PET and PVC sink. This simple method is often used as a last stage of separation after initial grading, shredding and flaking (Stessel, 1996).

Flotation

Flotation is a process that results in selected fine size particles floating to the surface of a slurry by means of attached bubbles; the common application is the removal of glass from ceramics and other contaminants. The key to this process is obviously the selective adhesion of air bubbles to the desired material a surface conditioning agent, which preferentially coats glass particles and makes their surfaces hydrophobic. Aeration causes the glass particles to float in the froth, while the non-glass contaminants sink and are rejected.

A relatively new application of this approach is the separation of different types of plastics. As most plastics are hydrophobic, their wetting characteristics can be selectively adjusted using surface conditioning agents. As the wetting characteristics are related to the ease of air bubble attachment, changing wetting characteristics allows certain plastics to float while others (even with equal densities) sink. This technique has been shown to be over 98% effective in commercial operation (Stessel, 1996).

Magnetic separation

The use of the force of a magnetic field gradient for the separation of ferrous metals from solid waste is one of the simplest and most developed material separation process in materials recovery. For magnetic separation to be effective waste must be processed (screening and shredding) to free metal from bags or containers and to dislodge attached or trapped non-metallic contamination, all of which otherwise reduce the efficiency of the operation (Stessel, 1996).

Note that the separation of ferrous metals from incinerator residue is often more efficient than from processed MSW due to the absence of organic contamination. However, oxidation and alloying of metals and non-metals during the combustion process means that ferrous material recovered from the incineration process does not meet all industry specifications.

Electromagnetic separation

Eddy current separation uses the principle of electromagnetic induction to separate conductive non-ferrous metals. By using modulating electromagnetic fields or the motion of the metal moving through a number of permanent magnets, eddy currents are generated in conductive metal particles. These eddy currents interact with the magnetic field and cause the metal particles to be repelled out of the process stream. Particle size reduction and effective pre-concentration of materials are necessary for this process to be effective (Vesilind and Rimer, 1989). Electro-magnets, located over the picking belt are used to separate out ferrous material in many MRFs (e.g. Dublin, Ireland). Eddy current separators, are also employed in many MRFs (e.g. Adur, UK). This technique is capable of separately removing both ferrous and aluminium material from a material stream. This means that both metals and plastic can be collected co-mingled and the three streams separated mechanically (plastic being the residue stream) at up to 5 tonnes per hour (Newell Engineering Ltd, 1993).

Electrostatic separation

Charged particles under the influence of electrostatic forces obey the laws of attraction and repulsion similar to those of permanent magnets. Electrostatic separators use an electric field generated by electrodes above a stream of particles as they flow onto a grounded metallic drum. The non-conductors (glass and organics) hold a static charge long enough to be attracted and held to the drum; conductors (metals) loose their charge quickly and are repelled from the drum and therefore separated (Vesilind and Rimer, 1989).

Feedstock for electrostatic separation require even more pre-processing than feedstock for electromagnetic separation. Here particle size must be small (<25 mm), the material must be dry and the level of paper kept to a minimum to avoid interference with the process. Due to these operating restrictions the actual application of electrostatic separators has been limited (Corbitt, 1998).

Detect and route systems

Detect and route systems separate the identification of a material from the final routing of that material into two operations. This approach depends on an array of sensors (visible light spectrophotometry, ultra violet, infra red and X-ray) acting upon individual objects. Therefore objects must pass each sensor separately, which is achieved by the configuration of the waste conveying system. Once the material has been identified air knives blow each object into the appropriate material collection bin (Corbitt, 1998). Coloured glass separation is possible using visible light spectrophotometry (Lewis and Newell, 1992), as is coloured plastic separation, while clear plastic separation (un-pigmented polymers, clear PET to translucent HDPE) is possible using near infra-red sensors. Opaque plastics need to be separated using X-ray sensors, although these are considerably more expensive than the other sensor types described. Appropriate configurations of visible light, ultra-violet and near infra-red sensors are able to efficiently sort most plastic streams.

The separation of plastic resin types is difficult due to the number of resins in common use. PVC can be separated from other plastic bottles using X-ray fluorescence to detect the chlorine atoms in PVC; polypropylene can be separated from clear HDPE, and coloured HDPE can be sorted into different colours by means of colour sensors. Recent technology for plastic sorting has mostly been focused on near infra-red (NIR) sensors linked to an air knife system. The identification speed of a NIR system often exceeds the speed with which the bottles can be processed to pass the sensor one at a time (Corbitt, 1998). It is claimed that this technique can operate at 20 bottles per second, carrying out 1000 spectral scans per second (Rhoe, 1998). NIR systems cannot reliably identify black plastics due to the infrared radiation being absorbed by the carbon filler. Fortunately this is not a major problem associated with current plastic bottle waste.

Together, an automated plastic bottle sorter can be assembled capable of handling up to 2 tonnes per hour (Magnetic Separations Systems Inc., 1995), but such systems are still not widely installed in MRFs on a global basis.

Elements of IWM

Roll crushing

This process is used in material recovery operations to grind brittle materials such as glass and flatten malleable materials such as metal cans and plastic bottles. This allows for subsequent separation by screening. Roll crushers were first employed for the reclamation of materials from incinerator residue and more recently are being used to process source separated recyclables, mainly glass containers and aluminium and steel cans. Once crushed this mixture can be sorted using screens and magnets, the material remaining being aluminium cans (Vesilind and Rimer, 1989).

Shredding

The first applications of shredders to MSW were aimed at improving final disposal, not facilitating materials recovery. Shredding breaks material into a homogenous mixture of uniform particle size. This results in volume reduction as large voids are removed, reduced odour as the material stays aerobic, and reduced litter problems as small pieces are not caught by the wind. Shredding MSW reduces the landfill volume required as shredded refuse compacts better within a landfill, and the total amount of daily cover (earth or rubble) is reduced, therefore shredding MSW becomes attractive for landfill sites where daily cover must be purchased. Leachate from shredded refuse is often produced more rapidly and is more concentrated than leachate from unprocessed MSW. Therefore when landfilling shredded waste, leachate collection and treatment systems are necessary (Vesilind and Rimer, 1989).

Shredded material is easier to handle than unprocessed material and this facilitates the identification and extraction of different fractions. Shredded waste also burns more readily, increasing its value as a fuel. Shredders are more often employed in MRFs that accept unsorted MSW, rather than facilities that accept co-mingled recyclable materials (Vesilind and Rimer, 1989).

Baling

Although a compression process, not a sorting technology, baling of sorted materials (paper, metal cans and plastic bottles) and MRF residues is a useful volume reduction technique. A bale is a predictable product that is easy to handle, an advantage when transporting recycled materials to reprocessing facilities or transporting residue materials to incinerators or landfill. Baled residues are less prone to methane generation, generally do not support combustion and produce less concentrated leachate (Corbitt, 1998).

The final disposal of baled solid waste (and MRF residues) is an easier materials handling exercise than uncompressed solid waste. Many operators of balefills claim that baling results in a more economical operation overall. Depending upon the original solid waste material, a balefill may require approximately 10% of the void space necessary to landfill the same amount of material using conventional landfill practices (Corbitt, 1998).

Central sorting at a Materials Recovery Facility (MRF)

The aim of MRFs is to separate materials that have enough value to make their recovery worthwhile. MRFs are relatively inexpensive when compared to most other waste treatment processes (e.g. incineration) and it is technically feasible to recover almost any fraction of the waste stream either manually or mechanically. Unfortunately this has often encouraged an approach where many materials are recycled because they can be recycled, not because it is environmentally or economically beneficial. There is little benefit to be gained from recycling materials that have no value. The economic costs and environmental burdens associated with such practices undermine the sustainability of the whole waste management system as the advantages of recovering and recycling a valuable material are often negated by the burdens of recovering and recycling a valueless one.

MRFs can be designed to accept and process unsorted household waste or co-mingled recyclables. The exact processing configuration of each facility depends on the input and the markets that exist for each recovered material. Thus, as emphasised in the previous chapter, there is a need to start by considering the end markets along with the waste stream composition, and then to design collection and sorting systems together to produce the materials that these markets require.

Not surprisingly, therefore, there is no standard MRF operation. While some collection schemes deliver unsorted waste to a MRF, others deliver all recyclables to the MRF in a mixed state, and other schemes will have already separated some recyclable materials out by this stage.

Blue box schemes, such as Adur and Sheffield (UK), usually involve a kerbside sort that separates out paper and glass (colour separated). Although such materials may need to be bulked up before sale and onward transport, no further sorting is necessary. Where materials such as glass are collected via glass banks, there may be no need to process it at a MRF at all, since it can be transported on to the materials processors directly.

Thus MRFs may process unsorted MSW, all dry recyclable materials, or just a restricted range of recyclable materials. Since the number of different fractions that can be collected separately is limited by practical considerations (Chapter 9), it is likely that at least some recyclables, in particular plastics, aluminium cans and steel cans, will be collected in a co-mingled fraction, for subsequent central sorting. The mechanical unit operations employed by a MRF will therefore depend on the process input and the desired final outputs.

Materials Recovery Facility (MRF) design

MRF design depends on process input (based upon the collection system), the desired process output (based upon the available material markets), the degree of mechanisation, the recovery efficiency of both manual and mechanical sorting processes and the economic viability of the whole process.

The material recovery efficiencies of several of the common separation techniques described above are presented in Table 10.2 and schematic diagrams of a simple MRF and an automated MRF are also presented in Figures 10.1 and 10.2.

Machine	Separates	Efficiency (%)
Electro magnet	Ferrous metal	60–90
Eddy current separator	Non-ferrous metal	60–90
Disc screen	Particulates	50–90
Trommel screen	Particulates	80–90
Vibrating screen	Particulates	60–90
Air classifier	Heavy and light fraction	60–90

Table 10.2 Material recovery efficiencies for separating equipment. Source: CalRecovery Inc. (1995)

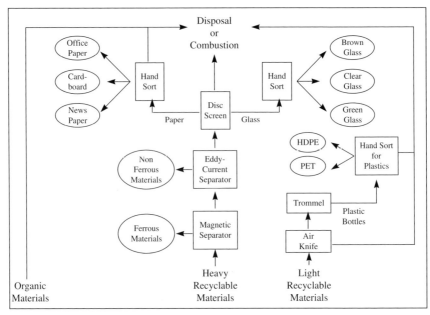

Figure 10.1 Generic MRF with manual sorting.

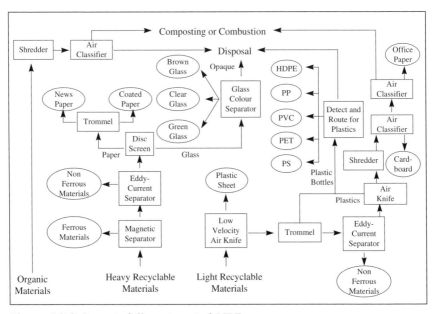

Figure 10.2 Generic fully automated MRF.

Advances in MRF technology

Single-stream processing of dry recyclables

Single-stream processing of co-mingled recyclables commonly requires a high level of manual sorting (see Box 10.1). As this is relatively costly in developed countries, few MRFs operate in such a way. Recently MRFs in California and Ohio have begun to use mechanical processes to separate a single flow of materials into paper, mixed glass, metal containers and glass containers (Egosi and Weinberg, 1998). The system still requires manual pre-sorting of large items (mainly cardboard) and non-recyclable items (mainly plastic film). In each facility the cost savings outweigh the additional processing costs of both labour and equipment.

The benefits of this single-stream processing are easy sorting instructions for residents, which results in increased diversion rates and no special collection trucks being required; rear-loading single-pass garbage and recycling trucks achieve more pick ups per hour.

The disadvantages of the system are that it collects both more recyclable and more non-recyclable material, due to incorrect sorting, resulting in an increase in the cost of residue disposal. This problem can be somewhat mitigated by more frequent education campaigns.

MRF	Hampshire, UK
Input	Single stream (co-mingled recyclables) 42,000 tonnes/year
Cost	£3.6 million (2.2 million euros), equipment £1.7 million (1.02 million euros)
Process	Hand sort for cardboard, plastic bags and other non-recyclable items Bounce adhesion belts (×2), separate paper from remaining recyclables Debris roll screen, separates items <50 mm, residue is compacted Rotating disk screen, removes remaining small items Vibrating tables, separates remaining paper Magnetic separator, removes steel cans Hand sort for plastics (three types)
Employs	78
Reference	INTEGRA (1998)

Box 10.1 Europe's largest co-mingled recyclables MRF.

Integrated waste processing

Some MRFs are now being designed not only to separate waste materials but also to produce and market raw materials ready for manufacturing processes. A 600,000 tonne/year facility in Crisp County, Georgia, USA, has been designed to produce separated green and clear PET, natural and mixed HDPE, ferrous and non-ferrous metal, newsprint, corrugated board, glass and compost (Fickes, 1998). All of these materials, with the exception of the metals will be ready for use by reprocessors (materials) and farmers or gardeners (compost).

The system operates two lines and begins with two bag openers, each able to open 65 tonnes of bags per hour. These feed two large trommel screens at the beginning of the separation process. Two low speed shredders are used to reduce the material to a size that ensures optimum operation of the plastics separator, which uses optical sensors and X-ray emitters to distinguish both coloured material and different polymer types and send specific materials to one of five processing lines. Each processing line feeds its material into a grinder that produces plastic flake, which is washed in float/sink tanks to remove labels and adhesive. The flakes are further separated by polymer type in the float/sink tanks based on the material's specific gravity. After drying, the clean flake plastic can be used as a raw material and is considerably more valuable than unprocessed recovered plastics.

The facility also operates an aggregate-separating system that removes rocks, stones, broken glass, coins, batteries and material <6 mm in diameter. This material is added to the rest of the process residues stream, which is baled and sent to an on-site balefill. Leftover organic material and unrecoverable paper flow into a mixing drum before being added to the source separated organic material for composting. Essentially, the majority of the end markets for each material produced by the MRF have been addressed by the design of the facility.

Sorting of mixed waste for Refuse-Derived Fuel (RDF)

Solid waste has been used to produce steam and electricity since around the beginning of the 1900s. The calorific value of most solid wastes is between one-quarter and one-half that of coal. The exact calorific value of most solid wastes is a function of the carbon content of the material. The ash content is generally low, between 20–40%. The amount of moisture in solid waste is highly variable and can be significantly changed due to processing, handling and storage (Manser and Keeling, 1996).

A wide range of MSW compositions can be burned without auxiliary fuel. However, since water and non-combustible material do not contribute to the calorific value of the waste, processing waste to minimise their moisture content and reduce their ash content can significantly enhance fuel quality and improves combustion efficiency.

Refuse-Derived Fuel (RDF) is produced by mechanically separating the combustible fraction from the non-combustible fraction of solid waste. The combustible fraction is then shredded, and may also be pelletised. RDF production thus forms part of a thermal treatment system, which aims to valorise part of the waste stream by recovering its energy content. The second stage, RDF combustion, can either occur on the same site, or the RDF can be transported for combustion elsewhere. In this book, production and combustion of RDF, even if they occur on the same site, will be treated separately. Since RDF production is a central sorting process, it will be discussed in this chapter. RDF combustion will be considered alongside other thermal treatment processes in Chapter 12.

A further reason for considering RDF sorting separately from thermal treatment is that the process need not only produce solid fuel; it can also produce an organic fraction, which can form the feedstock for biological treatment. As a result, in some cases, the RDF sorting process occurs in combination with a biological treatment process (e.g. at Novaro in Italy (ETSU, 1993)). Again, although the RDF sort may occur on the same site as biological treatment, it is considered here as a separate process.

There are two basic RDF processes, each producing a distinctive product, known as densified RDF (dRDF) and coarse RDF (cRDF) (also referred to as fluff or floc), respectively.

dRDF is produced as pellets, often similar in size and shape to wine corks. Prior to pelletising it is dried, so is relatively stable and can be transported, handled and stored like other solid fuels. It can either be burned alone, or co-fired with coal or other solid fuels.

Country	No. of RDF plants operating	Locations	Source
France	1 (dRDF)	Laval, Mayenne	1
Germany	1 (dRDF)	Herten	1, 2, 3
Italy	11 (5 dRDF)	Rome (2) Perugia Milan Modena Novaro Pieve di Corano Ceresara Tolmezzo Udine St Georgio	1
The Netherlands	1 (dRDF)	Amsterdam	1
Spain	1 (cRDF)	Madrid	3
Sweden	5 (cRDF)		3
Switzerland	1 (dRDF)	Chatel St Denis	3
UK	4 (dRDF)	Byker, Newcastle Polmadie, Glasgow Hastings Isle of Wight	1
Canada	1		1
South Korea	1	Seoul	3
USA	28 (6 dRDF)	Edin Prairie, MN Thief River Falls MI Northern Tier PA Yankton, SD Iowa Falls, IO Cherokee, IO	1, 3

Table 10.3 Locations of RDF sorting plants. Sources: (1) ETSU (1993); (2) Barton *et al.* (1985); (3) *Warmer Bulletin* (1993)

Elements of IWM

dRDF requires considerable processing, including drying and pelletising, and so has a relatively high processing energy requirement. As a result there has recently been interest in the alternative coarse RDF (cRDF). This comes in the form of a coarsely shredded product, that has been compared in appearance to 'the fluff from a vacuum cleaner' (Warmer Bulletin, 1993b). cRDF requires less processing, but as it has not been dried, cannot be stored for long periods. It is suitable for immediate use in on-site combustion for power generation and/or local heating. Depending on the level of processing, it can be suitable for combustion on conventional grates or in fluidised bed systems (ETSU, 1992); for more information on these combustion technologies see Chapter 12.

Status of RDF

The early development of RDF technology occurred mainly in the UK and to some extent in Italy, with plants built from the mid-1970s onward. Many of the early plants have since closed down, however, often due to difficulty finding markets for the dRDF fuel product. Lack of off-site markets has also led to the development of cRDF technology for on-site power generation. The current extent of RDF processing in Europe and elsewhere is shown in Table 10.3.

A simple rule for RDF systems is that rigorous waste processing results in a high-quality final fuel. This has been well proven by the Robbins Resource Recovery Facility Illinois, USA (APC, 1998). This facility receives 1600 tonnes of MSW per day from surrounding communities, which is processed to separate out aluminium, ferrous material, glass and compostable material. This recyclable material (25% of the incoming MSW) is sold to local recycling companies. The remaining material (75%) is shredded to form cRDF and burned in two circulating fluidised bed boilers to produce electricity (Studley and Moyer, 1997). The combination of efficient waste

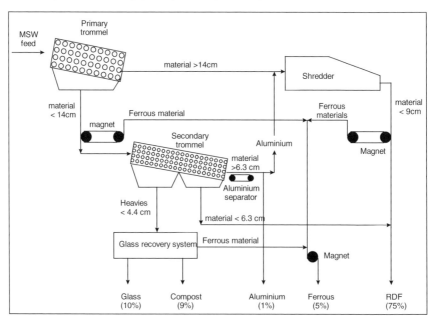

Figure 10.3 Automated cRDF processing at the Robbins Resource Recovery Facility Illinois. Source: Studley and Moyer (1997).

processing (Figure 10.3), high efficiency combustion boilers and state-of-the-art air pollution control devices has resulted in the facility achieving high material recovery rates, superior energy recovery efficiencies and low ash generation (APC, 1998).

RDF sorting processes

Details of the processing line in different RDF plants vary, but the basic dRDF process can be broken down into five distinct stages (ETSU, 1993). As shown in Figure 10.4, the production of cRDF is a simpler process (see also Figure 10.3), which omits either one or two of these stages.

Waste reception and storage

Mixed waste is delivered by the collection vehicles and tipped into a hopper or onto a tipping floor, where any unwanted 'rogue' items (e.g. car engines, logs, etc.) can be removed. This initial short-term storage stage acts as a buffer to provide the RDF production process with a steady feedstock level.

Waste liberation and screening

These processes free the waste from any refuse bags or containers, and provide the main fuel/non-fuel sort. Bag opening can involve the use of flail mills, shredders, spikes or ripping devices, though experience has shown that non-shredding devices have the advantage of not shredding or mixing the waste excessively, which can make separation more difficult.

Screening often involves a drum or rotary screen, which performs three functions. It completes the bag-emptying process, it removes the undersize (fines) fractions, and it separates the oversize (>500 mm) material from the fuel fraction. The fines fraction contains the high moisture content organic/putrescible material, as well as ash, dust and broken glass. The oversize fraction consists mainly of large pieces of paper, board and plastic film, and is usually landfilled along with other residues.

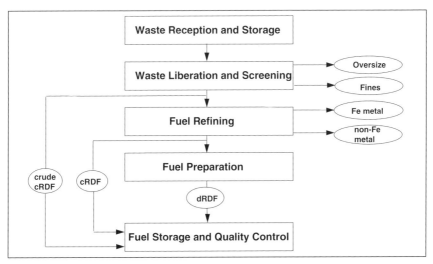

Figure 10.4 Stages in the production of Refuse-Derived Fuel (RDF).
Source: ETSU (1993).

The remaining fraction produced by this stage can be used as a crude cRDF, (cRDF type A, ETSU, 1992) though it will still contain metals and other non-combustible materials.

Fuel refining

Fuel refining involves size reduction, classification and magnetic separation. Size reduction using a shredder or hammer mill aids the separation into light and dense fractions. The density separation (classification) stage is necessary to separate the heavy fraction (metals, dense plastics) from the combustible light fraction (paper, plastic film), which will go on to form the dRDF product. Two main methods can be used to achieve this: air classification and ballistic separation, which rely on the behaviour of the object in an air stream and their 'bouncing behaviour', respectively. Magnetic separation can then be performed on the heavy fraction to remove both ferrous metal and in some plants also aluminium (by eddy current separation).

The light fraction, together with the remains of the magnetically sorted heavy fraction, can be used as a more refined form of cRDF (cRDF Type B; ETSU, 1992).

Fuel preparation

This stage represents the main difference between the cRDF and the dRDF processes. It involves the conversion of the fuel rich fraction (floc) into a dry, dense pellet form by re-shredding, drying, and then pelletising. Secondary shredding is needed to reduce the particle size of the fuel fraction to the size needed for the pelletising operation, and drying reduces the moisture content from about 30% to around 12%. Low moisture levels are needed for good storage and combustion characteristics. The dryers used are basic pneumatic conveying systems that operate on hot combustion gas from natural gas burners.

Once the combustible fraction is dry, organic and inert residues can easily be screened out, reducing the ash content of the product. Most of the chlorine, heavy metals and silicates in the product is contained within this inert residue. After this stage dRDF can be produced with a final ash content of 10% by weight and chlorine levels of 0.5%. In the absence of inert contaminants such as silicate, the calorific value of the material increases significantly.

dRDF can either take the form of pellets or briquettes, though most plants use a pellet mill to densify the product. Pellet mills are very energy intensive, consuming over 35 kWh per tonne of fuel produced, so the fuel needs to be cooled prior to storage, to remove the heat produced on compression. Pellet mills are also prone to damage from dense contaminants left in the fuel fraction, so that a further magnetic extraction stage and a ballistic separator are often used to remove both ferrous metal and other dense materials prior to the final pelletising stage.

Fuel storage and quality control

Once dried and in pellet form, dRDF can be stored before use; cRDF, in contrast, needs to be burned soon after production.

Elements of IWM

CHAPTER II

Biological Treatment

Summary

Biological treatment can be used to treat both the organic and non-recyclable paper fractions of solid waste. Biological treatment can be separated into two distinct processes – aerobic and anaerobic treatment – and therefore two main treatment types exist: composting (aerobic) and biogasification (anaerobic). Either can be used as a pre-treatment to reduce the volume and stabilise material for disposal in landfills or as a way to produce valuable products, such as compost and (from biogasification) biogas plus compost, from the waste stream. The inputs and outputs of each process are discussed, using available data. Further development of biological treatment depends on the further development of markets, and agreed standards, for the compost products.

Introduction

Biological treatment involves using naturally occurring micro-organisms to decompose the biodegradable components of waste. Aerobic organisms require molecular oxygen to use as an external electron acceptor in respiratory metabolism; this results in rapid growth rates and high cell yields. Anaerobic metabolism occurs in the absence of oxygen and does not involve an external electron acceptor. This fermentative metabolism is a less effective energy-producing process than aerobic respiration and therefore results in lower growth rates and cell yields.

If left to go to completion, biological processes result in the production of gases (mainly carbon dioxide and water vapour from aerobic processes and carbon dioxide and methane from anaerobic processes) plus a mineralised residue. Normally the process is interrupted when the residue still contains organic material, though in a more stable form, comprising a compost-like material.

The garden compost heap is the simplest form of biological treatment. With some care and regular turning (aeration), this can transform vegetable scraps and garden refuse into a rich and useful garden compost. Compost heaps that are not turned regularly become anaerobic in the middle of the heap and release methane, a potent greenhouse gas. This negates any benefit that composting has on the overall waste management system. Garden compost heaps can be a valuable method for treating part of the household waste at source, but are limited to more rural and suburban areas where space and gardens are plentiful. The alternative method to treat organic waste not composted at source (in particular from urban areas), involves centralised biological treatment plants.

241

Almost any organic material can be treated biologically. It is particularly suitable for many industrial wastes from such sources as breweries, fruit and vegetable producers and processors, slaughter-houses and meat processors, dairy producers and processors, paper mills, sugar mills, and leather, wool and textile producers (Bundesamtes für Energiewirtschaft, 1991). At the local community level, it is widely used to treat sewage sludges and organic wastes from parks and gardens.

Household waste is also rich in organic material, consisting of kitchen and garden waste. According to geography, this accounts for between 25% and 60% of MSW by weight (Chapter 8), with levels of organics particularly high in southern Europe and most developing countries. If one adds to this the non-recyclable paper fraction, which is also of organic origin and suitable for biological treatment, some 50–85% of MSW can be treated by such methods. Composting of biowaste and the non-recyclable paper fraction has been shown to have no negative effect on paper recycling schemes, the composting process or compost quality (Boelens et al., 1996). It has also been shown to have the potential to reduce the UK's total methane emissions by up to 3%, if all biowaste and non-recyclable paper fractions of MSW were to be source separated and composted (Hindle and McDougall, 1997).

The suitability of biological treatment for wet organic material contrasts markedly with other treatment methods, such as incineration and landfilling, where the high water content and putrescible nature of such material can cause problems, by reducing overall calorific value and increasing the production of leachate and landfill gas. This potential of biological treatment is being exploited in some countries, but almost ignored in others.

Category	Description
Mixed Wastes:	
MSW	Municipal Solid Waste – co-mingled solid waste collected from Households, Commerce and Institutions
HHW	Household Waste – co-mingled waste collected from households only
Centrally sorted waste:	
RDF sort fines	Putrescible material sorted mechanically from mixed waste during the production of Refuse-Derived Fuel (RDF)
Separately collected waste:	
Wet Waste	Household waste from which dry recyclables have been removed
Biowaste	This term is widely misused, ranging in meaning from garden waste only, to VFG material or to VFG plus non-recyclable paper. Here it refers to separately collected organic and non-recyclable paper waste only
VFG	Separately collected Vegetable, Fruit and Garden waste only
Greenwaste GW	Separately collected garden waste only

Table 11.1 The range of possible inputs to biological treatment plants

Numerous variants of biological treatment exist, differing according to the feedstock used (Table 11.1) and the process employed. Feedstocks range from highly mixed wastes, e.g. MSW, which require extensive treatment to remove the non-organic fractions prior to, or occasionally after, biological processing, to the separately collected and more narrowly defined Biowaste, VFG (Vegetable, Fruit and Garden) and green wastes.

Although there are many different types of biological treatment facilities available, there are two basic processes: aerobic and anaerobic treatment. In aerobic treatment, usually known as composting, organic material decomposes in the presence of oxygen to produce mainly carbon dioxide, water and compost. Considerable energy is released during the process, which is most often lost to the surroundings. Anaerobic processes are variously described as anaerobic fermentation, anaerobic digestion or biogasification. Throughout this chapter the term biogasification will be used. As the name implies this produces biogas, a useful product consisting mainly of methane and carbon dioxide, plus an organic residue which can be stabilised to produce compost, but differs somewhat from aerobically produced composts.

Biological treatment objectives

Both composting and biogasification can fulfil several functions, and it is necessary to identify the key objective(s) required of the process. They can either be considered as pre-treatments for ultimate disposal of a stabilised material (normally in a landfill) or as a treatment method.

Pre-treatment for disposal

Volume reduction

Breakdown into methane and/or carbon dioxide and water can result in the decomposition of up to 75% of the organic material on a dry weight basis (Bundesamtes für Energiewirtschaft, 1991). From wet biowaste to compost the weight loss is approximately 50%. As organics and paper represent the two largest fractions of the household waste stream this is a significant reduction. Additionally there is considerable loss of water, either by evaporation (in composting) or by pressing of the residue (biogasification). The moisture content of the organic fraction of household waste is between 60–70% (De Baere, 1999), whilst for compost made from biowaste it is around 30–40% (Fricke and Vogtmann, 1992) and for material from biogasification 25–45% (Six and De Baere, 1988; De Baere, 1993). The breakdown and moisture loss together result in a marked volume reduction in material for further treatment and disposal. Removal of water will also reduce the formation of leachate if the residues are subsequently landfilled.

Stabilisation

As a result of the decomposition processes that occur during biological treatment, the output materials are more stable than the original organic inputs, and thus more suitable for final disposal in a landfill. The cumulative oxygen demand of the organic material, a measure of biological activity and thus inversely related to stability, can decrease by a factor of six during biological treatment (Table 11.2). Similarly, the carbon/nitrogen (C/N) ratio, which gives a measure of the maturity of a compost (high C/N ratio indicates fresh organic material, low C/N ratio indicates mature, stable material), falls markedly during biological treatment processes.

	Fresh organic fraction	Windrow composting (after 6 weeks)	Biogasification (after 3 weeks)
C/N ratio	30	15	12
Cumulative oxygen demand (mg O_2/g organic matter over 10 days)	250–300	150–160	50–60
Pathogen destruction colonies/gram dry weight			
Faecal. Coliforms	3×10^3	2×10^2	0
Faecal. Streptococci	2×10^5	4×10^4	0

Table 11.2 Compost quality. Source: OWS (1995)

Standard	Exposure (days)	Temperature (°C)	Comments
Austria S2022	6 (or 2 × 3 days)	65	Moisture 40% by mass and 35°C for beneficial bacteria build-up
Belgium VLACO	21	55 (or 14 days at 60)	
Denmark	14	55	1 hour at 70°C/lime to pH 12 for 3 months
France	4	60	
Germany LAGA			Inoculated test organisms must be inactivated satisfactorily
Italy	3	55	
The Netherlands	14	50	Moisture 35%, 2× turning
Switzerland	21	55 (or 7 days at 60)	
UK code of practice for agricultural use of sewage sludge	5	40	With 4 hours at 55°C and maturation

Table 11.3 Sterilisation requirements for selected European countries. Source: AFR Report 154 (1997) with additions

Sterilisation

Both composting and biogasification are effective in destroying the majority of pathogens present in the feedstock. Composting is a strongly exothermic process, and temperatures of 60–65°C are built up and maintained in composting piles or vessels over an extended period of time, sufficient to ensure the destruction of most pathogens and seeds (Table 11.3). Biogasification processes are only mildly exothermic, but may be operated at temperatures of 55°C (thermophilic process) by the addition of heat. The combination of this temperature and anaerobic conditions is sufficient to destroy most pathogens (Table 11.2), though if lower process temperatures are used (mesophilic process), further heat treatment during the final aerobic stabilisation stage may be required to produce sanitary residues.

Valorisation

In contrast to the above, the main objective of most biological treatment is to produce useful products (biogas/energy, compost) from organic waste, i.e. to valorise part of the waste stream. An extensive range of organic materials have been studied with respect to their methane-forming potential under anaerobic conditions (Gunaseelan, 1997).

Biogas production

Biogasification produces a flammable gas with a calorific value of around 6–8 kWh (21.6–28.8 MJ) per cubic metre (German Government report, 1993), which can be sold as gas, or burned on-site in gas-engine generators to produce electricity. Some of the biogas will be burned to provide process heating on site, and some electricity will be consumed, but there can be a net export of either gas or electrical power from the plant. There will normally be a market for this product, at least for the electricity. As the gas can be stored between production and use for power generation, electricity export into the national grid can be timed to coincide with peak consumption times, and thus highest energy prices. This economic advantage is increased further in countries where additional premium prices are paid for electricity generated from non-fossil fuel sources (e.g. under the UK Non-Fossil Fuel Obligation (NFFO) scheme).

Compost production

Both composting and biogasification produce a partly stabilised organic material that may be used as a compost, soil-improver, fertiliser, filler, filter material or for decontaminating polluted soils (Ernst, 1990). Alternatively the material can be considered as a residue and landfilled. The only point that determines whether the material is a useful product, and hence of value, or a residue to be disposed of at a cost, is the presence of a market.

Markets for compost differ widely across Europe. In southern Europe, the lack of organic matter in the soil creates a large need for additional organic material. There is therefore a strong market for compost made from any feedstock, provided that the compost is safe for use, even though the level of contamination may be high. The same compost, if produced in Holland or Germany, however, would be considered only for landfill cover material or as a residue for disposal. As it would not meet the relevant quality guide-lines, there would be no market for such a product, though markets do exist for higher quality composts. Since the main determinants of compost quality are the composition of the feedstock (Fricke and Vogtmann, 1992) and the process used, production of compost for sale in such countries may require the use of restricted feedstocks involving separate collection (biowaste, either with or without paper; VFG; green waste), or the use of more sophisticated processing techniques. There is a grey area, however, around the distinction between product and residue, as in many cases, waste-derived compost is freely distributed.

The need to define the objective of any biological treatment process bears repeating. If the aim is to produce a quality compost for sale, then a restricted input (e.g. VFG, or biowaste) is preferable, and the necessary source-separated collection schemes must be put in place. If, however, the objective is to maximise diversion from landfill, while still producing a quality compost, the biowaste definition can be widened to include paper as well as organic materials *so long as there is a market for the resulting compost*. If the objective is to pre-treat waste prior to final disposal, then treating a mixed waste stream will be effective. In biogasification, where there are two possible products, biogas and compost, it is also necessary to decide which should be optimised.

Overview of biological treatment

Use of biological treatment for dealing with MSW varies considerably across Europe (see Figure 11.1). It is extensively used in Southern European countries, e.g. Spain, Portugal, France and Italy, which correlates with the generally high organic content of municipal wastes from these areas (Chapter 8). The feedstocks used for biological treatment also varies geographically: Southern Europe generally treats MSW, whilst countries such as Germany, Austria, The Netherlands, Belgium, and Luxembourg treat more narrowly defined feedstocks, which are separately collected (Table 11.4).

Country	Main feedstocks
Austria	Biowaste, VFG
Belgium	MSW, green waste
Denmark	MSW, green waste
France	MSW, biowaste, greenwaste
Germany	Biowaste, greenwaste
Italy	MSW
Luxembourg	MSW, green waste
The Netherlands	VFG
Spain	MSW
Sweden	MSW, biowaste
Switzerland	VFG, biowaste, MSW (plants >1000 tonnes/year)
Turkey	MSW

Table 11.4 Main organic waste feedstocks for biological treatment across Europe. Source: Thome-Kozmiensky (1991), with additions

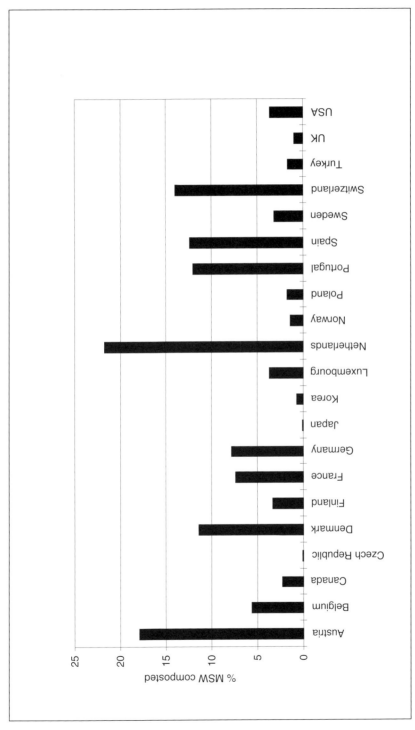

Figure 11.1 Biological treatment of MSW. Source: OECD (1997).

By far the most common biological treatment process world-wide is composting. Often large numbers of small plants exist in countries that exploit this treatment method. Switzerland alone has over 140 plants with a capacity of over 100 tonnes/year, of which 40 plants have capacities of over 1000 tonnes/year, and 15 plants over 4000 t/year (Schleiss, 1990). In the USA there are more than 262 operational composting facilities and a further 55 are either at pilot scale or at the permitting, design, construction or planning stages (Goldstein and Block, 1997). There are 14 composting facilities in the USA, processing either MSW or a combination of MSW and other organic material (Glenn, 1997; see Table 11.5 below); a further 26 facilities are at the permitting, design, construction or planning stages.

Anaerobic treatment is a more limited application, although the number of operational full-scale facilities world-wide is increasing rapidly. A list of the biogasification facilities treating MSW, either as source-separated material or mixed waste is shown in Table 11.6.

Facility	Process	Capacity (tonnes/year)
Arizona, Pinetop-Lakeside	Aerated windrow (Bedminster)	4500 (+ 2000 biosolids)
Florida, Sumter County	Windrow	18,000
Iowa, Buena Vista County	Windrow	10,000
Michigan, Mackinac Island	ASP	3000
Minnesota, Fillmore County	Aerated windrow	2000
Minnesota, Lake of the Woods County	Windrow	2000
Minnesota, Pennington County	Windrow	9000
Minnesota, Swift County	Windrow	2000
Minnesota, Truman	In-vessel (OTVD)	36,500
Nebraska, Lexington	In-vessel (Agranom)	36,500
New York, East Hampton	In-vessel (IPS)	Not available
Tennessee, Sevierville	Aerated windrow (Bedminster)	84,000 (+ 22,000 biosolids)
Wisconsin, Columbia County	Digester/Windrow	25,000
Wisconsin, Portage	Digester/Windrow	5500 (+ some biosolids)

Table 11.5 Operational MSW composting facilities in the USA. Biosolids are organic residues from waste water treatment processes. Source: Glenn (1997)

Country	Location	Process	Feedstock	Capacity (tonnes/year)
Austria	Boheimkirchen	Arge Biogas	Biowaste, manure	5,000
	Kainsdorf	Entec	Biowaste, Manure, OIW	14,000
	Lustenau	Kompogas	Biowaste	8,000
	Mayerhofen	Arge Biogas	Biowaste, manure	2,500
	Salzburg	Dranco	Biowaste	20,000
	Wels	BTA one stage/ Linde-KCA	Biowaste	15,000
Belgium	Brecht	Dranco	Biowaste	12,000
	Mons	Valorga	Mixed waste	35,000
China	Shilou	Eco-Tec	Mixed waste	17,000
Denmark	Arhus	C.G. Jenson	Biowaste, Manure, OIW	116,600
	Grirdsted	Kruger	Biowaste	40,000
	Helsingor	BTA multistage/ Carl Bro	Biowaste, OIW	20,000
	Nysted	Kruger	Biowaste, Manure, OIW	82,000
	Sinding	Prikom/HKV	Biowaste, Manure, OIW	52,700
	Studsgard	Prikom/HKV	Biowaste, Manure, OIW	130,000
	Vaarst-Fjellerad	NNR	Biowaste, Manure, OIW	55,000
	Vegger	Jysk	Biowaste, Manure, OIW	18,000
Finland	Vaasa	CiTEC	Mixed waste	15,000
France	Amiens	Valorga	Mixed waste	85,000
Germany	Baafler	Kruger	Biowaste, Manure, OIW	60,000
	Baden-Baden	BTA one stage	Biowaste	5,000
	Berlin	Eco-Tec	Biowaste, OIW	30,000
	Bottrop	Eco-Tec	Biowaste	6,500
				(Continued)

Table 11.6 Biogasification plants treating MSW, 2500 tonnes/year or larger. Source: IEA Bioenergy (1998)

Country	Location	Process	Feedstock	Capacity (tonnes/year)
Germany (contd)	Braunschweig Dietrichsdorf-	Kompogas	Biowaste	20,000
	Volkenschwand	BTA one stage	Biowaste, OIW	13,000
	Ellert	Entec	Biowaste	5,000
	Engelskirchen	Valorga	Biowaste	35,000
	Erkheim	BTA one stage	Biowaste, OIW	11,000
	Finsterwalde	Schwarting UHDE	Biowaste, Manure	90,000
	Ganderkesee	ANM	Biowaste	3,000
	Gross Muhlingen	DSD	Biowaste, Manure, OIW	42,000
	Herten	IMK	Biowaste	18,000
	Kaiseserslautern	Dranco	Biowaste	20,000
	Karlsruhe	BTA one stage	Biowaste	8,000
	Kaufbeuren	BTA	Biowaste, OIW	6,000
	Kempten	Kompogas	Biowaste	10,000
	Kiel	Eco-Tec	Biowaste	20,000
	Kirchstockach	BTA multi stage	Biowaste	25,000
	Munchen/Eitting	Kompogas	Biowaste	20,000
	Munster	BTA one stage	Biowaste	20,000
	Nordhausen	Hasse	Biowaste	16,000
	Osnabruck	Bioscan	Biowaste	19,000
	Regensburg	TBW/Biocomp	Biowaste	13,000
	Saarland	BTA one stage	Biowaste	20,000
	Sagard	BRV	Biowaste, Manure, OIW	20,000
	Schwabach	BTA one stage/ATU	Biowaste	12,000
	Simmern	Kompogas	Biowaste	10,000
	Singen	DUT	Biowaste, OIW	25,000
	Zobes	DSD	Biowaste, Manure, OIW	20,000
India	Kalyan	Eco-Tec	Mixed waste	55,000
	Kanpur	Entec	Biowaste	220,000
	Lucknow	Entec	Biowaste	145,000
	Pune	Paques	Mixed waste	150,000
Italy	Bellaria	Ionics Italba	Mixed waste	4,000
	Campo-Basso	Valorga	Mixed waste	48,000
	Chieri	Valorga	Mixed waste	60,000
				(Continued)

Table 11.6 (continued) Biogasification plants treating MSW, 2500 tonnes/year or larger. Source: IEA Bioenergy (1998)

Country	Location	Process	Feedstock	Capacity (tonnes/year)
The Netherlands	Breda	Paques	Biowaste	10,000
	Lelystad	Biocel/ Heidemij	Biowaste	35,000
	Tilburg	Valorga	Biowaste	52,000
Poland	Rzeszow	Eco-Tec	Mixed waste	50,000
Spain	La Coruna	Eco-Tec	Biowaste	113,500
Sweden	Boras	Projectror	Biowaste	9,000
	Borlange	BKS Nordic	Biowaste	9,000
	Helsingborg	NSR	Biowaste, Manure, OIW	80,000
	Kil	CiTEC	Biowaste	3,000
	Kristianstad	Kruger	Biowaste, Manure, OIW	50,000
	Uppsala	Projectror/BioMil	Biowaste, Manure, OIW	30,000
Switzerland	Aarburg	Dranco	Biowaste	11,000
	Baar	BRV	Biowaste	6,000
	Bachenbulach	Kompogas	Biowaste, Garden waste	10,000
	Frauenfeld	rom-OPUR	Biowaste, OIW	12,000
	Islikon	rom-OPUR	Biowaste	2,500
	Nieder-Uzwil	Kompogas	Biowaste	6,000
	Ottlefingen	Kompogas	Biowaste	10,000
	Rumlang	Kompogas	Biowaste, Garden waste	4,000
	Samstagern	Kompogas	Biowaste, Garden waste	10,000
Thailand	Bangkok	Eco-Tec	Mixed waste, manure	14,000
UK	Ashford	WMC	Mixed waste	40,000
USA	Greensboro, NC	Duke Engineering	Garden waste	30,000
	Waimanalo, HI	Unisyn Biowaste Tech.	Biowaste, Manure, Garden waste	19,000

Table 11.6 (continued) Biogasification plants treating MSW, 2500 tonnes/year or larger. Source: IEA Bioenergy (1998)

Biological treatment processes

A classification of the types of biological treatment processes is presented in Figure 11.2. Each consists of a pre-treatment stage followed by a biological decomposition process.

Table 11.7 below compares the land usage of composting and biogas processing. Whilst there is likely to be an effect of plant capacity on land usage (larger plants will have proportionately less free space at any time than smaller plants), it can be seen that generally composting is a more space-intensive process than biogasification. This is because biogas plants are built vertically, whilst composting plants are built horizontally. Also, in composting, a greater percentage of the input is produced as compost, which requires maturing, and so occupies space for some time.

Pre-treatment

Pre-treatment has two basic functions: the separation of the organic material from other fractions in the feedstock, and the preparation of this organic material for the subsequent biological processing. Clearly, the amount of pre-treatment will depend on the nature of feedstock – the more narrowly defined the incoming material, the less separation will be required, although pre-treatment for size reduction, homogenisation and moisture control will still be needed.

Where the plant input is mixed waste, such as MSW, the non-organic material (plastic, glass, metal, etc.) needs to be removed at this stage (unless the overall objective is pre-treatment prior to landfill). In some plants, part of this material can be recovered for use as secondary materials. In the Duisburg–Huckingen composting plant in Germany, for example, incoming mixed waste is passed under a magnet to remove ferrous metals, and then along conveyor belts where glass bottles, non-ferrous metals and plastic items are hand-picked and recovered (Ernst, 1990; Figure 11.3).

Recently there has been an increased level of criticism aimed at mixed waste composting, supported by experience from Germany, France and the UK where large quantities of low-

Process	Space (m^2/tonne capacity)
Composting	
Windrow	1.45
Drum	0.6
Tunnel reactor	0.5
Biogasification	
Dry 1-stage	0.12
Dry 1-stage	0.4
Wet 2-stage	0.32
Dry 1-stage	0.23
Wet 2-stage	0.57
Dry 1-stage	0.14

Table 11.7 Space requirements for biological treatment plants. Source: Bergmann and Lentz (1992)

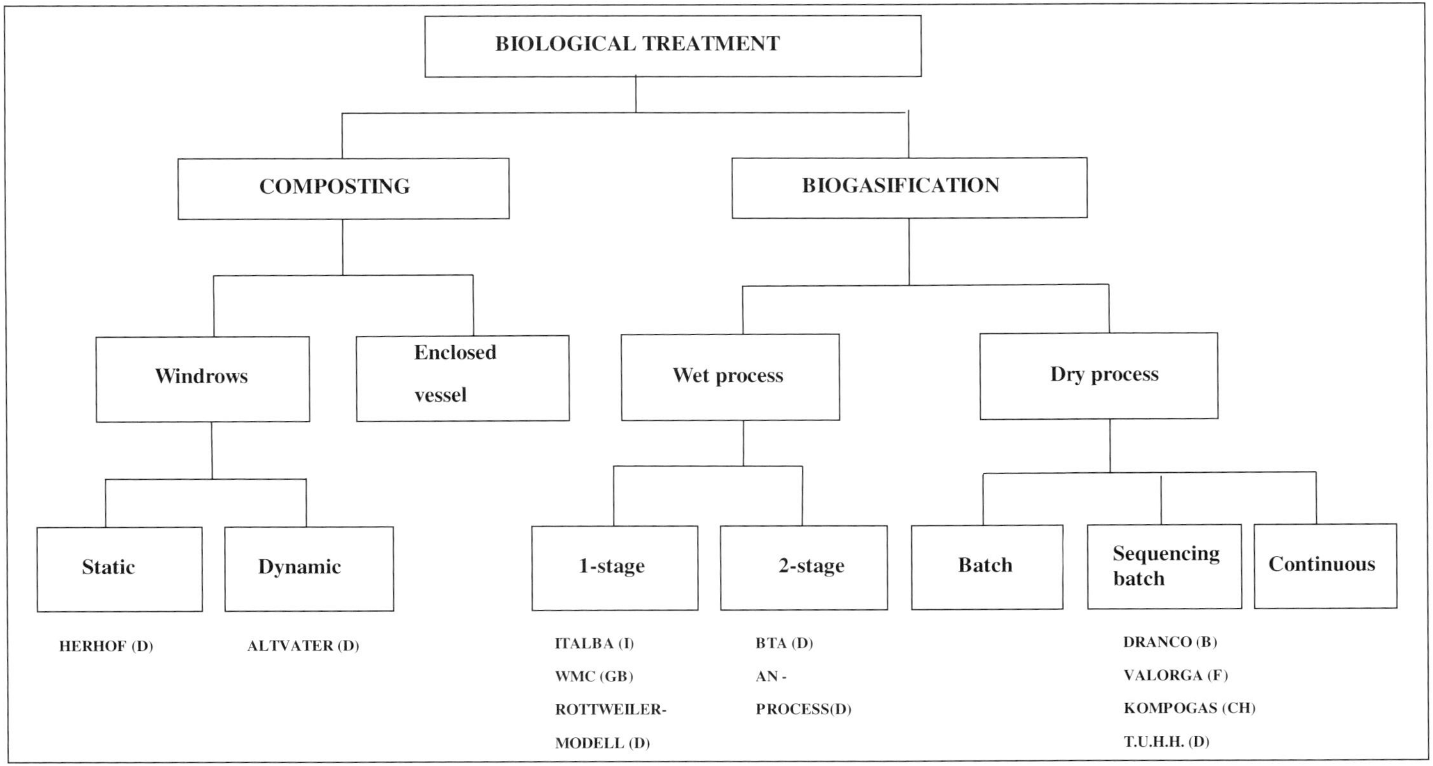

Figure 11.2 A classification of biological treatment processes.

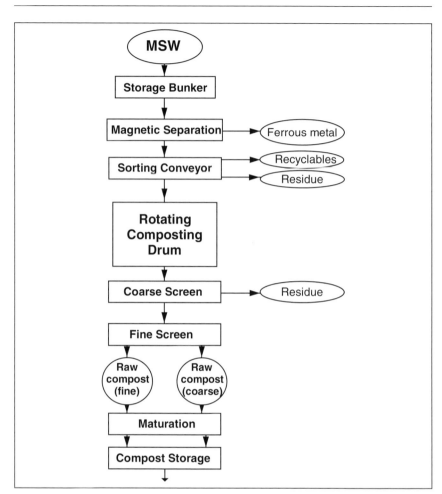

Figure 11.3 Simplified flow chart for composting process for co-mingled MSW (Duisburg–Huckingen Plant). Source: Ernst (1990).

quality compost containing high levels of heavy metals were produced (ENDS, 1997). A report by the consultancy DHV of The Netherlands, for the European Commission, stated that 'The preferred method of biological treatment now emerging in Europe is based on source-separated feedstock' (DHV, 1997). This view has been supported by representatives of Herhof, a German manufacturer of composting plants, who say 'Only by source-separated collection of the biowaste fraction can a compost with low contents of heavy metals and other contaminants be achieved' (Grüneklee, 1997). In the UK this approach is also now supported by the Government, a report for the Department of the Environment, Transport and the Regions states that 'Uniform and uncontaminated compost is required if markets are to be secured. Only wastes from source-segregated household collections and civic amenity sites will be suitable' (DETR, 1997). The future of composting may well rely on the development of collection systems that require householders to source separate biowaste.

Preparation for the actual composting or biogasification usually involves some form of screening to remove oversize items, size reduction and homogenisation. For biogasification it is normally necessary to produce a pumpable feedstock. Size reduction and mixing are achieved either by shredding the feedstock in a mill, or by the use of a large rotating drum. Shredding the feedstock removes the need for a screening stage prior to processing, but means that nuisance materials are also shredded. This makes them much more difficult to separate from the compost in the later refining stage. A drum achieves some degree of size reduction and homogenisation as it rotates, but does not shred nuisance materials. These can then be removed intact by a screen (which can be incorporated into the rotating drum) prior to the biological process, so that the later refining stages can be simplified.

Nuisance materials can therefore be removed either before or after the biological treatment stage. There is an advantage in removing them as early as possible, since the longer they are in contact with the organic material, the greater the likelihood of cross-contamination. Additionally, by removing material that would not be degraded, biological treatment processes operate more efficiently.

The amount of nuisance material removed in the pre-treatment stage depends on the waste used as feedstock. Even where the feedstock arises from source-separated collection of biowaste, separation of nuisance material during pre-treatment is advisable, especially if the

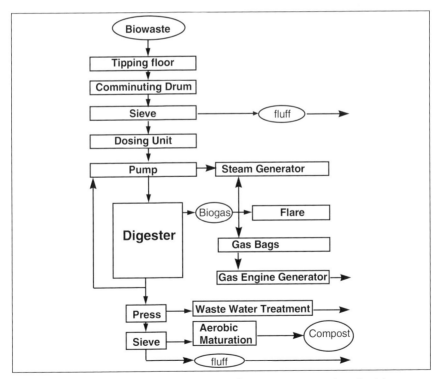

Figure 11.4 Flow chart for a dry biogasification process at a plant in Brecht, Belgium. Source: De Baere (1993).

feedstock comes from urban areas (see Chapter 9). A biogas plant in Brecht, Belgium, for example, using a feedstock of separately collected VFG plus paper (Table 11.1) discards up to 19% of its input during the pre-treatment stage (De Baere, 1993) (Figure 11.4).

Aerobic processing – composting

Biological treatment can be described as the biological decomposition of organic wastes under controlled conditions; for composting these conditions need to be aerobic and elevated temperatures are achieved due to the exothermic processes catalysed by microbial enzymes. Three main groups of micro-organism are involved in the composting process: bacteria, actinomycetes and fungi. Initially bacteria and fungi predominate, and their activity causes the temperature to rise to around 70°C in the centre of a pile. At this temperature only thermophilic (heat-tolerant) bacteria and actinomycetes are active. As the rate of decomposition and hence temperature subsequently falls, fungi and other heat-sensitive bacteria become active again (Lopez-Real, 1990). Temperature, therefore, is one of the key factors in composting plants that needs to be constantly monitored and controlled. Actual operating temperatures within the composting material are controlled in most cases to maximise both stabilisation (high temperatures inhibit this process) and sterilisation (for which high temperatures are necessary); a compromise is often the end result.

To maintain the high rate of decomposition, oxygen must be constantly available. In the simplest process type, as with a garden compost heap, this is achieved by regular turning of the composting material in long piles or windrows. The alternative method is forced aeration, whereby air is forced through a static pile using small vents in the floor of the composting area. Air can either be forced out of the vents, or drawn down through the composting pile by applying a vacuum to the vents. The former method aids dispersion of the heat from the centre of the pile to the outside, making for a more uniform process. The latter helps in controlling odours as the air passing through the pile is effectively filtered to reduce many odours before release. Aeration also helps remove carbon dioxide and volatile organic compounds, such as fatty acids, and buffers the pH of the material.

Percolation of composting piles by air depends on the structure and water content of the input material. The water content for aerobic composting needs to be over 40%, otherwise the rate of decomposition will start to fall, but if it is too high the material will become waterlogged and limit air movement. If the input material is too wet, water-absorbing and bulking agents such as woody garden waste, wood chips, straw or sawdust can be added to improve the structure, and increase the air circulation.

Windrow composting is the most common technology used, being least capital intensive. However, when it is open to the elements, control over moisture content, temperature and odour emissions is limited. One way round this is to have the entire area enclosed, and the exhaust air filtered. In The Netherlands, open air composting is generally not allowed for plants with a capacity exceeding 2000 tonnes/year, but due to the cost of building enclosed facilities, Fricke and Vogtmann, (1992) recommend this only for plants in excess of 12,000 tonnes/year. Further control over both composting conditions (moisture and temperature), odours, bioaerosols and emissions are possible in more advanced technologies, using a variety of enclosed vessels (boxes and drums) for totally enclosed processing. Open or semi-enclosed windrows are often still used in these systems, for the maturation stage. Table 11.8 summarises a number of factors affecting the choice of composting technology.

Table 11.8 Technical and operating parameters of different composting systems. Source: Panter *et al.* (1996)

	Heaps	Windrows	Aerated static pile	Shed hanger system	Vertical closed vessel	Tunnel systems
Odour control	low	low	low/medium	medium	variable	high
Temp control	low	medium	medium	medium/high	variable	high
Pathogen kill	low	low	medium	medium/high	variable	high
Weather tolerance	low	low	medium	high	high	high
Ease of operation	medium	medium	medium	medium	medium/low	medium/high
Operation & management staffing	medium	medium	medium	low	medium/low	low
Refits	low	medium	high	medium	high	low
Operation & management costs	medium	medium	medium	medium	medium/low	medium/low
Operator environment	low	low	medium/low	medium/high	high	high
Capital costs	low	low		high	high	high
Noise	low	high	medium/high	low	low	low
Process reliability	low	medium	low	medium/high	low	high
Feedstock versatility	low	low	low	medium/high	low/high	high
Process tolerance	low	medium	medium	medium/high	low	high
Product reliability	low	medium	medium	medium/high	low/high	high

The duration of the composting process varies with the technology employed, and the maturity of the compost required. The assessment of compost stability is not easy, as currently there is incomplete understanding of all the processes involved. There are several methods used:

1. **Carbon-based analysis** – compost maturity can be assessed by the carbon/nitrogen (C/N) ratio of the material, which falls from around 20 in raw organic waste to around 12 in a mature compost after some 12–14 weeks. This method suffers from a lack of sensitivity and the results vary depending on the C and N content of the starting material. Application to soils of immature composts with high residual microbial activity and high C/N ratios can result in uptake of the nitrogen from the soil by the compost, which will reduce, rather than enhance the soil fertility.
2. **Enzyme assays** – different enzymes change in concentration during the composting process. Further research is required to develop an accurate measure of stabilisation using such assay techniques.
3. **Respiration measurements** – respiratory activity falls as composting proceeds. It has been proposed that a compost may be considered stable when its oxygen uptake is less than 40 mg/kg dry matter per hour at 20°C.
4. **Phytotoxicity assays** – the presence of phytotoxins (organic materials toxic to some plants) can be assessed using cress seed emergence tests and assays for individual phytotoxins such as acetic, propionic, butyric and valeric acids and phenolic acids.
5. **Humification indicators** – measurement of humic substances, especially humic acid carbon to fulvic acid carbon ratio, may provide an effective measure of stabilisation. Humic acid content increases as composting progresses.
6. **Molecular size determination** – molecular sizes rise as humic substances are formed. This test requires specialised equipment and expert operators.

Before marketing compost, further maturation and refining are often needed. Additional maturation may be necessary to break down complex phytotoxins, which may still be present in the compost. Refining involves size classification of the compost particles and the removal of nuisance materials by sieving, ballistic separation or air classification, ready for the chosen end

Country	Total	Plastic	Metal	Glass	Stone
Austria VFG	<0.5%>2 mm	<0.2%>2 mm 0>20 mm			
Austria MSW		<3%>4 mm	<0.5%>6.3 mm 0>6.3 mm iron	<2%>2 mm	
Denmark	<5%				
Germany RAL GZ-251	<0.5%>2mm				<5%>5 mm
The Netherlands	<0.2%>2mm			<0.2%>2 mm	<2%>5 mm

Table 11.9 Inert contaminant limits of selected European countries.
Source: DHV Environment and Infrastructure (1997)

Feedstock	Total N (%)	Total P (%)	Total K (%)	pH	Glass (%)	Metal (%)	Plastic (%)
Greenwaste (GW)	1.2	0.2	0.64	8.8	0.07	0.43	0.13
VFG	1.1	0.24	0.75	8.5	0.17	0.04	0.06
GW + VFG	1.3	0.22	0.82	8.7	0.07	0.04	0.19
Commercial	4.4	0.37	1.24	7.3	0.36	0.00	0.26
Mixed (all of above)	1.5	0.35	0.62	8.6	0.09	0.01	0.17

Table 11.10 The characteristics of compost from different feedstocks
Source: Shields (1999)

Feedstock	Cd (mg/kg)	Cr (mg/kg)	Cu (mg/kg)	Pb (mg/kg)	Hg (mg/kg)	Ni (mg/kg)	Zn (mg/kg)
GW	0.67	20.9	51.1	118.2	0.17	18.7	198
VFG	0.55	20.3	84.3	79.9	0.16	25.4	185
GW + VFG	0.86	17.1	50.3	126.6	0.22	19.6	206
Commercial	0.37	5.5	31.6	17.8	0.05	5.3	117
Mixed (all of above)	0.67	42.4	76.9	103.9	0.25	16.4	267

Table 11.11 The characteristics of compost from different feedstocks (continued). Source: Shields (1999)

Feedstock	Weed seeds (viable seeds/litre)	Salmonella (spp in 25 grams)	E. coli (colony forming units/gram)
GW	1	Absent	9,018
VFG	6	Absent	<10
GW + VFG	2	Absent	100,352
Commercial	1	Absent	<10
Mixed (all of above)	7	Absent	19,674

Table 11.12 The characteristics of compost from different feedstocks
Source: Shields (1999) (continued)

use. Nuisance materials may include oversized material, stones, metal fragments, glass, plastic film and hard plastic, the limits for some of these inert materials in selected European countries are presented in Table 11.9. Oversize organic fragments can be recycled into the composting process, but the rest of the residue will need to be incinerated or landfilled.

Final compost characteristics depend on a combination of the process (mainly temperature and time) and the composition of the original feedstock. Tables 11.10–11.12 present values for a number of parameters that can be expected from composting a range of different feedstocks.

Dry-stabilization

A new process, which takes advantage of the composting process is the dry stabilization system. Developed in Germany by the Herhof Umwelttechnik engineering company the technology is applied to the restwaste fraction of the household waste stream. In Germany due to the German Packaging Ordinance (Chapter 9) the restwaste stream generally contains low levels of recoverable material and is often sent directly to landfill. The restwaste contains material that is considered unrecoverable, but still has a significant calorific value. The process begins with the restwaste being composted in enclosed vessel composters. The heat generated by the composting process is recovered and then used to dry the restwaste material itself. This not only makes the material more manageable but further increases its calorific value. The dry material is separated into a light (high-energy) fraction and a heavy inert fraction. The inert fraction is passed through magnetic and eddy current separators to recover ferrous and non-ferrous metals, which are sold. The residual inert material is washed and sold as sand/gravel for road construction.

The light fraction (the dry stabilite) has a calorific value of between 15 and 18 MJ/kg (Warmer Bulletin, 1998b), which compares favourably with coal, wood and paper (8–16, 10–16 and 13–14 MJ/kg, respectively). When this fraction is incinerated it produces 80% less ash than conventional fuels. Only 2–3% of fly ash remains at the end of the process. The heavy metal content of the dry stabilite is also lower than that of MSW and again compares favourably with coal. Incineration with energy recovery will displace the use of fossil fuels and will therefore make a positive contribution to reducing fossil fuel CO_2 emissions.

The process is currently being used in Asslar in Lahn-Dill, Westerwald Kreis in Rheinland-Pfalz and in Dresden in Germany. This novel approach, a combination of composting and incineration with energy recovery, appears to be a major step in significantly reducing the environmental burdens of the restwaste fraction of MSW on the whole waste management system.

Anaerobic processing – biogasification

Anaerobic treatment of organic material is basically a two-stage process: large organic polymers are fermented into short-chain volatile fatty acids. These acids are then converted into methane and carbon dioxide (Figure 11.5). Both processes occur at the same time, in single-phase systems. The separation of the acid producing (acidogenic) bacteria from the methane producing (methanogenic) bacteria results in a two-phase system. The main attraction of anaerobic treatment is the concurrent waste stabilisation and production of methane gas (an energy source). The retention time for solid material in an anaerobic process can range from a few days to several weeks depending on the chemical characteristics of the solid material, the amount of pre-processing it has undergone and the design (single-stage, two-stage multi-stage, wet or dry, temperature and pH control) of the biogasification system itself.

Conditions for biogasification need to be anaerobic, so a totally enclosed process vessel is required. Although this necessitates a higher level of technology than some forms of composting, containment allows greater control over the process itself and also of emissions such as noxious odours. Greater process control, especially of temperature, allows a reduction in treatment time when compared to composting. Since a biogas plant is usually vertical, it also requires less land area than a composting plant (Table 11.7).

Biogasification is particularly suitable for wet substrates, such as sludges or food wastes, which present difficulties in composting as their lack of structural material will restrict air circula-

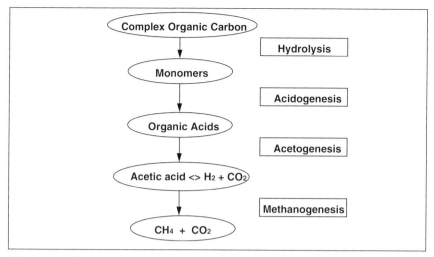

Figure 11.5 Metabolic stages in the biogasification of organic wastes.

tion. The anaerobic process has been used for some time to digest sewage sludges (Noone, 1990) and organic industrial wastes (BfE, 1991), and this has been extended to fractions of household solid waste (Coombs, 1990).

The various biogasification processes can be classified according to the solids content of the material digested, and the temperature at which the process operates. Dry anaerobic digestion may be defined as taking place at a total solids concentration of over 25% (De Baere et al., 1987); below this level of solids the process is described as wet digestion. With regard to temperature, processes are either described as mesophilic (operating between 30 and 40°C) or thermophilic (operating between 50 and 65°C). It has been well established that different anaerobic micro-organisms have optimum growth rates within these temperature ranges (Archer and Kirsop, 1990). In contrast to aerobic processing (composting) the biogasification process is only mildly exothermic. Thus heat needs to be supplied to maintain the process temperature, especially for thermophilic processes. The advantage of the higher temperature is that the reactions will occur at a faster rate, so shorter residence times are needed in the reactor vessel.

'Wet' anaerobic digestion

In its simplest form, this process consists of a single stage in a completely mixed mesophilic digester, operating at a total solids content of around 3–8% (De Baere et al., 1987). To produce this level of dilution, considerable water has to be added (and heated), and then removed after the digestion process. This method is routinely used to digest sewage sludge and animal wastes, but has also been used to treat household waste in Italy and Germany.

The single-stage wet process can suffer from several practical problems; however, such as the formation of a hard scum layer in the digester, and difficulty in keeping the contents completely mixed. The major problem with the single-stage process is that the different reactions in the process cannot be separately optimised. The acidogenic micro-organisms are fast growing and act to lower the pH of the reaction mixture, whereas the methanogens, which grow more slowly, have a pH optimum around 7.0. This problem has been solved by the development of the two-stage process. Hydrolysis and acidification are stimulated in the first reactor vessel, kept at a pH of around 6.0. Methanogenesis occurs in the second separate vessel, operated at a pH of 7.5–8.2. Variations of the two-stage wet (mesophilic) digestion process have been developed and implemented in Germany. The whole process can be run with a retention time of 5–8 days (De Baere et al., 1987).

'Dry' anaerobic digestion

Several processes have been developed that digest semi-solid organic wastes (over 25% total solids) to produce biogas in a single stage, either as a batch process or a continuous process. The processes can be either mesophilic, or thermophilic, and can use organic material from mixed wastes such as MSW, or separated biowaste.

The dry fermentation process means that little process water has to be added (or heated), which favours thermophilic operation. No mixing equipment is necessary and crust formation is not possible due to the relatively solid nature of the digester contents. This anaerobic process usually takes from 12–18 days, followed by several days in the post-digestion stage for residue stabilisation and maturation (De Baere et al., 1987).

Maturation and refining

The residues of both wet and dry biogasification processes require extensive maturing under normal aerobic conditions. This period can be significantly reduced by effective aeration. The maturation process facilitates the release of entrapped methane, elimination of phytotoxins (substances that are harmful to plant growth, such as volatile organic acids) and reduces the moisture content to an acceptable level. These residues contain high levels of water; even the dry process residue contains around 65% water, compared to German maximum recommended water levels for compost of 35% and 45% for bagged and loose compost, respectively (ORCA, 1992b). Excess water can be removed by filtering or pressing, to produce a cake-like residue; further drying can be achieved using waste heat from the gas engines if the biogas is burnt on site to produce electricity (De Baere et al., 1987). Some of the waste water can be recirculated and used to adjust the water content of the digester input; the rest represents an aqueous effluent requiring treatment prior to discharge.

The digested residue, initially anaerobic, will also still contain many volatile organic acids and reduced organic material. This needs to be matured aerobically to oxidise and stabilise these compounds, in a process similar to the maturation of aerobic composts, prior to sale as compost, or disposal as a residue.

The production of odours is a sensitive issue with regard to the neighbours of a biological treatment facility. Odour production, measured as the total amount of volatile organics produced per tonne of biowaste during composting and the final aerobic maturation after anaerobic digestion, is compared in Table 11.13.

In composting systems many of the compounds listed in Table 11.13 are volatized as soon as intensive aeration begins. This can result in odour problems. In anaerobic systems these compounds are effectively broken down by the anaerobic bacteria. After the anaerobic sludge is dewatered, the pressed cake contains few of these compounds so overall emissions, and therefore odours, are far less than from aerobic systems.

Compounds (g/tonne biowaste)	Aerobic composting	Maturation after anaerobic digestion
Alcohols	283.6	0.033
Ketones	150.4	0.466
Terpenes	82.4	2.2
Esters	52.7	0.003
Organic sulphides	9.3	0.202
Aldehydes	7.5	0.086
Ethers	2.6	0.027
Total volatile organic compounds	588.5	3.017
Ammonia	158.9	97.6
Total	747.4	100.617

Table 11.13 Emissions of volatile compounds from composting and maturation after anaerobic digestion. Source: De Baere (1999)

Compost markets

It is the presence or absence of a viable market that determines whether the composted output from biological treatment represents a valued product or a residue for disposal. Consequently much effort has been put into the definition and development of markets for waste-derived composts both in Europe and the USA by the Organic Reclamation and Composting Association (ORCA) and the Solid Waste Composting Council (SWCC), respectively. Broadly speaking there are three main areas of compost application: agriculture, horticulture and landscaping.

Within these areas compost can fulfil one or more of four basic functions:

1. **Soil conditioner or improver** – by adding organic matter to the soil, compost will improve the structure of the soil and replace the organic material lost during sustained intensive cultivation.
2. **Soil fertiliser** – the actual value of compost as a fertiliser will depend on its nutrient content in general and of its nitrate and phosphate content in particular. This is normally much lower than for inorganic fertilisers, and because these nutrients are bound to the organic matter, their release is slow and sustained. This is an advantage that is becoming increasingly important in countries such as Denmark, Belgium, The Netherlands and Germany where strict limitations on nitrate application are being implemented to reduce ammonia emissions and possible groundwater contamination (ORCA, 1992a).
3. **Mulch** – compost can be applied to the soil surface to reduce evaporation losses and weed growth.

4. **Peat replacement** – the use of peat is facing growing public opposition, being seen as the exploitation of an irreplaceable natural biotope. In both the UK and Germany some sectors of the trade have specified that no peat be used in products for home gardening. Whilst the use of waste-derived composts instead of peat may be limited in the potted plant industry due to very strict phytosanitary regulations in Europe, to control the spread of plant diseases, there appears to be a market as a peat replacement in the home gardening and landscaping sectors (ORCA, 1992a). Table 11.14 shows the current situation in the UK. It is clear that peat utilisation, by local authorities and landscapers, as both a soil improver and growing medium is decreasing, but peat utilisation as a growing medium by amateur gardeners (97% of all peat used) has not changed. This is potentially a very large market available to compost producers if product quality and performance can be guaranteed.

Sector:	Soil improvers		Growing media		Total	
	1996	1997	1996	1997	1996	1997
Amateur gardening						
Peat	160,000	99,800	2,031,900	2,108,400	2,191,900	2,208,200
Alternatives	330,300	400,000	115,200	94,800	445,500	494,800
Totals	490,300	499,800	2,147,100	2,203,200	2,637,400	2,703,000
% peat	33%	20%	95%	96%	83%	82%
Local authority						
Peat	11,500	3,500	16,200	7,900	27,700	11,400
Alternatives	139,800	160,300	1,000	1,100	140,800	161,400
Totals	151,300	163,800	17,200	9,000	168,500	172,800
% peat	8%	2%	94%	88%	16%	7%
Landscaping						
Peat	30,900	16,000	26,300	32,600	57,200	48,600
Alternatives	486,200	598,900	900	1,400	487,100	600,300
Totals	517,100	614,900	27,200	34,000	544,300	648,900
% peat	6%	3%	97%	96%	11%	7%
Total peat	202,400	119,300	2,074,400	2,148,900	2,276,800	2,268,200
Total alt.	956,300	1,159,200	117,100	97,300	1,073,400	1,256,500
Total	1,158,700	1,278,500	2,191,500	2,246,200	3,350,200	3,524,700
% peat	17%	9%	95%	96%	68%	64%

Table 11.14 Amateur gardening, local authority and private sector landscape sectors: results of a survey of producers for 1996 and 1997 in the UK – summary of material use (in m^3) by sector. Source: DETR (1998)

Elements of IWM

	Sale of compost (tonnes)	Sale of compost (%)
Landscaping	560,619	25.3
Agriculture	419,250	19.0
Hobby gardening	353,930	16.1
Special crops	220,915	9.9
Communities	108,053	5.0
Commercial gardening	140,692	6.4
Soil improver	256,240	11.6
Other	146,993	6.7
Total	2,221,445*	100%

Table 11.15 Utilisation of quality-controlled compost in Germany in 1996. Source: BGK (1998). *The total amount of compost corresponds to approximately two-thirds of the total amount of compost produced in Germany in 1996, as not all composting facilities are subject to quality control

As well as fulfilling different functions, composts from biological treatments come in different forms. Many processes produce more than one grade of compost (e.g. coarse and fine) at the final refining stage. The essential marketing step is to match these products to the market requirements. In some cases, new markets may need to be developed; for example, where composting plants produce a novel product, such as the very fine textured and uniformly graded compost produced by the lumbricomposting process. The market potential can be assessed by considering both current and potential future usage of composts.

Surveys of compost consumption show that in Switzerland, of the 100,000 tonnes of compost produced each year, 46% is used in agriculture and vineyards, 30% in horticulture and tree nurseries, 13% in hobby gardening and 11% in recultivation (Schleiss, 1990). A more detailed analysis of German usage is given in Table 11.15.

A report published by the UK Department of the Environment (1997) suggests that there is a significant market in the UK for waste-derived composts and digestates (Table 11.16).

Sector	Market size (Mt/year or Mm^3)
Agriculture	0–148
Land reclamation	0.5
Forestry	2.0
Professional horticulture and landscaping	0.5–1.0 Mm^3
Retail and domestic	0.5–1.0 Mm^3

Table 11.16 Compost market potential assessment for the UK. Source: DOE (1997)

Elements of IWM

Penetration of new market areas, such as agriculture, will depend on effective marketing of waste-derived compost as a quality product, i.e. that it is safe and fit for use, and gives clear benefits compared to competing products at an affordable cost (ORCA, 1992a). Whilst these are general pre-conditions, there are additional specific compost quality requirements that will vary between different compost uses (Table 11.17).

The failure to gain widespread acceptance of waste-derived composts, especially in the agricultural sector has most likely been due to concerns over its safety and quality. Failure of many early plants to completely separate visual contaminants (e.g. plastic film) from the final compost reinforced the idea of waste-based composts as inferior products. The very use of the word 'waste' also raises concerns over safety, in particular the possible presence of pathogens, although these should be effectively destroyed in the biological treatment process. More recently the level of heavy metals in waste-derived composts has become a concern. The high levels of some heavy metals in some fractions of household waste (Table 8.8) can result in contamination of the final compost, if biological treatment of mixed wastes is used. What are needed, if waste-derived composts are to become fully accepted and used more extensively, are:

1. widely accepted quality standards, which reassure potential users that the compost is both safe and fit for use, especially with repeated applications
2. plant growth studies demonstrating a commercial benefit from the application of compost or compost-based products.

Market outlet	Impurities (glass, plastic)	Maturity	Organic material	Particle size	Salinity	Humidity
Agriculture	1 xxx	3	2 xxx	4	6	5
Market gardening	1 xxxx	1 xxxx	3 xxx	5	3 xxx	6
Produce farming	1 xxxx	2 xxxx	3 xxx	5	4 xxx	6
Viniculture	1 xxx	2	2	4	6	5
Arboriculture	1 xxx	2	2 xxx	4	6	5
Mushroom growing	2 xxx	1 xxxx	4	5	6	2 xxx

Table 11.17 Market requirements for compost quality in France (numbers 1–6 give quality criteria in descending order of importance; xxx/xxxx: customers sensitive/very sensitive to this criterion). Source: ANRED (1990)

Elements of IWM

Compost standards

Compost market development would be facilitated by the application of consistent quality standards across Europe, but present standards vary widely between countries both in approach and detail. ORCA have published a review of compost standards for 12 European countries (ORCA, 1992b). Of these, nine countries (Austria, Belgium, Denmark, France, Germany, Greece, Italy, The Netherlands and Switzerland) have implemented or proposed standards, Sweden and the UK are in the process of drafting standards while Spain uses standards relating to fertilisers. In countries such as Germany these criteria take the form of marketing standards, whereas in other countries they actually comprise a legally defined standard (Table 11.18). A more recent set of data describing the limit values for heavy metals in composts in a number of European countries is presented in Table 11.19.

The objective of the standards is to protect land from contamination and to ensure that the composts marketed are fit for use. Since there are many uses for compost, however, different compost quality criteria need to be applied for each separate application. Several countries, such as the The Netherlands and Austria, define several different grades of compost with different maximum levels of contaminants for each (Table 11.19). Belgium also specifies which grades can be used for different applications such as growing food or fodder crops.

Most standards relate to the physical and chemical properties of the compost, though there is normally more emphasis on what should not be in the compost (i.e. contaminants) rather than what should be in the compost (e.g. nutrients). Heavy metal levels come in for close scrutiny, but as Tables 11.18 and 11.19 demonstrate, limit levels vary widely between countries. To a large extent this reflects differences in the interpretation of the available scientific data on the heavy metal levels that constitute a significant health or environmental risk. Another cause for variability, however, is the use to which the compost may be put. Many of the most restricted heavy metal limits refer to composts that can be freely applied; some of the more relaxed standards are supplemented by restrictions on their level or time of use, frequency of application, application during wet weather, soil type or proximity of water supply plants (ORCA, 1992b). Measured levels for contaminants such as heavy metals will also depend on the analytical methods used. Whilst most of the country standards include details of the analytical methods required, some, such as Switzerland, do not. Clearly this lack of uniformity can only hinder the development of free markets for compost across Europe.

Not all standards systems even take the same basic approach. Since the quality of a compost is largely determined by the feedstock used and the processing method, some standards set criteria for these rather than the quality of the resulting compost itself. Criteria for some of the high-quality composts specify that unseparated household waste cannot be used as a feedstock. Criteria for compost processing methods are commonly used to determine microbiological safety. Several standards include both the temperature that must be achieved and its duration for destruction of pathogens during aerobic composting (Table 11.3).

The French compost standard takes yet another approach. Rather than set criteria in terms of compost/composting conditions considered safe and fit for use, the criteria reflect the engineering capabilities of existing plants. In this case, some plants should be capable of meeting the requirements, but elsewhere the problem of compost not meeting quality criteria seems widespread.

	Legal definition (LD), Marketing Standard (MS)	Number of grades of compost	Minimum processing requirements	Heavy metal limits	Limits on other characteristics	Analytical methods	Quality control procedures
Austria	LD	2	No	Yes	Yes	Yes	Yes
Belgium	LD	2	Yes	Yes	Yes	Yes	No
Denmark	LD	1	Yes	Yes	Yes	Yes	Yes
France	LD & MS	4	Yes	Yes	Yes	Yes	Yes
Germany	Three MS	1 (for each standard)	No	Yes	Yes	Yes	Yes
Greece	LD	1	No	No	Yes	No	No
Italy	LD	2	Yes	Yes	Yes	Yes	Yes
The Netherlands	LD	3	No	Yes	No	Yes	No
Spain	LD	1	No	Yes	Yes	Yes	No
Sweden	None	0	No	No	No	No	No
Switzerland	LD	1	No	Yes	No	No	No
United Kingdom	None	0	No	No	No	No	No

Table 11.18 Summary of criteria specified in existing compost standards. Source: ORCA (1992b)

Standard	Cd	Cr^{tot}	Cr^{VI}	Cu	Hg	Ni	Pb	Zn	As
EU 488/98 Ecolabel for Soil Improvers	1.0	100		100	1	50	100	300	10
Austria ÖN S2200 1993									
Class 1	0.7	70		70	0.7	42	40	210	
Class 2	1.0	70		100	1.0	60	150	400	
Class 3	4	150		400	4	100	500	1000	
Federal Draft									
Farming	1.0	70		150	0.7	60	150	500	
Land remediation	3.0	250		500	3.0	100	250	1200	
Belgium									
Farming	5.0	50		100	5.0	50	600	1000	
Parks and Gardening	5.0	200		500	5.0	100	1000	1500	
Vlaco (Flanders)	1.5	70		90	1.0	20	120	300	
Denmark									
Before 1/6/2000	0.8				0.8	30	120		
After 1/6/2000	0.4				0.8	30	120		

(Continued)

Table 11.19 Limit values for elements in compost, in current standards. in mg/kg dry matter. Source: Centemero *et al.* (1999) with additions

Standard	Cd	Cr^{tot}	Cr^{VI}	Cu	Hg	Ni	Pb	Zn	As
Finland Decision 46/94	3			600	2	100	150	1500	50
France Fertiliser Law	1.5	300		600	1.0	100	100	1500	
Germany Federal Bio AbfV (1998)									
Class 1	1.0	70		70	0.7	35	100	300	
Class 2	1.5	100		100	1	50	150	400	
Italy Fertiliser Law (3/98)	1.5		0.5	150	1.5	50	140	500	
Luxemburg Draft (as RAL)	1.5	100		100	1.0	50	150	400	
The Netherlands									
BRL K256/02 VFG	1.0	50		60	0.3	20	100	200	15
BRL K256/02 high quality VFG	0.7	50		25	0.2	10	65	75	5.0
Spain Draft amendment	10	400	0	450	7	120	300	1100	
UK OWCA	10	1000		400	2	100	250	1000	

Table 11.19 (continued) Limit values for elements in compost, in current standards, in mg/kg dry matter. Source: Centemero et al. (1999) with additions

In the USA, the Environmental Protection Agency (EPA) has taken yet a different approach, based on a risk assessment of soil to which compost has been added. This approach is related to the US (and UK) approach towards application of sewage sludge to fields, and does have a certain scientific logic.

In conclusion, criteria are needed to provide reassurance that marketed composts are fit and safe for use, but such criteria should be set uniformly across Europe on the basis of good scientific data, considering all aspects of compost usage. Such standards should define the quality and quantity of compost that can be used for different applications, ranging from horticulture and agriculture to the reclamation of derelict land and erosion control.

CHAPTER 12

Thermal Treatment

Summary

Thermal treatment can be regarded as either a pre-treatment of waste prior to final disposal, or as a means of valorising waste by recovering energy. It includes both the burning of mixed MSW in municipal incinerators and the burning of selected parts of the waste stream as a fuel. These different methods reflect the different objectives that thermal treatment can address. This chapter describes the various thermal treatment processes and their application worldwide.

Introduction

Thermal treatment of solid waste within an Integrated Waste Management system can include at least three distinct processes, Mass burn, Refuse-Derived Fuel burn and Paper and Plastic fuel burn. The most well known is mass-burning, or incineration, of mixed Municipal Solid Waste (MSW) in large incinerator plants, but there are two additional 'select-burn' processes whereby combustible fractions from the solid waste are burned as fuels. These fuels can be separated from mixed MSW either mechanically to form Refuse-Derived Fuel (RDF), or can be source-separated materials from household collections such as paper and plastic, which have been recovered but not recycled (Chapter 10).

These three methods reflect the different objectives that thermal treatment can address (Table 12.1). The process can be considered as an Energy from Waste (i.e. valorisation) technique or as a pre-treatment to final disposal. Although their objectives may differ, all methods of burning solid waste will be dealt with together in this chapter, due to the similarity of the underlying physical processes and issues involved.

Thermal treatment objectives

Burning of solid waste can fulfil up to four distinct objectives.

1. **Volume reduction** – depending on its composition, incinerating MSW reduces the volume of solid waste to be disposed of by, on average, 90%. The weight of solid waste to be dealt with is reduced by 70–75%. This has both environmental and economic advantages since there is less demand for final disposal to landfill, as well as reduced costs and environmental burdens due to transport if a distant landfill is used.

2. **Stabilisation of waste** – incinerator output (ash) is considerably more inert than incinerator input (MSW), mainly due to the oxidation of the organic component of the waste stream. This leads to reduced landfill management problems since the organic fraction is responsible for landfill gas production and the organic compounds present in landfill leachate.

3. **Recovery of Energy from Waste (EfW)** – this represents a valorisation method, rather than just a pre-treatment of waste prior to disposal. Energy recovered from burning waste is used to generate steam for use in on-site electricity generation or export to local factories and district heating schemes. Combined heat and power (CHP) plants increase the efficiency of energy recovery by producing electricity as well as utilising the residual heat. Often viewed as a 'renewable resource' (van Santen, 1993), burning solid waste can replace use of fossil fuels for energy generation. As a large part of the energy content of MSW comes from truly renewable resources (biomass), there should be a lower overall net carbon dioxide production than from burning fossil fuels, since carbon dioxide is absorbed in the initial growing phase of the biomass.

4. **Sterilisation of waste** – whilst this is of primary importance in incineration of clinical waste, incineration of MSW will also ensure destruction of pathogens prior to final disposal in a landfill.

Current state of thermal treatment

The prevalence of thermal treatment, and the actual approach that it takes, reflects the relative importance attached to the different objectives listed above, and varies from country to country (Figure 12.1). In 1996 there were approximately 2400 large-scale waste incineration facilities in operation around the world, 150 were under construction and a further 250 facilities were planned. Globally, around 2800 incineration facilities are predicted to be operational by the year 2005 (Helmut Kaiser Consultancy, 1996). Countries with an acute shortage of landfill capacity, such as Switzerland, The Netherlands and Japan, incinerate a high proportion of MSW principally for volume reduction, with some energy recovery. By comparison, countries with plentiful and currently cheap landfills, such as the UK and Spain, have little MSW incineration (12% and 4% of MSW, respectively).

Volume reduction, for both environmental and economic reasons, and sterilisation of waste have historically been important objectives for incineration. These were the prime reasons for the MSW incinerators built in the UK, for example, in the 1960s. It is also likely that the future will see more emphasis on using incineration for stabilising wastes for subsequent landfilling. This will also increase the proportion of MSW incinerated. Due to growing concern over the production of landfill gas and the organic compounds in leachate from landfills receiving untreated MSW, stabilisation of waste prior to disposal is becoming an important additional objective in some countries. Landfill gas and leachate arise principally from the organic fraction of MSW, which can be effectively converted to gases and mineralised ash by incineration.

Concern over emissions from landfill sites led Germany to pass the National Technical Directive for the future management of communal waste (TA Siedlungsabfall, 1993), which states that only inert materials can be landfilled. Inert materials are defined as those that have less than 1% (for normal community waste landfills) or 5% (for 'special' community landfills) weight loss on incineration. Note that the directive does not make incineration mandatory for material going to landfill, but at present there is no other technology available that can achieve this level of inert-

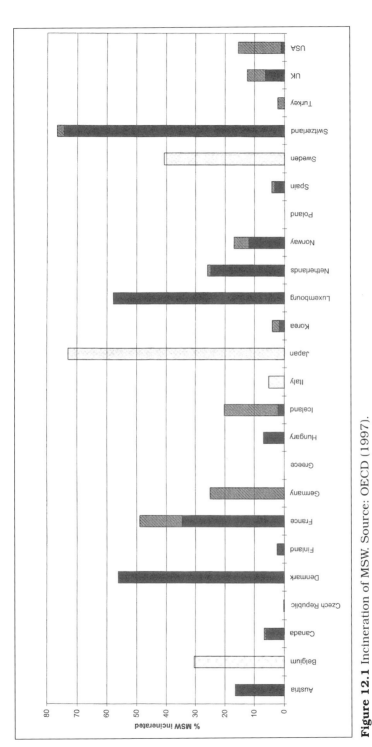

Figure 12.1 Incineration of MSW. Source: OECD (1997).
Key: black – % of incineration with energy recovery; hatched – % of incineration without energy recovery; grey – no data on % of incineration with energy recovery.

Mass-burn incineration of MSW and residues from other waste treatment options	Reduces volume of waste for final disposal Produces a stable residue that will produce little or no gas on landfilling. Produces a sanitised residue for landfilling Energy recovery possible, but limited by high moisture content of MSW, presence of non-combustible material and corrosive contaminants. High level of airborne emissions from combustion, requiring extensive gas-cleaning equipment to meet emission standards.
Burning Refuse-Derived Fuel (RDF)	Uses the combustible part of the waste stream for energy production. Either loose cRDF or pelletised dRDF can be burned. cRDF must be burned as produced. dRDF can be stored and transported, but needs more energy to produce. RDF has a higher calorific value than MSW, so energy recovery is higher. RDF contains less non-combustible material than MSW, so less ash is produced. Combustion characteristics of RDF are more consistent than MSW, so combustion can be better controlled.
Burning source-separated paper and plastic as fuel	Can use the paper and plastic collected separately from households that is in excess of recycling capacity. Due to low moisture content can be stored and transported. High calorific value, low ash content and consistent combustion advantages apply to this fuel as they do to RDF.

Table 12.1 Thermal treatment options

ness. This may require the building of new incinerators to meet the new legal requirements. Austria is also considering similar legislation. At the time of writing, recent EC legislation (EC Landfill Directive 1999/31/EC) states that not later than July 2003 Member States will reduce biodegradable municipal waste going to landfill to 75% of the total amount (by weight) of biodegradable municipal waste produced in 1995 (EC, 1999a). This figure drops to 50% by 2009 and to 35% by 2016. This legislation is likely to result in an increase in the amount of material incinerated throughout the EC countries.

Scandinavian countries have generally exploited incineration of MSW for energy recovery (e.g. Denmark and Sweden use 56% and 42% of their MSW, respectively, in this way) usually via district heating schemes (van Santen, 1993). Energy recovery has been taken further in the development of the Refuse-Derived-Fuel process (RDF). RDF technology has been developed extensively in the UK and Sweden, though plants have also been set up in France, Germany, Switzerland and the USA (see Chapter 10 for details of the sorting process). The other process discussed in this chapter, the burning of source-separated paper and plastic as an alternative fuel, has not been fully developed. However, as the amount of paper and plastic collected by materials recovery schemes across Europe increases and exceeds reprocessing capacity, interest in this area is likely to grow.

Mass-burn incineration of MSW

This form of thermal treatment can be divided into several distinct stages: the incineration process, energy recovery, emission control and treatment of solid residues.

The incineration process

Municipal Solid Waste incineration processes are dominated by the so-called 'mass burn' technologies. These systems accept solid waste with little pre-processing treatment other than the removal of recyclable material and bulky items. A typical mass burn incinerator uses a single large furnace with an inclined moving or roller grate system.

A schematic diagram of a generic modern incinerator is shown in Figure 12.2. Mixed waste for incineration is delivered into a reception hall or tipping bunker, from which it is fed into the furnace feed hopper, usually by mechanical grab. The reception area can be a source of local

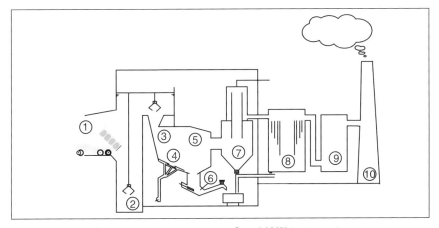

Figure 12.2 Schematic representation of an MSW incinerator.
Key: 1 – reception hall; 2 – waste pit; 3 – incinerator feed hopper;
4 – combustion grate; 5 – combustion chamber; 6 – quench tank for
bottom ash; 7 – heat recovery boiler; 8 – electrostatic precipitator; 9 – acid
gas scrubbing equipment; 10 – incinerator stack.

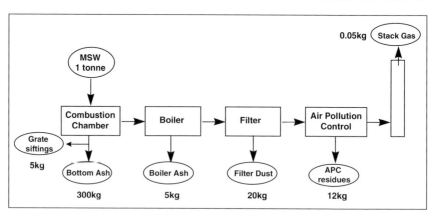

Figure 12.3 Solid mass flow in an MSW Incinerator. Source: IAWG (1997).

nuisances such as noise, odour and litter, which need to be controlled. The majority of MSW incinerators have a furnace with a moving grate design, also known as a 'stoker' type incinerator. The moving grate keeps the waste moving through the furnace as it burns and deposits the unburned residue (bottom ash) into a quench tank. Primary air for combustion is pumped through from under the grate, and secondary air is introduced over the fire to ensure good combustion in the gas phase. An approximate solid mass flow for waste in a mass-burning system is shown in Figure 12.3.

The key factors for high levels of combustion and destruction of organic pollutants in the incoming waste are temperature (high), residence time (long) and turbulence (high) (Vogg, 1992). The EC Directives on Incineration (1989a, b), for example, requires a residence time (for the gas in the combustion chamber) of at least 2 seconds at temperatures in excess of 850°C, in the presence of at least 6% oxygen to ensure maximum oxidation of dioxins and other organic pollutants. Furnace design and operating efficiency are the crucial factors that determine the levels of pollutants in the crude gas entering the flue gas-cleaning system (Table 12.2).

The hot gases then enter the energy recovery boiler, where they cool rapidly. If no energy is recovered, the flue gases must be cooled using air or water sprays before they enter the gas cleaning systems.

	Old facility	Modern facility
Dust (mg/Nm^3)	6500	1700
C in dust (%)	2.7	1.4
PCDD (ng/Nm^3) (dioxins)	270	55
PCDF (ng/Nm^3) (furans)	1100	110
TEQ (ng/Nm^3)	25	2.5

Table 12.2 Crude gas values of two different waste incinerator plants showing the effect of furnace design on combustion efficiency. Source: Vogg (1992)

Some systems have added to the objectives of volume reduction and sterilization the recovery of energy for electricity generation or local heating and/or power combinations. The energy is recovered from the flue gases using a boiler system to generate hot water or steam. Hot water may be used for heating and steam for industrial purposes or electricity generation. Combined Heat and Power plants (CHP) can be fitted into MSW incineration processes. The different incineration processes are described more fully below.

Grate incinerators

There are different designs for the grates used in municipal waste incineration. Grate systems are used to transport the refuse through the furnace and to promote combustion by agitation and mixing with combustion air. Once the refuse is in the grate, it passes through a drying stage where the most volatile compounds are burnt. The refuse moves further down the grate and continues to burn slowly until it gets to a burnout stage before the ash is discharged at the end of the grate. Ash temperatures are a limitation to the operation of grate systems. If the temperature is high enough for the ash to melt and form slag, the slag reduces the air supply as it blocks up the grates. Overfire air is injected above the grates to provide enough air to combust the flue gas and the particulate material it contains. Some of the most common grates systems used in Municipal Solid Waste incineration are:

1. **Roller systems**: which consist of drums placed to form an inclined surface. The drums rotate slowly in the direction of the refuse movement.
2. **Reciprocating systems**: the grates are placed above each other. Alternated grate sections slide back and forth while the adjacent sections remain in place.
3. **Reverse reciprocating**: the grates move forward and then reverse direction.
4. **Continuous**: two sections are used in cascade. This process is cheap and reliable, but does not agitate the waste, and needs large amounts of air, which in turn produces large amounts of flue gas.
5. **Rocking grates**: the grates are placed across the width of the furnace. Alternate rows are rocked to produce an upward-forward motion. This motion agitates and moves the waste forward through the incinerator.

Fluidised bed incinerators

Fluidised beds are simple devices consisting of a vessel lined with heat-resistant material, containing inert granular particles. Gases are blown through the inert particles at a rate high enough to cause the bed to expand and act as an ideal fluid. Normally bed design restricts combustion to the immediate area of the bed. This maintains an area above the bed for separating the inert particles from the rising gases and other components. The hot gases leave the fluidised bed and enter heat-recovery or gas-cleaning devices. Because of the close contact between combustion gases and the waste being burned, excess air for normal incineration is usually limited to approximately 40% above the stoichiometric air requirements for combustion of the waste. Fluidised beds are subject to problems caused by low ash-fusion temperatures and materials with low melting points. Such materials (i.e. aluminium and glass) have to be removed before entering the process. These can be avoided by keeping the operating temperature below the ash-fusion level or by adding chemicals that raise the fusion temperature of the ash (Cheremisinoff, 1987). An advantage of this technology is that reagents that capture halogens (chlorides and fluorides) can be added to the process, reducing the final discharge of acid gases. This

	Grate incinerator	Fluidised bed incinerator
Structure		
Orientation	Horizontal	Vertical
Moving parts?	Yes	No
Max capacity of single stream	*ca.* 1200 tonnes/day	*ca.* 350 tonnes/day
Combustion		
Mixing	Mild agitation	Turbulent
Rate	Slow	Rapid
Burn out	Often incomplete	Complete
Air ratio	1.8–2.5	1.5–2.0
Load	200–250 kg/m^2/h	400–600 kg/m^2/h
Fuel size	75 cm	50 cm (shredding required)
Combustion residue		
Unburnt carbon	3–5% by weight	0.1% by weight
Volume	Larger	Smaller
State	Wet	Dry
Iron recovery	Difficult	Easy
Fly ash		
Volume	Smaller	Larger
Unburnt carbon	3–7% by weight	1% by weight
Flue gas		
Volume	Larger	Smaller
NOx control	By added chemicals	By air ratio control
Operation Stop	Few hours (unburnt fuel remains)	Few minutes (no unburnt fuel)
Restart after 8 h stop	1 h	5 minutes
Restart after weekend stop	2 h	30 minutes

Table 12.3 Comparison of Mass-burn (grate) and fluidised-bed systems for MSW incineration. Source: Patel and Edgecumbe (1993)

process has been developed and implemented particularly in Japan. A comparison of moving-grate and fluidised-bed incinerators is given in Table 12.3.

Rotary combustors or rotary kilns

This process consists of a horizontal, cylindrical, shell lined with heat-resistant material that is mounted on a slight slope, and is the most commonly used design for the incineration of waste. The rotation of the shell mixes the waste with the combustion air. The range of the combustion temperatures is between 820 and 1650°C. Residence times vary from several seconds to hours, depending on the waste and its characteristics. Rotary kilns are effective when the size or nature of the waste prevents the use of other types of incineration equipment.

Multiple-chamber incinerators

This configuration provides complete burnout of combustion products, which decreases the concentration of particulate material in the flue gas. The combustion of the solid waste takes place in a primary chamber and the unburned gaseous products are combusted in a secondary chamber. The secondary chamber provides the required residence time and supplementary fuel for the combustion of the gaseous emissions.

Multiple-hearth furnace

The diameter and number of hearths employed by this process depend on the waste material, on the required processing time and the type of thermal processing employed. A normal incineration process usually requires a minimum of six hearths. The waste material enters the furnace by a feed port in the furnace top. Rabble arms and teeth, attached to a vertical shaft, rotate counter-clockwise to spiral the waste across the hearths and through the furnace. The waste drops from hearth to hearth through passages located either along the edge of the hearth or near to the central shaft. Although the rabble arms and teeth all rotate in the same direction, additional agitation of the waste is obtained by reversing the angles of the rabble-tooth pattern and the rotational speed of the central shaft. Burners and combustion air ports are located in the walls of the furnace. Each hearth contains temperature sensors and controllers. The construction materials of the hearths vary in grade to suite waste requirements. This technology is used mostly for the incineration of sludges.

Pyrolysis or starved air

This technique is employed when the waste material has a high calorific content. The waste has to be organic and able to maintain combustion. Normal incineration requires 40–100% excess air over the stoichiometric value. Pyrolysis is theoretically a zero-air indirect-heat process. However, in practical applications it is an air-starved process in which combustion occurs with air levels less than stoichiometric requirements. If the heat content of the waste is not high enough to maintain combustion, sub-stoichiometric burning will not succeed. During pyrolysis waste organic compounds are distilled or vaporised to form combustible gas, by heating them in an oxygen-free atmosphere. The process is endothermic and requires an external heat source. Heat for the process can be provided by the partial combustion of the pyrolysis gas within the furnace and by the combustion of elemental carbon (Cheremisinoff, 1987). The unoxidised portion of the combustible gas may be used as fuel in an external combustion chamber, with the resulting energy recovered by conventional waste-heat-boiler technology. Carbon levels in the furnace ash are higher for pyrolysis than for normal incineration. Pyrolysis is generally considered unsuitable for handling large volumes of mixed, untreated MSW at the time of writing, although Deutsche Babcock Alagen GmbH have been operating a 50,000 tonnes/year plant in Günzburg, Barvaria since 1985 (Warmer Bulletin, 1999b).

Energy from Waste plants (EfW)

Energy can be recovered from the combustion of waste. The heat from the flue gases is used for heating water or for steam generation. The hot water is then used for other industrial or space heating applications. The steam can be used for heating or for electricity generation. Heat recovery can be achieved by two means: by waterwall combustion chambers or by

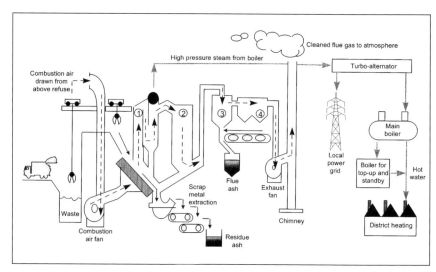

Figure 12.4 Schematic of a generic Energy from Waste facility.
Key: 1 – boiler; 2 – economiser; 3 – flue gas lime scrubber; 4 – bag filters.

waste-heat boilers. The heated water turns to steam, which can generate mechanical energy in a steam turbine, which is then converted into electricity (Figure 12.4).

In Amsterdam, 15% of the city's electricity is obtained from an EfW plant (EEWC, 1997). The energy efficiency of waste can be compared to that obtained from other fuels. In treated waste where glass, metal and wet organic materials have been removed the energy potential is just under half that of coal (EEWC, 1997). The recovery of heat has other beneficial side-effects. By recovering heat from the flue gases, the flue gas volume is reduced. Therefore, the required capacity of gas-cleaning equipment is reduced.

Conventional energy recovery involves passing the hot flue gases through a boiler, whose walls are lined with boiler tubes. Water circulated through these tubes is turned to steam, which can be heated further, using a superheater, to increase its temperature and pressure to make electricity generation more efficient. The thermal efficiency of modern boilers is around 80%. If the steam is used to generate electricity, however, the overall energy recovery efficiency (from calorific content of fuel to electricity generated) is around 20% (RCEP, 1993).

If wet feedstocks are used (e.g. high in food and garden waste), much of the gross calorific content of the waste is used up evaporating this moisture. This latent heat contained within the flue gas is not normally recovered. Recent advances in energy recovery techniques, in particular the flue gas condensation (FGC) process, however, allow some of this latent heat to be recovered also, so increasing the overall efficiency of energy recovery. An additional benefit of FGC systems is the relatively high rate of removal of acid gases from the flue gas during the process, leaving less to be removed by subsequent emission control equipment.

Emission control

The majority of modern MSW incinerators produce less particulate and gaseous pollutants than their predecessors, which had few environmental controls. With the introduction of tighter controls on air emissions by regulatory bodies world-wide, multi-stage pollution control systems are becoming more common. Emissions from MSW incinerators are controlled by a combination of measures that use both the pollution prevention approach and end-of-pipe solutions.

The operation of the combustion process plays an important role in the formation of some pollutants. Carbon monoxide, NOx, hydrocarbons and other volatile organic compound emissions can be minimised by optimising the combustion process. This is achieved in Europe by regulations that state that the furnace temperature must remain above 850°C for 2 seconds in the presence of at least 6% oxygen. Because the formation of pollutants cannot be prevented or completely controlled, the use of end-of-pipe equipment is needed. This is the case for gaseous streams containing dust, acid gases (HCl, SO_2, HF), heavy metals and dioxins. The combustion of MSW can produce trace quantities of dioxins and the industry has pioneered the use of dry scrubber/ fabric filter systems, activated carbon filters and carbon injection for dioxin control. To control dioxin emissions good combustion practices and proper temperature control of the flue gases are essential and can achieve reductions of dioxins to exceedingly low levels. The main gaseous emissions from the incineration process are described below.

Carbon dioxide (CO_2)
Carbon dioxide is one of the two main products of the incineration process. The other main product is water. In low concentrations CO_2 has no short-term toxic or irritating effects; it is abundant in the atmosphere, necessary for plant life and is not considered a pollutant (Tchobanoglous et al., 1993). Nevertheless, due to the high increase in global concentrations of CO_2, it has been recognised as one of the gases responsible for global warming (IPCC, 1996).

Carbon monoxide (CO)
Carbon monoxide is formed by the incomplete combustion of carbon due to the lack of oxygen to complete the oxidation to CO_2. This gas is very toxic; it reacts with haemoglobin in the blood causing a decrease of available oxygen to the organism. This lack of oxygen produces headaches, nausea and eventually death. Carbon monoxide in the flue gas is used to monitor the incomplete combustion of other emissions, such as unburned hydrocarbons and to provide information on the combustion performance.

Hydrochloric acid (HCl)
Hydrochloric acid results from the high concentration of chlorine-containing materials in MSW (some types of plastics like polyvinyl chloride). Chlorine easily dissolves in water to form HCl. Its presence in the gaseous stream may increase the acidity of local rain and groundwater, which can damage exposed unprotected metal surfaces, erode buildings and may affect the mobilisation of heavy metals in soils (Clayton, 1991).

Hydrogen fluoride (HF)
Hydrogen fluoride is more toxic and corrosive than HCl, although its presence in the emissions from MSW incinerators occurs in much smaller quantities. It is formed due to the presence of trace amounts of fluorine in the waste.

Sulphur oxides (SOx)
The emission of SOx is a direct result of the oxidation of sulphur present in MSW, but other conditions such as the type of incinerator used and its operating conditions may also influence

production although to a lesser degree. Approximately 90% of SOx emissions are as SO_2 and 10% are as SO_3. In the atmosphere most of the SO_2 is transformed into SO_3 (Benitez, 1995). SO_2 may lead to the production of H_2SO_3 and H_2SO_4 (sulphurous and sulphuric acids, respectively) in the atmosphere increasing the acidity of rain. Its effects on human beings depend on concentration. At high concentrations, it can produce eye, nose and throat irritation and other respiratory problems. Particulates that carry SOx on their surface intensify harmful respiratory effects, as this particulate matter can penetrate deep into the lungs (Benitez, 1995).

Nitrogen oxides (NOx)

Nitrogen oxides are predominantly formed during the incineration process; however they oxidise to NO_2 in the atmosphere. NOx is formed from two main sources: thermal NOx and fuel NOx. In thermal formation the oxygen and nitrogen in the air react, the free oxygen atoms produced in the flames by dissociation of O_2 or by radical attack, themselves attack the nitrogen molecules and begin a chain reaction. Fuel NOx production is formed during reactions between oxygen and nitrogen in the fuel. These reactions are highly sensitive to temperature, and are also affected by the degree that the fuel mixes with air (Benitez, 1995). Nitrogen oxides are important because they participate in several processes in atmospheric chemistry. They are precursors of the formation of ozone (O_3) and peroxyacetal nitrate (PAN): photochemical oxidants known as smog and which contribute to acid rain formation.

Particulates

Particulates are formed during the combustion process by several mechanisms. The turbulence in the combustion chambers may carry some ash into the exhaust flow. Other inorganic materials present in the waste volatise at combustion temperatures. This material condenses downstream to form particles or deposits on ash particles. Organic material can also be emitted through pyrolitic reactions near the fuel bed. This material can also be carried away and condensed downstream. The main component of fly ash is chemically inert silica but it may also contain toxic metals and some toxic organic substances (Benitez, 1995).

Heavy metals (Hg, Cd, Pb, Zn, Cu, Ni, Cr)

Municipal solid waste contains heavy metals and metallic compounds in the combustible and incombustible fractions. During the incineration process, metals may vaporise directly or form oxides and chlorides in the high temperatures in the combustion zone. They condense over other particles and leave the incineration process in the flue gas (Tchobanoglous et al., 1993).

Dioxins and furans

Polychlorinated dibenzo-p-dioxins and polychlorinated dibenzofurans (PCDD/PCDF) have been detected in the emissions from Municipal Solid Waste incinerators (Olie et al., 1977). They immediately became a major issue in the debate over the place of incineration within the municipal waste management strategy. Dioxins can be formed in all combustion processes where organic carbon, oxygen and chlorine are present, although the processes by which they are formed during incineration are not completely understood or agreed upon.

Dioxin is the generic name given to a family of over 200 chlorinated organic compounds. Their molecular structure is very similar; 12 carbon atoms form two benzene rings, which are connected by two oxygen atoms. They differ from each other by the number of chlorine atoms and their spatial arrangements. Furans are similar to dioxins, but the difference relies on the location of the chlorine atoms, which are situated at positions 1–4 and 6–9. There are 75 dioxin isomers and 135 furan isomers and all are collectively named 'dioxins'.

The concern over dioxins and furans is due to a number of animal studies that have shown that for some species they are highly toxic at very low levels of exposure (Tosine, 1983; Oakland, 1988). The extrapolation of these animal data to humans, albeit contentious, has helped these compounds acquire their notoriety. Nevertheless, they have known adverse health effects on humans such as chloracne, which was clearly documented in the Seveso incident (Porteous, 1994).

Even though the relationship between dioxins and human cancer at everyday levels of exposure is still under debate and not convincingly demonstrated, research by a group of scientists from the World Health Organisation's (WHO) International Agency for Research on Cancer (IARC) concluded that there is enough evidence to classify one dioxin (2,3,7,8-TCDD) as a known human carcinogen. This verdict was based on the fact that TCDD has been shown to exert carcinogenic and other effects in animals by binding to an intracellular complex called the Ah receptor, which is also found in humans. These scientists believe that the overall cancer risk is increased by a factor of only 1.4 in a highly exposed worker, a risk comparable to that from environmental tobacco smoke. Heavy cigarette smoking increases the risk of lung cancer by a factor of around 20. TCDD exposure seems not to be linked to any particular type of cancer. Again there is not enough evidence to conclude that dioxins have carcinogenic effects at low ambient concentrations.

Three pathways were proposed to explain the presence of dioxins in incinerator emissions (Hutzinger et al., 1985):

1. Dioxins are present in the incoming feed and are incompletely destroyed or transformed during combustion.
2. Dioxins are produced from related chlorinated precursors such as polychlorinated biphenols (PCB), chlorinated phenols and chlorinated benzenes.
3. Dioxins are formed via de novo synthesis from chemically unrelated compounds such as polyvinyl chloride (PVC) and other chlorocarbons, or are formed by the burning of nonchlorinated organic matter such as polystyrene, cellulose, lignin, coal and particulate carbon in the presence of chlorine donors.

It has now been verified that dioxins are present in all emissions, flue gases, bottom ashes, fly ashes and scrubber water. Although each of the three pathways described above do occur in large-scale incinerators, pathways 2 and 3 are more significant than pathway 1. This is due to a combination of modern combustion and flue-gas-cleaning technologies and the fact that (for thermodynamic reasons) dioxins are destroyed at temperatures above 800°C at residence times of 2 seconds, the required operating conditions for waste incineration now.

There is general agreement that the most significant way in which dioxins are formed is when flue gases are transported down cooling tubes between temperatures of 250 and 450°C (Hutzinger et al., 1993). Both fly ash (with its constituents, organic carbon, chlorides of alkali and earth metals, metal activators and catalysts) and dioxin/furan precursors in the gas phase play a role in the formation of dioxins (Stieglitz et al., 1989). In addition to this, parameters such as oxygen, water vapour and temperature have to be considered (Fielder, 1998). Dioxin generation occurs mainly in dedusting equipment, especially electrostatic precipitators (Vogg, 1995). There is evidence that both reactions in the gas phase and reactions on particle surfaces play a role in the formation of dioxins (Hutzinger and Fielder, 1991), and there are indications that the mechanisms of these reactions are different.

No significant effect of the PVC content of waste on the formation of dioxins has yet been established (Chang, 1996), although some studies have related the inputs of dioxins and furans in the waste to the outputs (Obermeier, 1990).

In summary, there are several factors that affect the formation and emission of dioxins and furans: the composition and properties of the waste, the combustion conditions, the composition of the flue gas, the amount of particulate matter carried in the flue gas, the flue gas residence time and temperature, the cleaning equipment used to control particulate matter and its operating temperature and the method used to control acid gases combined with the one used to control particulate matter (Kilgroe, 1995). It is not yet known which of these factors has the most influence on dioxin formation.

Despite de novo synthesis, however, the levels of dioxin emitted from a well-operated MSW incinerator are considerably less than the levels in the input, as shown in Figure 12.5. Overall modern incinerators act as net destroyers of PCDD/PCDF.

Historically, incineration was the biggest source of dioxin emissions to air in the UK. Her Majesty's Inspectorate of Pollution estimated that municipal waste incinerators released between 460 and 580 g TEQ of dioxins compared with a total UK release of 560–1100 g TEQ in 1995 (HMIP, 1995). This figure had dramatically fallen to just 4.7 g TEQ by 1998 (UK

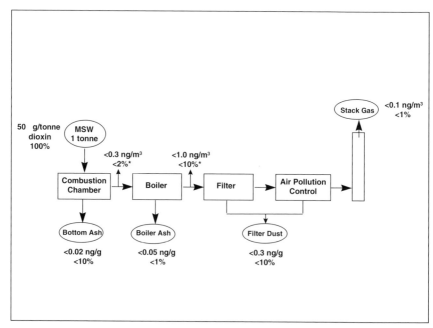

Figure 12.5 Dioxin balance for the incineration of Municipal Solid Waste. * = raw gas concentration. Source: IAWG (1997).

Environment Agency, 1999). This reduction was due to the new emission controls that came into force at the end of 1996 (these included a 1 ng/m^3 limit on dioxins) and resulted in the closure of a number of old municipal incinerators and the upgrading of the gas-cleaning technology of the remaining incinerators.

Gas-cleaning equipment

The technologies employed to carry out the necessary flue gas cleaning are described below.

Electrostatic precipitators (ESP)

Electrostatic precipitators are used for particle control. They use electrical forces to move the particles flowing out of the gas stream onto collector electrodes. The particles are given a negative charge by forcing them to pass through a corona or ionising field. The electrical field that forces the charged particles to the walls comes from discharge electrodes maintained at high voltage in the centre of the flow lane (Steinsvaag, 1996). When particles are collected they must be carefully removed to avoid them re-entering the gaseous stream. This is achieved by knocking them loose from the plates and by intermittent or continuous washing with water.

ESPs act only on the particles and not on the entire gaseous stream. The removal efficiency for particles is more than 99% with a low pressure drop. The performance of the equipment is affected by particle size and other physical characteristics such as gas stream temperature, flue gas volume, moisture content, gas stream composition, particle composition, and particle surface characteristics (Clayton, 1991).

ESPs come in one or two stage designs and can have different configurations: plate wire, flat plate and tubular, These configurations can be wet or dry depending on the method of dust collection. Plate wire precipitators are widely used in solid waste incineration facilities because they are well suited to handle large volumes of gas. In this configuration, gas flows between parallel plates of sheet metal where weighted high voltage wire electrodes hang. The gas must flow sequentially over each wire within each flow path. This distribution allows parallel operation of many flow lanes. The removal of the particles captured by the collecting walls is done by banging on the plates. The banging can dislodge some of the particles back into the gas stream, reducing the efficiency of the process. Some of these released particles are re-captured in other sections downstream of the same cleaning process, but the particles re-entering the stream from the last section are not recaptured and escape the cleaning unit (Benitez, 1995; Steinsvaag, 1996).

Wet precipitators

The configurations described above can be cleaned using a continuous or intermittent flow of warm water mixed with detergents. Re-entrance of particles caused by banging operations is minimal, but washing makes the device more complex, and water treatment has to be considered including the cost of disposing of the sludges formed (Steinsvaag, 1996).

Two-stage precipitators

Two-stage precipitators separate the charging and collection stages to optimise the electrical conditions for each stage. Due to the fact that charging requires high current density and electric field and collection requires high electrical field but much less current a discharge electrode (ioniser) is placed before the collector electrodes.

Fabric filters

In fabric filtration, the gas flows through a number of filter bags placed in parallel, leaving the dust captured by the fabric. Extended operation of a fabric filter requires periodic cleaning of the cloth surface. After a new fabric goes through a number of cycles of use and cleaning, it forms a residual cake of dust that becomes the filter medium, which is responsible for the highly efficient filtering of small particles that characterises fabric filters. They are widely accepted for control of particulate emissions. The type of cloth fabric limits the temperature of operation of fabric filters: cotton is the least resistant (355 K) while fiberglass is the most resistant (530 K). This temperature requirement makes the use of a cooling system necessary for the gas before it enters the equipment, but it is also necessary that the temperature of the exhaust gas stream is maintained above the dew point because liquid particles block the pores in the fabric very quickly. The major difference between different configurations of fabric filters is the cleaning method used during operating cycles.

Shaker cleaning

Here the bag is suspended from a motor-driven hook or framework, which oscillates, and such motion creates a sine wave along the fabric, which dislodges the collected dust. The dust falls into a hopper below the compartment.

Reverse-air cleaning

This method was developed as a less intensive way to clean the bags. Gas flow to the bags is stopped in the compartment being cleaned and a reverse flow of air is directed through the bags. This gently collapses the bags and the shear forces developed remove the dust from their surface.

Pulse-jet cleaning

This form of cleaning forces a burst of compressed air down through the bag expanding it violently. The fabric reaches its extension limit and the dust separates from the bag. Bags are mounted on wire cages to prevent their collapse while the dusty gas flows through them. The top of the bag and cage assembly is attached to the equipment structure, whereas the bottom end is loose and tends to move in the turbulent gas flow.

Scrubbers (wet, dry, semi-dry)

Scrubbers are used to control acid gases leaving the incinerator.

Dry scrubbing

This process works by injecting an alkaline reagent ($Ca(OH)_2$) into the flue gas as a fine dry powder in an up-flow reaction tower. This equipment is usually fitted after a boiler and before the particulate control system. In the most common design, the gases enter the bottom of the reaction tower and the reagent is injected pneumatically through one or more nozzles. Cyclone effects are used to promote good mixing of the powder and the gas. Most of the reaction product and the unreacted $Ca(OH)_2$ is carried in the gas stream and removed in the particulate control system along with dust and fly ash. A small amount of material is deposited in the bottom of the tower and removed by a conveyor. Gas residence time in the tower is generally longer than in a semi-dry or wet system (Loader, 1991; Pritchard, 1995).

The system requires cooling of the gas to temperatures below 180°C before entering the process. This is necessary because acid gas removal efficiencies are improved below this temperature. Volatile metals such as mercury can be captured because they precipitate out at lower temperatures (and the reformation rates for dioxin and furans are also reduced).

Dry injection systems are recommended to be used in combination with bag filters instead of ESPs. The performance of acid gas removal is enhanced by the bag filter's cake, which provides an additional reaction surface.

Semi-dry scrubbing

Here the alkaline reagent is introduced as a wet slurry, but the residues from the process are dry. The flue gases leaving the incinerator enter the top of the reaction tower at around 230°C. Flow spreaders ensure good mixing and an even flow downwards (Loader, 1991; Pritchard, 1995). The reagent is injected as a spray of fine droplets, and the acid gases are absorbed into the aqueous phase on the surface of the droplets and neutralised. The water evaporates on its movement downwards leaving only a dry powder consisting of calcium chloride, sulphate, sulphite and unreacted calcium hydroxide. Some of this material is deposited in the bottom of the tower, but most is removed from the flue gas by the particle control devices. This equipment is slightly more complicated than the dry injection system and requires a slurry make-up system.

Semi-dry systems can be used directly on incinerator flue gases at high temperatures, but it is preferable to reduce the gas temperatures below 220°C to improve abatement performance for acid gases and volatile metals and reduce the reformation of dioxins and furans (Clayton, 1991). These systems are often used in combination with an ESP to increase the efficiency of the particulate control device. If a bag filter is used, acid gas removal is also enhanced as the flue gases pass through the filter cake, which contains alkaline fly ash plus any unreacted hydroxide (Loader, 1991; Pritchard, 1995).

Wet scrubbing

Wet scrubbing removes acid gases by direct contact with a washing solution. Particulate matter is also removed to some extent by sticking to water droplets or wet surfaces. These processes are defined as wet systems because they produce a wet effluent. There are many more designs for this equipment than there are for dry or semi-dry systems. The principle remains the same, however; the flue gases pass through a reaction tower in which the acid gases are removed by direct absorption into an aqueous washing solution. The flow inside the tower can be upwards or downwards. The washing solution varies; it can be water for highly soluble compounds such as HCl and HF and some metal compounds, whereas for SO_2 removal an initial water wash stage can be followed by washing with an alkaline solution of NaOH (Pritchard, 1995).

Nitrogen control

Process optimisation and combustion controls such as flue gas recirculation are used to control nitrogen oxide emissions. Nevertheless, it may be necessary to use flue gas treatment to ensure low emission of NOx. In such cases, two technologies are used: selective catalytic reduction (SCR) and selective non-catalytic reduction (SNCR) (Clayton, 1991; Loader, 1991; Pritchard, 1995).

Elements of IWM

SCR requires the injection of ammonia into the flue gas. After this stage the gas moves through a catalyst bed of base metals (copper, iron, chromium, nickel, etc.). This technique reduces NOx emissions by up to 90% (Tchobanoglous et al., 1993).

SNCR uses the same principle of ammonia injection but no catalyst is used. The ammonia is injected directly into the furnace as a gas. The removal efficiencies achieved are between 50 and 80% (Tchobanoglous et al., 1993).

Treatment of solid residues

The bottom ash, once it has been cooled in the quench tank, is usually passed under an overhead magnet to recover any ferrous metal. Around 90% of the ferrous metal can be recovered this way (IFEU, 1992). The remaining bottom ash, and the fly ash and residues from gas cleaning then require disposal.

MSW contains inorganic pollutants, of which heavy metals form an important group (as listed in Table 12.4 and in Table 8.8) The chemical nature of these may be modified by heat, but they are not destroyed. They will therefore leave the incinerator either in the air emissions, bottom ash, filter dust or sludge from gas cleaning.

The distribution of common inorganic pollutants between these different outputs is shown in Table 12.4. Due to the volume reduction involved in incineration, considerable concentration of these materials will occur in the resulting residues, especially in the fly ash. This has both advantages and disadvantages. Concentration of pollutants means that the residues may need to be handled as hazardous wastes, necessitating disposal in special hazardous waste landfills, or further treatment. The converse of this is that being concentrated, there is less material to render inert and treatment to remove mobile heavy metals (e.g cadmium, mercury, copper, zinc) for re-use is possible (Vogg, 1992). Processes involving solidification, washing or melting techniques to render flue dusts safe for final disposal are under investigation (Vogg, 1989).

Fraction	Stack gas	Bottom ash	Filter dust	Salts and sludges from gas cleaning
Cadmium	0.04%	11%	85%	3.6%
Chlorine	0.12%	9%	15%	76%
Chromium	0.01%	94%	5.8%	0.27%
Copper	0.01%	95%	4.9%	0.53%
Fluorine	1.5%	69%	3.0%	26%
Lead	0.01%	75%	24%	0.9%
Mercury	2.1%	7%	5.1%	86%
Nickel	0.04%	87%	13%	0.61%
Sulphur	0.47%	50%	10%	40%
Zinc	0.05%	49%	51%	0.7%

Table 12.4 Distribution of heavy metals and other elements in MSW incinerator residues. Source: IFEU (1992)

Elements of IWM

Bottom ash generally has a low carbon content (1–2%), making this material suitable for disposal to landfill, either directly or after further processing. In Japan, for example, ash melting in a carbon arc furnace is used to reduce the volume of the bottom ash. This produces a vitrified slag with half the volume of the original ash, but the process is very energy intensive, using up to 1000 kWh per tonne of ash. Ash recycling is common in several European countries: Denmark recycles 82%, The Netherlands recycles 90%, while France and Germany recycle 45 and 50%, respectively (International Ash Working Group, 1997 and Warmer Bulletin, 1999c). Recent initiatives in the UK have raised the level of bottom ash recycling to 27%, the limiting factor here being the development of markets rather than technical restrictions, although concern over the leachability of bottom ash has persuaded ash recyclers to concentrate on bound applications such as asphalt, concrete blocks and cement sub-bases for the time being (ENDS, 1999).

Burning of refuse-derived fuel (RDF)

Refuse-derived fuel (RDF) is produced by a specific process designed to select the combustible from the non-combustible fraction of mixed MSW (described in Chapter 10). RDF consists mainly of the lighter materials in MSW (paper and plastic), which are separated out and shredded to produce 'floc' or non-densified cRDF. This can then be pelletized to improve handling, producing densified RDF (dRDF). Pelletizing the RDF is costly however, with a high energy consumption, so attention has begun to move to the use of floc RDF, handled by bulk handling techniques (Ogilvie, 1992). A similar waste-derived fuel is used in some industries, notably the cement industry, where shredded MSW can be used as a partial replacement for coal (Warmer Bulletin, 1993b).

Coarse cRDF can be burned in a fluidised bed incinerator (ETSU, 1993), and densified dRDF can be burned in either a fluidised bed or on a conventional grate.

The benefits of using RDF, and cRDF in particular, have been summarised as follows (RCEP, 1993):

1. cRDF has a higher calorific value than raw MSW and is more uniform in combustion characteristics
2. cRDF contains much less non-combustible material than MSW, so there is less ash left for disposal
3. overall efficiency is higher since the combustion characteristics can be tailored more precisely to the fuel specification.

The air emissions from the Robbins Resource Recovery Facility, Illinois, where MSW is processed to form high-quality cRDF, which is then combusted on site, are presented in Table 12.5. The emissions from both of the facility's chimney stacks are well within the required limits due to effective combustion of the cRDF and a high-efficiency air pollution control system. This air pollution control system treats flue gases using a dry scrubber, a fabric filter baghouse and a selective non-catalytic reduction system (SNCR).

Elements of IWM

Pollutant	Units	Emissions – Unit A	Emissions – Unit B	Permit limit RDF facility
Arsenic	(mg/m^3)	0.0002	0.0002	0.01
Cadmium	(mg/m^3)	ND	0.0003	0.04
Chromium	(mg/m^3)	0.0048	0.0056	0.120
Lead	(mg/m^3)	0.0036	0.0309	0.490
Mercury	(mg/m^3)	0.0158	0.0029	0.080
Nickel	(mg/m^3)	0.0031	0.0031	0.100
Particulates	(mg/m^3)	52.97	197.97	350
Total dioxins/furans	(ng/m^3)	2.1	4.9	30
HCl	ppm	4.6	6.2	25
SO_2	ppm	2.1	4.9	30

Table 12.5 Robbins Resource Recovery Facility stack emissions test results. ND = not detected, all concentrations measured at 7% oxygen. Source: Weaver and Azzinnari (1997)

Burning of source-separated paper and plastic

Source-separated fuel is similar in some ways to floc RDF (above), but would arise from source-separated collections for materials recycling, rather than produced from a mechanical screening process (Chapter 10). Current kerbside collection schemes for dry recyclable materials have demonstrated that high recovery rates for most materials can be achieved (e.g. IGD, 1992). If such schemes were to be extended country-wide, large amounts of materials would be collected (Chapter 9). Whilst there is likely to be recycling capacity for metals and most glass, the amounts of paper and plastic collected are likely to be well in excess of what can be recycled in an environmentally and economically sustainable way. In the case of paper it will always be necessary to inject virgin pulp into the system to compensate for the degradation of fibre length with each successive use. There will always be excess paper in the waste stream over recycling requirement. In the case of plastic, recycling capacity is low because it is still a costly operation, resulting in a premium being paid for recycled material over virgin. In the case of very lightweight plastic packaging items (e.g. yoghurt pots, carrier bags, sweet wrappers), recycling is unlikely to be environmentally sustainable because of the environmental burdens associated with washing and transport.

A logical use of this source-separated paper and plastic that cannot or should not be recycled is as a fuel, since it has a high calorific value, and should be relatively free of material unsuitable for the incineration process, since it is produced by a positive sort at source (compared to a negative sort from mixed MSW as in the case of dRDF). This alternative fuel could either be burned in power stations along with conventional fuel (using coal-burning grate technology), in small dedicated boilers in industrial plants to provide steam or heating, or in cement kilns. Trials are underway to assess the feasibility of such fuel usage (EPIC, 1997). The Association of

Plastic Manufacturers in Europe (APME) believes that emissions data from trial burns in Switzerland prove that waste plastic is a cleaner fuel than existing fossil fuels (Cosslett, 1997).

Emission limits

Strict emission limits for solid municipal waste incineration are currently in force in all EC countries as a result of the EC Directive for new and existing MSW incineration plants (E.C., 1989a, b) (Table 12.6). Old plants that were equipped only with electrostatic precipitators (e.g. in the UK, Clayton et al., 1991) were not able to meet these standards, and therefore were forced to either upgrade their emission control, to include at least gas scrubbing equipment, or to shut down. It is possible that some plants failed the requirement for a minimum residence time of 2 seconds at over 850°C, and in these cases retrofitting (modifying the facility after initial installation) would not be an option.

The improvement in incinerator emissions after retrofitting new gas-cleaning technology is presented in Table 12.7. Retrofitting existing facilities to meet higher standards is sometimes a more attractive option than building a completely new incinerator, although retrofitting costs

Regulated emission	EC Directive for new plants >3 tonnes/h (1989) (7 day average value)	Germany Limit values Ordinance 17. BImSch V (Dec. 1990)	The Netherlands (1990)
HCl	50	10	10
HF	20	1	1
SOx	300	50	40
NOx	–	200	–
CO	100*	50 (100)*	50
Organics†	20	10	10
Dust	30	10	5
Heavy metals	Cd, Hg 0.2 Ni , As 1.0 Pb, Cr, Cu, Mn 5.0	Cd, Tl 0.05 Hg 0.5 Class III (Sb, As, Pb, Cr, Co, Cu, Mn, Ni, V, Sn)	Cd 0.05 Hg 0.05 Class III 1.0
PCDD + PCDF	–	0.1 ng I-TE	0.1 ng
Values relate to	11% O_2 dry flue gas	11% O_2 dry flue gas	11% O_2 dry flue gas

Table 12.6 Emission values for MSW incineration plants (mg/m^3).
*1 h average value; †total carbon

	Limit value TA Luft Clean Air Act	17th emission Regulation	Plant emissions before retrofitting (mg/m³)	Plant emissions after retrofitting (mg/m³)
Total dust	30	10	50	<0.3
Organic substances (total carbon)	20	10	–	<2
Gaseous inorganic chlorine compounds (HCl)	50	10	35–40	<1
Gaseous inorganic fluorine compounds (HF)	–	1	1	0.9
Sulphur dioxide and trioxide (SO_2 and SO_3)	100	50	300	<1
Carbon monoxide (CO)	100	50	20–30	11.2
Cadmium, thallium (Cd, Th)	–	0.5	–	<0.001
Mercury (Hg)	0.05	0.05	0.2	<0.02
Total of all other heavy metals (Sb, As, Pb, Cr, V, Sn, Co, Cu, Mn, Ni)	100	0.5	–	0.02*
Dioxins/furans (PCDD/PCDF)	–	0.1	8	
Total TEQ in ng/m³	–			0.0059

Table 12.7 Effect of improved gas cleaning equipment on performance of MSW incineration (the table gives limit values and actual values before and after retrofitting of incineration plant at Zirndorf, Fürth, Germany. Notes: the '<' symbol represents below level of detection, *most heavy metals are below level of detection). Source: Warmer Bulletin (1993c).

may be higher than just the difference in cost between a new plant meeting existing standards and a new plant built to achieve higher performance (EC, 1997a). This is because retrofitting often requires modifications and changes in design of the additional plant in order for it to fit existing equipment, which itself may also require modification. Retrofitting costs range from approximately 25% to 40% of the new build costs of an incinerator (EC, 1997b).

For the burning of RDF and source-separated materials the position is less clear. In several countries (e.g. Germany, Belgium, The Netherlands and the UK), these materials are legally considered to be waste, and are therefore subject to the same emission controls as mass-burn incinerators. Since they involve burning only part of the waste stream, both the variety and levels of many pollutants may be expected to be lower than for mass-burn incineration.

Public acceptability

During the 1980s public concern over emissions from incinerators was very high. This period saw the growth both of the NIMBY (Not In My Back Yard) syndrome, and of genuine concern of the environmental effects of airborne emissions from the incinerators then operating. To some extent, the public debate has failed to separate issues relating to incineration of hazardous chemicals from those relating to the incineration of Municipal Solid Waste. Also, typically at that time, dust-collecting electrostatic filters were the only emission abatement techniques used in most European countries, giving rise to high levels of HCl, heavy metals and dioxins in stack gases released. More stringent emission regulations, both on a national and European Community scale now mean that all new facilities require sophisticated gas-cleaning equipment to achieve the regulated levels. Emission levels from such incinerators may be sufficiently low to reduce concern over their environmental effects.

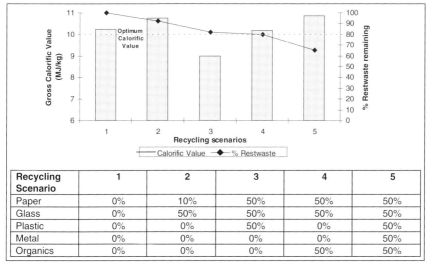

Recycling Scenario	1	2	3	4	5
Paper	0%	10%	50%	50%	50%
Glass	0%	50%	50%	50%	50%
Plastic	0%	0%	50%	0%	50%
Metal	0%	0%	0%	0%	50%
Organics	0%	0%	0%	50%	50%

Figure 12.6 Variation in the calorific content of restwaste with different recycling scenarios. Source: DTI (1996).

Public concern and pressure group focus has turned instead to the perceived incompatibility of incineration with materials recycling (ENDS, 1992c). If incinerators attempt to maintain a certain throughput for profitability, this will obstruct schemes to remove certain materials from the waste stream. This may be true for existing incinerators. For new incinerators, the design phase should match proposed incinerator capacity to the waste generated from an area, after a given level of materials recovery has occurred. Figure 12.6 shows that recycling metals and glass and composting of organics has in fact a very positive effect on the calorific value of the residue. The application of Energy from Waste incineration can complement materials recycling by utilising collected paper and plastic that is in excess of recycling capacity, or when due to market variability the prices of recycled paper and plastic fall below those which make recycling economically feasible. The decision-making process must also take into account the environmental burdens associated with each of the available processes. Flexibility and informed decision making is the key to economically and environmentally sustainable waste management systems.

Elements of IWM

CHAPTER 13

Landfilling

Summary

In this chapter landfilling is considered as a waste treatment process, with its own inputs and outputs, rather than as a final disposal method for solid waste. Landfilling essentially involves long-term storage for inert materials along with relatively uncontrolled decomposition of biodegradable waste. The use of landfilling is described and landfilling methods are discussed, including techniques for landfill gas and leachate control, collection and treatment.

Introduction

Landfilling stands alone as the only waste disposal method that can deal with all materials in the solid waste stream. Other options such as biological or thermal treatment themselves produce waste residues that subsequently need to be landfilled. Consequently, there will always be a need for landfilling in any solid waste management system. Landfilling is also considered the simplest, and in many areas the cheapest, of disposal methods, so has historically been relied on for the majority of solid waste disposal. In several European countries (UK, Eire, Spain), the USA and virtually all developing countries, landfilling continues to be the principal waste disposal method, although as land prices and environmental pressure increase, it is becoming more difficult to find suitable landfill sites, and so this position shows signs of changing.

Not all cases of 'landfill' actually involve filling of land. Although the filling of exhausted quarries and clay pits occurs in many countries (the UK in particular), above-ground structures are also common. In countries such as Japan 'landfilling' can also take the form of 'sea-filling', where material is used to construct man-made islands in Tokyo Bay and Osaka Bay.

The concept of landfilling as a final disposal method for solid waste can be challenged. A landfill is not a 'black hole' into which material is deposited and from which it never leaves. Like all the other waste options discussed in this book, landfilling is a waste treatment process, rather than a method of final disposal (Finnveden, 1993, 1995). Solid wastes of various compositions form the majority of the inputs, along with some energy to run the process. The process itself involves the decomposition of part of the landfilled waste. The outputs from the process are the final stabilised solid waste, plus the gaseous and aqueous products of decomposition, which emerge as landfill gas and leachate. As in all processes, process effectiveness and the amounts and quality of the products depend on the process inputs and the way that the process is operated and controlled. The same applies to landfilling: what comes out of a landfill depends on the quantity and composition of the waste deposited, and the way that the landfill is operated. This chapter considers both the objectives of the landfilling process and the methods used to achieve them.

Landfilling objectives

The principal objective of landfilling (Table 13.1) is the safe long-term disposal of solid waste, both from a health and environmental viewpoint; hence the term 'sanitary landfill' which is often used. Sanitary landfill describes an operation in which the wastes to be disposed of are compacted and covered with a layer of soil at the end of each working day (WHO, 1993). As there are emissions from the process (landfill gas and leachate), these also need to be controlled and treated as far as possible.

To a limited extent, landfilling can also be considered as a valorisation process. Once collected, the energy content of landfill gas can be exploited, so landfilling could be argued to be an Energy from Waste technology (Figure 2.4). Gendebien et al. (1991) estimated that there are potentially 730 billion (10^9) cubic metres of landfill gas produced annually from domestic solid waste, and that this would be equivalent in energy terms to 345 million tones of oil. If allowed to diffuse freely from landfill sites, this landfill gas can present a serious risk both to the environment (methane is a potent greenhouse gas) and the health and safety of local residents (methane is highly explosive). Collection and control of landfill gas is therefore needed for safety and environmental reasons. Once collected it makes sense to utilise the energy content of landfill gas where it is produced in commercially exploitable quantities. It is not designed as an Energy from Waste technology, however, since conditions in the landfill are relatively uncontrolled, and a large part of the gas often escapes uncollected.

Landfill can deal with all waste materials	
Essentially a waste treatment process with the following outputs:	Landfill gas
	Leachate
	Inert solid waste
The waste treatment process parameters can be optimised, e.g.	Dry containment
	Leachate circulation
	Lining technology
	Landfill gas and leachate collection
Can be used to reclaim land (or sea)	
Should avoid groundwater catchment and extraction areas	

Table 13.1 Landfilling: key considerations

Landfill can also be considered as a valorisation method when it reclaims land, either from dereliction (e.g. exhausted quarries) or from the sea (e.g. as practised in Japan), and returns it to general use. This is not always the case, however, since in many countries such opportunities are not available, and landfills need to be sited on otherwise useful land. Thus while landfilling can be a method for land generation, in most instances it consumes land.

So, although landfilling constitutes a means of valorising waste in two limited ways, its prime objective is the safe disposal of solid waste residues, whether direct from households or from other waste treatment processes.

Current landfilling activity

Reliance on landfill for solid waste disposal varies geographically around the world (Figure 13.1). Countries such as the UK have traditionally used landfilling as the predominant disposal route, partly because of its geology and mineral extraction industry, which has left many empty quarries that can be filled with waste. Such sites, however, may not always be in suitable locations for minimising their environmental burdens (see below). Conversely, countries such as The Netherlands, where the lack of physical relief and high water table have meant that large void spaces are not available, have had to develop alternative disposal routes.

Landfilling – basic philosophy

Archaeologists can gather information on the development of human culture by the excavation of ancient settlements and the surrounding dump sites. Low population densities and a nomadic hunter-gatherer existence meant that health problems due to solid waste arisings were minimal. As settled, agricultural-based society evolved the disposal of solid waste became more of a problem, the solution to which (at least initially) was open dumps. An open dump is an area of land where uncontrolled deposition of waste materials occur. No distinction between household and hazardous material is made and the waste is often set on fire to reduce its volume. Pollution of surface and groundwaters is common as well as migration of methane from the site. Open dumps cause severe litter problems, noxious odours and the breeding of disease carriers. The act of scavenging (see below) on such sites has a very adverse effect on human health.

Today, open dumps occur mainly in developing economies and are a result of limited technical and financial resources (US EPA, 1998). This leads to inadequate storage at the point of waste generation and inefficient or deficient collection systems. Open dumping is an unsatisfactory method of final disposal as it is an uncontrolled system not based on an engineering design. To control or eliminate the adverse consequences of open dumps they must be controlled by isolating them using fencing, managing surface water, groundwater, leachate and landfill gas. Closure is also necessary by capping or even moving the contents to a sanitary landfill.

Historically, landfilling has consisted of dumping waste in deep earthen pits. Over time, and with the percolation of rain water, the degradable fractions of the waste decompose and the resulting products are diluted and dispersed into the underlying soil. On a small scale, this 'dilute and disperse' method of operation can be effective, since soils have a natural capacity to

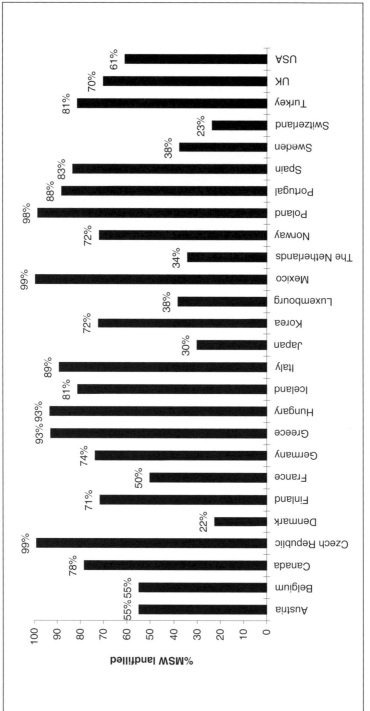

Figure 13.1 Percentage of MSW landfilled in OECD countries. Source: OECD (1997).

further decompose organic material and to adsorb many inorganic residues. Such sites will still produce landfill gas, however, which will diffuse into the atmosphere, and can cause safety concerns if it is allowed to accumulate. With increasing urbanisation, increased waste generation and increased difficulty in locating suitable and publicly acceptable sites, landfills have increased in size over time. As a result, dilution and dispersion is no longer an effective way of dealing with the landfill site emissions. Leachate produced by large unlined sites can pose a serious risk to groundwater supplies. The US Environmental Protection Agency, for example, has estimated that in the USA, around 40,000 landfill sites may be contaminating groundwater (Uehling, 1993).

To address the problems of landfill gas emissions and groundwater contamination, most modern landfills are operated on a containment, as opposed to a dilute and disperse, basis. Sites are lined with an impermeable layer or layers of mineral (e.g. clay) and/or synthetic (e.g. geotextile) liner, and include systems for collecting and treating both the resulting landfill gas and the leachate.

In Germany, for example, 94% of all municipal waste landfills are lined and collect and treat leachate prior to release into public sewage systems (UBA, 1993). The quantity of leachate produced in a landfill depends on, amongst other things, the amount of water that percolates into the site from rainfall and groundwater, so there has been a further tendency to seal the landfill by capping and make the whole structure water-tight. Since 1993, for example, all landfills in the USA have to be kept sealed and dry, with plastic membranes isolating them from percolating rain and groundwater (Uehling, 1993). Dry containment will also reduce the initial production of landfill gas, since a high moisture level is necessary for biodegradation. The methanogenic micro-organisms, for example, require a moisture content of over 50% to be active (Nyns, 1989).

The dry containment method of operating a landfill has been described as long-term storage of waste rather than waste treatment or waste disposal (Campbell, 1991), and does have some significant drawbacks. There will always be pockets of moisture within waste, and it is generally accepted that all lining and capping systems will eventually leak, so rain and/or groundwater will eventually enter the site. Thus the decomposition of the organic fraction of the waste will eventually occur, with resulting emissions of landfill gas and leachate. Since pipes and pumps buried within the waste eventually block and fail, there will be less chance of collecting and treating these emissions if they occur in the distant future. In place of such dry containment, therefore, some experts are advocating almost the opposite: the acceleration of the decomposition process by keeping the waste wet. This can be achieved by recirculating the leachate collected within the landfill to keep conditions suitable for microbial activity. In this way most of the gas and leachate production will occur in the early years of the landfill's life, while gas and leachate collection systems are operating effectively. By the time that the lining system eventually fails, the leachate should be very dilute and so reduce the risk of groundwater pollution. Operating the landfill in this way as a large 'bioreactor' also means that the gas is given off at higher rates, so making energy recovery economically more attractive. Further research with the wet bioreactor method is required, however, before it replaces dry containment as the preferred landfilling technique.

Elements of IWM

Landfill siting

No discussion of landfilling can neglect the problem of finding suitable, and publicly acceptable, sites. Along with local residents' concerns over traffic, noise, odour, wind-blown litter (and resultant effects on local property values), groundwater pollution has also recently become an important issue in selecting suitable sites. Since all landfill sites are likely to leak eventually, new landfills should not be placed within the catchment areas of groundwater abstraction points, where the contamination of drinking water could occur. Ideally, they should not be sited at all on major aquifers, where the potential for groundwater percolation would be greatest. Ideally, any new landfills should be located over minor aquifers or non-aquifers (Harris, 1992). On these grounds, the UK National Rivers Authority (now part of the UK Environment Agency) produced a national Groundwater Vulnerability Map, which will be used to assess the ground-water contamination potential of any new landfill developments. What has already become clear, however, is that most suitable void space for landfilling in the UK occurs above major aquifers. This is because most quarries, which form the bulk of suitable void space, have been developed to extract chalk, sandstone and limestone, which form the majority of the major UK aquifers. Thus, although the UK may have an abundance of void space for landfilling, in terms of groundwater protection, it is not necessarily in the right place. As in other countries, it may be necessary in future to consider greater use of above-ground sites located on non-aquifers (Harris, 1992). Some of the main factors that must be considered when evaluating a potential landfill site are listed in Table 13.2.

Considerations	Comments
Haulage distances	Minimum haul distance is desirable
Location restrictions	Both available land area and site access
Soil conditions and topography	Local soil suitable for use as daily cover
Climatic conditions	Average rainfall, flash floods, average temperature
Surface-water hydrology	Surface-water management system
Geological and hydrogeological conditions	Groundwater protection
Existing land-use patterns	Site security
Local environmental conditions	Leachate and landfill gas management systems
Potential uses for the completed site	Restoration of amenity value
Public attitudes to landfill	Public consultation if necessary

Table 13.2 Landfill siting considerations

Landfill site design and operation

The structure of one form of lined landfill site for containment of leachate is shown in Figure 13.2. The bottom liner of the site can either be a plastic (often butyl rubber or HDPE) or a layer of another low-permeability material such as clay. Whilst the permeability of the synthetic material is lower, they are vulnerable to mechanical puncture and so can then act as a point source for leaking leachate. By comparison, clay barriers (often a number of metres thick) are not subject to such localised failure, though they act as a diffuse source of leachate over the whole area of the landfill site (Campbell, 1991).

As well as a choice of material types, there is also a range of options in the way liners are laid down. As liner systems increase in complexity their costs will increase, but the risk of failure decreases, so there is less likelihood of expensive remediation work following leakages. The simplest form of barrier consists of a single liner, normally with a leachate collection system above the liner. Rather than rely on one type of liner material, a single composite liner system has two or more liners of different materials in direct contact with each other. In this design it is common to have a leachate collection system above a plastic liner, on top of a low permeability clay layer. The liner system shown in Figure 13.2 is a double liner system: two liners with a leachate collection system above the upper (primary) liner, and a leachate detection system between the two layers. The leachate detection system has a high permeability to allow any leachate that has leaked through the primary liner to be drawn off. Again, each layer in a double liner system may either be a single liner or a composite of two or more materials (Deardorff, 1991). A detailed cross-section of the liner systems required in Germany for materials containing different amounts of Total Organic Carbon (TOC) is presented in Figure 13.3.

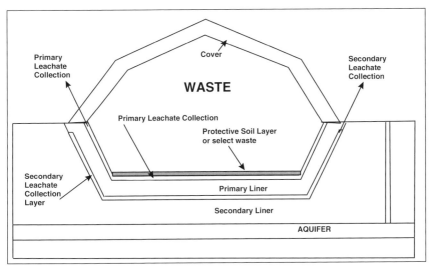

Figure 13.2 Simplified plan of a landfill with a double liner system. Adapted from Rowe (1991).

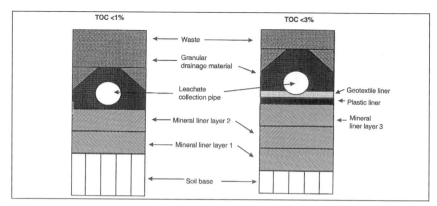

Figure 13.3 Landfill liner systems required in Germany for materials containing different levels of Total Organic Carbon. Source: TA Siedlungsabfall (1993).

Once the liner system has been installed, a cover of clay, soil or other inert material is normally applied to protect it from mechanical damage. Waste is then deposited and compacted, and layers of inert material (soil, coarse composted material) are normally added to sandwich the waste. The actual working face of the landfill is kept small and the fresh waste is covered by landfill cover material at the end of every day to reduce the nuisance from wind-blown material, and to keep off rodents, birds and other potential pathogen-carrying vermin.

Landfill leachate

The leachate collection system normally consists of a network of perforated pipes, from which the leachate can be either gravity drained or pumped to a leachate treatment plant. The most significant influence on leachate quantity is the amount of rainfall, which will vary seasonally. Leachate production begins shortly after the process of landfilling begins and may continue for a period of hundreds or possibly thousands of years. This is demonstrated in Figure 13.4, which presents an estimation (based on the model of Baccini et al., 1992) of the time when different compounds in leachate will no longer be considered harmful to the environment.

A storage sump or pool is often used so that surges in leachate production can be flow balanced before entering the treatment process. Landfill management practices greatly affect leachate quality. Acceleration of the early phases of decomposition is needed to produce low concentrations of organic matter and heavy metals in the leachate (Carra and Cossu, 1990). This can be facilitated by having a low waste input rate, moisture control (by leachate recirculation) or by having a composted bottom layer of waste. Leachate treatment can be carried out on or off site by physical, chemical or biological methods. One of the most common methods of leachate treatment is the use of aerated lagoons.

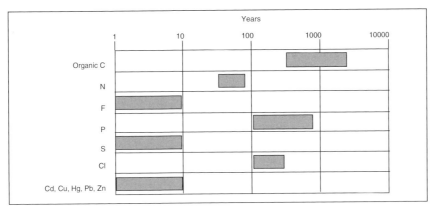

Figure 13.4 Estimated times when compounds in leachate become harmless to the environment. The grey boxes represent the time periods when the environmental burden of each compound in landfill leachate becomes negligible. Source: Baccini *et al.* (1992).

Landfill gas

The anaerobic breakdown of organic material (Chapter 11) within a landfill results in the production of gas, the composition of which varies over time, as shown in Figure 13.5.

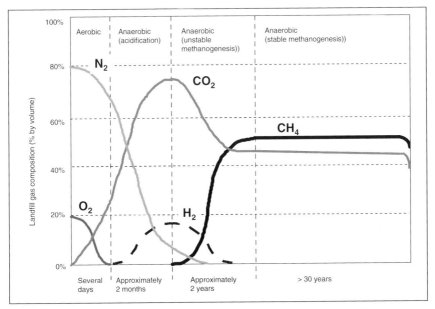

Figure 13.5 The composition of landfill gas over time. Source: Ramke (1991).

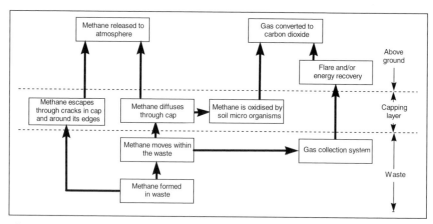

Figure 13.6 Pathways of methane release. Source: AEA (1999).

The key processes and pathways that impact on methane release from landfills are shown in Figure 13.6. Landfill gas is collected using a system of either vertical or horizontal perforated pipes. Since the gas will migrate horizontally along the layers of waste, vertical collection pipes are likely to collect gas more effectively. The density of pipes will vary across the landfill, with the greatest density needed at the periphery to prevent the migration of the gas laterally from the site. Pumped extraction of gas is needed for efficient collection, and thus less odour and emission problems. Once collected, the gas can either be flared off, to destroy the methane and organic contaminants, or used as a fuel. As produced, landfill gas is saturated with water vapour, and contains many trace impurities. This leads to a highly corrosive condensate, so if the gas is going to be used in a gas engine for energy recovery, gas cleaning is normally required. Similarly, if the gas is to be piped elsewhere for use as a fuel, in many cases it is puri-fied to remove the contaminants and the carbon dioxide, the latter to increase its calorific value.

The rate of gas production also depends on how the landfill is managed. From a commercial point of view, gas utilisation appears to be profitable from about 1 year after the waste is land-filled, and can be expected to continue to be so for no more than 15–20 years (Carra and Cossu, 1990). In the USA there were approximately 150 landfill gas recovery with energy gen-eration schemes operational in 1997; this was expected to rise to over 200 schemes by the end of 1998 (Thorneloe et al., 1997). Approximately 100 schemes were operational in the UK by 1997, generating a total output of 190 MW (ETSU, 1996a). More than 25% of all Canadian landfill gas is recovered for energy generation, while in Germany approximately 65% of municipal waste landfills recover energy from landfill gas (Warmer Bulletin, 1997b).

Waste inputs

The above measures to ensure safe disposal of landfilled material are based on containment, collection and treatment of emissions. As with all processes, however, the emissions will depend on the inputs, i.e. on what waste is placed in the landfill in the first instance. Restricting the type of waste entering the landfill can ensure that fewer emissions are produced. This strategy has gained ground in several countries. The German T.A. Siedlungsabfall (1993) ordinance, for example, defines the characteristics of waste that can be deposited in each of two classes of landfill. Class 1 landfills will only accept waste with a total organic carbon (TOC) content of $\leq$ 1% (See Figure 13.3). This means that the material will have to be incinerated before landfilling, as biological treatment cannot achieve this low level of TOC. Class 2 landfills will accept up to 3% TOC, but there are stricter requirements on the construction of such sites. Landfilling of untreated mixed Municipal Solid Waste will therefore no longer be possible in Germany by 2005. A key policy development that builds on the concept of the landfill ban is the EC Landfill Directive published in 1999. Under the directive, biodegradable Municipal Solid Waste landfilling must be reduced to 75% by 2006 (compared to 1995 levels), dropping to 50% by 2009 and 35% by 2016. Member states that landfill over 80% of their MSW may postpone these targets by a period not exceeding 4 years (EC Landfill Directive, 1999a). This directive will result in many changes throughout the waste management industry in Europe. Germany has gone one step further than this in a strategy unveiled by their environment ministry in August 1999, which states that all landfilling of German household waste should be phased out by 2020. This calls for treatment techniques to be developed so that all domestic wastes can be recovered 'fully and environment-compatibly' within two decades. Currently, some 60% of the 30 million tonnes of domestic waste arising annually in Germany is landfilled (ENDS, 1998).

In marked contrast to this restriction of landfill inputs, the UK has consistently argued in favour of the co-disposal of certain forms of industrial and potentially hazardous waste with Municipal Solid Waste. The rationale is that the difficult-to-treat materials are diluted and can be decomposed along with the normal solid waste. In a landfill, however, there is little control over this process, and the potential for serious groundwater pollution should such landfills leak is significant, so it is hard to view this as an environmentally sustainable method for the future.

Scavenging

The *Oxford Concise Dictionary* defines a scavenger as 'a person who seeks and collects discarded items'; given this, scavenging has been an ongoing practice in waste management systems since waste management began. Scavenging occurs at collection sites and landfill sites in developing countries and also in some developed countries. The practice is most prevalent in developing countries where social problems such as poverty, lack of training, education and jobs, and homelessness force the poorest of the poor into surviving on the discarded material of the rest of society. In developing countries scavengers are unfortunately part of most waste management systems. Table 13.4 shows how widespread this existence is for many people in developing countries.

Scavenging of waste can be split into two main types: scavenging for survival, where the target materials are food, shelter and clothing, and scavenging for revenue-generating material such as plastic, paper, glass metal and fabrics. Whatever type of scavenging activity is being carried out, the danger and ill effects associated with working and/or living on a landfill are significant. The immediate physical dangers of working on a landfill include working alongside heavy machinery, material slides, fire/explosions, cuts or abrasions from contaminated sharps, rodent bites, exposure to H_2S from landfill gas and contamination of drinking water by leachate. Long-term exposure to such conditions results in higher disease levels amongst scavengers (from a detailed review by Cointreau-Levine, 1997).

The situation with scavengers is by no means ideal, but must be addressed by waste planners in areas where scavenging is an issue, when the planners are designing new waste management systems. Outright banning of scavengers from landfills is not the answer as this would result in the scavengers having no means of earning a living or even just surviving, and would therefore increase the social problems of the area.

Location	Country	Source
Manila, open dumps	Philippines	Adan (1982)
Manila, open dumps	Philippines	Camacho (1995)
Cape Town, hazardous landfill	South Africa	Charters (1996)
Various sites (focusing on health issues)	Egypt (Calcutta), India (Bangalore, Bombay and New Delhi), Nepal (Khatmandu), Thailand (Bangkok)	Cointreau-Levine (1997)
Various sites	China, India, Indonesia, Malaysia and Vietnam	Furedy (1991)
Olongapo City, Cabalan landfill	Philippines	Gendzelevich (1996)
Rio de Janeiro, Itaoca landfill	Brazil	Gomes *et al.* (1995)
Cape Town	South Africa	Kahn (1996)
Harare City	Zimbabwe	Keeling (1991)
Manila, open dumps	Philippines	Torres (1991)
Boipatong, landfill	South Africa	Van Zyl (1996)
Rio de Janeiro, beaches (summer)	Brazil	Wells (1995)
Dar Es Salaam	Tanzania	Yhdego (1991)
Guatemala City, dump	Guatemala	Zabalza (1995)

Table 13.4 Studies on different aspects of landfill scavenging

The following recommendations for landfill managers were made by Boswell and Charters (1997) to minimise the negative impact and maximise whatever positive benefits can be gained from the practice of scavenging.

1. Separate high-value loads from the general waste stream.
2. Separate activities of scavenging from the working face of the landfill.
3. Formalise the relationship between the scavengers and the site operator.
4. Improve operating standards on landfills, including compaction, daily cover, management of surface water and leachate, control of landfill gas and prevention of fires.
5. Operate close supervision and operational controls at the landfill working face.
6. Provide protective clothing for both landfill workers and scavengers.
7. Separate scavenger settlements from waste management facilities and landfills.

Very few people choose to become scavengers, but when the situation forces people to choose between starvation, crime and scavenging, many in the developing countries resort to life on the landfills.

Elements of IWM

CHAPTER 14

Materials Recycling

Summary

The reprocessing of recovered materials into recycled materials is outside the boundary of the waste management system modelled in this book. Recovered material that is reprocessed can, however, be used to replace virgin materials, and this may result in overall savings in energy consumption and emissions.

In this chapter the manufacturing and recycling processes used for each material are briefly described to allow the reader to balance the advantages and disadvantages of materials recycling processes against virgin material production processes. The mantra that 'Recycling is best' is not always true, as the efficiency of virgin materials production continues to improve at a rapid pace and each recycling process must be considered in the context of its specific location, the availability of raw materials, the exact process to be used and the availability of markets.

This is presented as a final option within the computer LCI model, so that savings associated with the production of recovered material can be considered in the overall balance.

Introduction

According to the boundaries defined in this book, materials recycling processes lie outside the waste management system (Figure 14.1). In this system, materials destined for recycling cross the system boundary as recovered secondary materials at the exits of Materials Recovery Facilities, RDF sorting plants, biological treatment plants, mass-burn incinerators or transfer stations for mono-material bank-collected material. These materials then enter the industrial processing system for each particular material. This was chosen as the waste management system boundary because at this point the recovered and sorted material has generally re-acquired value (the secondary materials are sold on to the reprocessors), and so has ceased to be 'waste' (as defined in Chapter 1). Recovered materials, rather than recycled materials, thus form one of the outputs of this waste management system.

One problem with this system definition is that it is based on economic criteria, and thus is not necessarily absolute. As with any commodities, the market value of recovered materials fluctuates with supply and demand; in the case of recovered materials it can actually fall to a negative value. A solution to this definition problem is to include the recycling of the recovered materials within the waste management system boundary. In such a case, this material would leave the system as recycled, rather than recovered material (i.e. as metal ingots, granules of recycled plastic resin, etc.). The energy consumption and emissions associated with transporting the material to the reprocessors, and the processing stages themselves would then also need to be allocated to the waste management system.

Including recycling industries within the IWM system increases the complexity of the model. For glass, steel and aluminium, recycling is part of the production process for the virgin material,

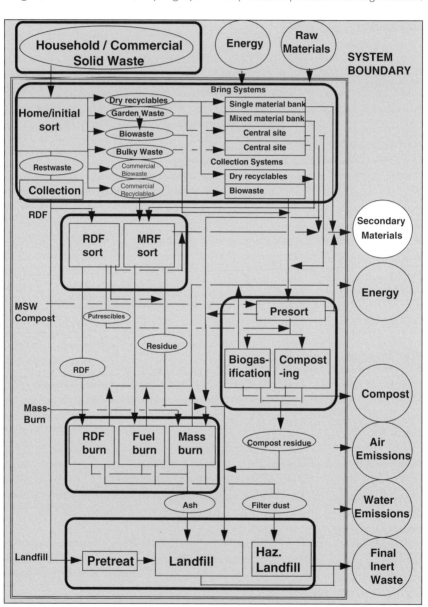

Figure 14.1 Materials recycling in relation to Integrated Waste Management.

so is difficult to separate out. The output of such an enlarged system is recycled material. Provided that a market exists for this material (i.e. produced at a competitive cost compared to virgin material alternatives) it would then replace virgin materials. The costs and environmental impacts of the virgin material production would thus be saved. To include these savings within the present model requires that the virgin material process, from cradle to grave, also be included within the system boundary. The method used to achieve this in the model is outlined in Figure 14.2.

In Chapter 22 (Model guide, Materials Recycling) a set of data on the environmental burdens and economic costs of reprocessing the recovered materials is provided, to give an indication of possible savings (or costs) compared to the use of virgin raw materials. This information is included in the computer model as a final option, to allow calculation of the overall environmental burdens and economic costs for an enlarged waste management system that includes the reprocessing stage and produces recycled, rather than just recovered materials.

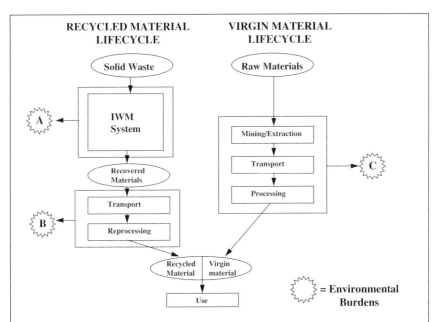

Figure 14.2 Relationship of materials recycling processes to an IWM system.

Materials manufacturing and recycling processes

Transportation

The first stage in the conversion of all recovered materials into recycled materials is the transport from the sorting or collection facility, to the reprocessing facility. The distances involved clearly depend on the relative locations of the waste management scheme and the reprocessing plants, so there is a strategic need to locate such plants within easy reach of large potential sources of recovered materials. Brief details of the subsequent reprocessing stages are given below for each material.

Paper and board manufacturing and recycling

Paper manufacture relies on the fact that wet cellulose fibres bind together with hydrogen bonds when dried under pressure. Basically paper recycling reverses this process by wetting, agitating and then separating the cellulose fibres. Estimates suggest that the maximum number of times paper fibres can be reprocessed is four, so virgin fibre will always be needed (Warmer Bulletin, 1997c). Approximately 95% of the base material used in paper and board manufacture is fibrous (AFPA, 1997c) and a large percentage (90%) originates from wood, although crops such as hemp, jute, flax and bamboo are also used. The filler (mineral additives used to improve opacity, strength and smoothness) content of some grades of paper approaches 30%. Many tree species, both hard and soft wood, are used to produce wood pulp. The wood used for pulp is taken from the parts of the tree that are left after it has been used for other commercial purposes such as construction and furniture making and also from forest thinning. The paper industry uses approximately 12% of timber world-wide (PPIC, 1999) and 94% of the wood used by the European paper and board industry comes from managed forests in Europe; the other 6% comes from North America (PPIC, 1999).

There are two major processes involved in the conversion of wood to pulp: mechanical processes and chemical processes. Mechanical pulps are formed by the mechanical separation of the fibre from the wood matrix. The process results in a high (90–95%) pulp yield. These pulps are used where opacity and good print quality are needed. The presence of large amounts of lignin, however, results in reduced inter-fibre bonding (resulting in low tensile strength) and poor light stability. Mechanical pulps are bleached using alkaline hydrogen peroxide or sodium hydrosulphite that maintain a high pulp yield and do not remove lignin. Brightness levels of 80% are common. The process of thermomechanical pulping (TMP) steams wood chips at 120°C before the fibre is mechanically extracted in a pressurised container. The resulting fibres are longer and less damaged and therefore give a stronger product than mechanical pulps with only a slight loss of opacity. Softwoods are the preferred raw material to produce optimum strength in the final product. Chemical treatments can be added to the TMP process to improve the strength of the final product. Such pulps are known as chemithermomechanical pulps (CTMP). The increase in strength is balanced by an associated loss of yield and opacity.

Semichemical pulps are produced by mild chemical digestion of wood chips prior to mechanical separation, and yields are 70–80%. The main use for this type of pulp is corrugated media. Hardwoods are usually the raw materials for semichemical pulps.

Chemical pulps are produced by chemical digestion of both lignin and hemicellulose. Little mechanical energy is needed to separate the fibres from the wood matrix. This results in undamaged fibres that are both long and strong. Chemical pulps are mainly used for strength and performance in a variety of paper and board products. The Kraft process is the principal chemical pulping process: mixtures of sodium sulphide and sodium hydroxide are the pulping chemicals and yields are 45–55%. Higher yield pulps contain more lignin and are used in bags and other products where strength is important. Lower yield pulps are bleached to remove virtually all of the lignin producing high brightness products (>90%), and are used where permanence and whiteness are needed in addition to strength. Historically, elemental chlorine was used as a bleaching agent, but environmental concerns over the effluents (which contained dioxins) from these processes resulted in replacement by chlorine dioxide and the development of dioxin control processes.

Modern pulp mill processes maintain a white water system that is as closed as possible, so as much water as is compatible with efficient machine operation is recycled. Since 1975 water usage has been reduced by 70% (PFGB, 1999) due mainly to better 'housekeeping'. The result of this is that 36 m^3 water was used per tonne of pulp produced in 1992 (Warmer Bulletin, 1997c). Fibre recovery from discharged water is undertaken using filtration, flotation and sedimentation. Pulp mills often use settling tanks and biological treatment methods followed by further settling to treat liquid effluents. These treatment processes themselves result in the production of sludges that require appropriate disposal.

Over the past 10 years energy consumption has been reduced by 30% (PFGB, 1999), which has largely been due to improvements in energy efficiency, paper machine performance, improved boiler efficiencies and a move to Combined Heat and Power (CHP) boilers, which are at least twice as efficient as conventional fossil fuel electricity power stations.

In 1997, the waste paper 'recovery rate' in Europe was 48.9%, the world average was 37% and the leaders of waste paper recovery in the developed countries included the USA (45%), Canada (45%) and Japan (53.1%) (CEPI, 1999). Waste paper reprocessing varies according to the type of recycled paper product, which will in turn determine the type of waste paper that is used as the process feed stock. Waste paper is graded into numerous categories (11 in the UK; 23 in the USA; five main grades in Germany, with 41 sub-grades: the Confederation of European Paper Industries list consists of five main grades and 64 sub-grades) according to quality (Cathie and Guest, 1991). The higher quality grades (UK grades 1–4) (paper mill production scrap, office and writing papers), which need little cleaning, are used to make printing and writing papers, tissues and wrapping papers, and are known as pulp substitute grades since they are used to replace virgin pulps. Newsprint (UK grade 5) and other papers needing de-inking are reprocessed for further use in the production of newspaper and hygiene papers. The lower (bulk) grades (UK grades 6–11) are mainly used for the production of packaging papers and board.

The details of the process stages will vary according to whether pulp substitute grades, newsprint or bulk grades are treated, but the basic steps are shown in Figure 14.3. After an initial soaking, the waste paper is pulped to separate the fibres, screened to remove contaminants, de-inked, thickened and washed. During these refining processes both nuisance materials and some fibres are removed from the system; such losses have been estimated at 15% for newsprint reprocessing (Shotton, 1992). Therefore, the input of one tonne of recovered paper will result in the production of approximately 850 kg of recycled paper.

The rejects, effluents and sludges generated by the recycling process include inks and solid pigments, adhesive particles, small plastic particles and wax, short cellulose fibres, paper filler

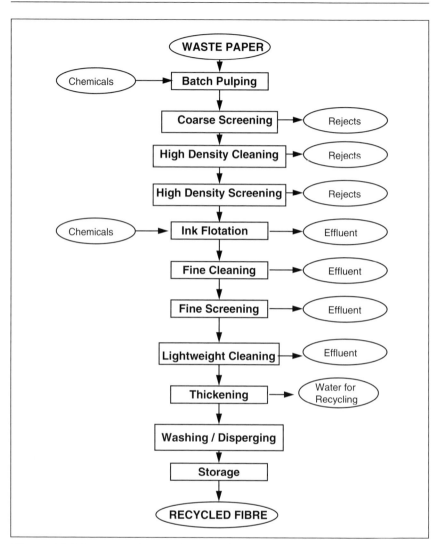

Figure 14.3 Stages in the production of recycled paper fibres. Source: Porteous (1992).

and coating particles and large solid materials such as grit, wire (paper clips and staples) and ceramics. Treatment and disposal of these wastes tend to be more complicated and costly than treatment and disposal of effluents and sludges from virgin pulp mills due to the increased variability and contamination of the raw waste paper feedstock.

Glass

Glass was formed naturally from common elements in the earth's crust long before humans began experimenting with its composition. Most glass is now manufactured by a process in

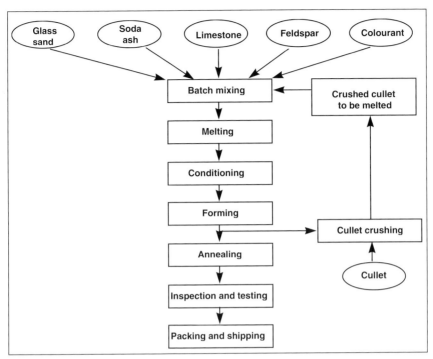

Figure 14.4 Glass manufacture – showing the input of recycled glass cullet.

which raw materials are converted at high temperatures (1420–1600°C) to a homogeneous melt that is then formed into products (see Figure 14.4). Raw materials are selected according to purity, supply, pollution potential, ease of melting and cost. Sand is the most common ingredient, of which purity and grain size are important. Container-glass manufacture tends to use sand between 590 and 840 μm for the best compromise between the high cost of producing fine sand and melting efficiency. Transport costs are often three to four times the cost of the sand, so manufacturing plant siting should be close to a source of good raw materials. Common colourants for glass include iron oxides, chromium, copper, cobalt and nickel. Colour separation of recycled glass is necessary to avoid colour quality concerns upon remelting.

Cullet, or broken glass is used as a batch material to enhance glass melting. The input of recovered cullet to the furnace lowers the temperature needed to melt the virgin raw materials, thus leads to considerable energy savings (Ogilvie, 1992) and it reduces the amount of dust and other particulate matter that accompanies a batch made exclusively from virgin materials. Certain glass-forming operations generate as much as 70% waste glass, which must be recycled as cullet. More efficient manufacturing operations, such as the container industry, may purchase cullet from recycled glass distributors. Typically between 10 and 50% of a glass batch is comprised of cullet, but operations at 70–80% cullet are not uncommon. For container glass, a 10% increase in cullet use reduces the melting energy by 2.5%, particulate emissions by 8%, NOx emissions by 4% and SOx emissions by 10% (Gaines and Mintz, 1994).

The first stage of glass reprocessing usually consists of a manual sort to remove gross contaminants (plastic bottles, ceramics, lead wine bottle collars) followed by automatic sorting to remove ferrous contaminants and low-density materials (paper labels, aluminium bottle tops). The former is achieved by magnetic extraction, the latter by a combination of crushing, screening and density separation techniques. Around 5–6% of the recovered glass input is removed in this way (Ogilvie, 1992). The crushed cullet is then ready for mixing with virgin raw materials, prior to melting in the furnace and blowing or moulding of the final glass products. Recycled glass cullet is not only made into new containers such as bottles and jars, it is also used for secondary markets such as fibre glass and 'glasphalt', paving asphalt using crushed cullet replacing stone aggregate. Since the use of recovered glass cullet is integrated within the normal glass production process, in consideration of the environmental burdens, glass reprocessing will be considered up to the production of finished glass containers.

Ferrous metal manufacture and recycling

Steel is essentially an alloy of iron and carbon. It contains less than 2% carbon, less than 1% manganese and small amounts of silicon, phosphorous, sulphur and oxygen. Steel is made by smelting iron ore in a furnace to produce pig iron, which is added to melted down scrap steel before being further purified. Iron ore is the fourth most common element in the earth's crust. The abundance of the raw materials necessary for steel production (iron ore, coal, limestone and scrap steel) and the relatively low manufacturing costs have resulted in the widespread use of steel we see today. There are two main types of furnace used in the production of steel. The basic oxygen furnace that produces sheet steel uses a minimum of 25% scrap steel, while the electric arc furnace uses almost 100% scrap steel. The mechanical properties of steel can be varied over a wide range by changes in composition and heat treatment. Mild steel contains less than 0.15% carbon, medium steel contains 0.15–0.3% carbon, while hard steel contains greater than 0.3% carbon. Stainless steel is a high alloy steel containing greater than 8% alloying elements such as chromium, nickel or silicon. Galvanised steel is steel coated with zinc to protect against atmospheric corrosion even when the coating is scratched, as the zinc is preferentially attacked by carbonic acid forming a protective coating of basic zinc carbonates. The steel used for can manufacture is coated in a thin layer of tin (tinplate); this resists atmospheric oxidation and attack by many organic acids.

Since the 1960s steelmaking has undergone many changes, essentially due to marked increases in energy costs, the need for better production flexibility and the impact of world-wide competition. The competition of other metals and materials and the greater availability of scrap has adversely affected the once exhaustive market for steels. Using the basic oxygen process, almost pure oxygen is blown onto the surface of the molten iron and conversion to steel occurs ten times faster than with the old open-hearth process. The drawback of this process is that it is limited to the utilisation of 30% scrap. This amount of scrap in the steel mix is often barely adequate to utilise the scrap produced at the manufacturing plant. Electric furnace processes were initially used to produce special steels for which the open-hearth process was not suitable. At operating temperatures of 3400°C the furnace is highly versatile (comparable to the open-hearth furnace) but most use high percentages of scrap. Common grades of steel can be produced in up to 300 tonne batches in under 3 hours.

Steel-making process	Energy use, MJ/tonne of raw steel
Open hearth	14.24
Basic oxygen process	14.42
Electric process	5.99

Table 14.1 Energy required per tonne of raw steel produced. Source: Electric Power Research Institute (1986)

Energy requirements for the three different steelmaking processes are shown in Table 14.1. When the energy requirement is weighted according to the different proportions of raw materials used by the three basic processes, the total amount of energy per tonne of raw steel is approximately 14 MJ per tonne for the open-hearth and the basic oxygen process compared to approximately 6 MJ for the electric process. The main reason for this large difference is that electric steel is almost always made from scrap, therefore energy is not required to reduce iron oxide to elemental iron.

It is important to appreciate that currently about 30% of the output of steel products is not readily recoverable as scrap. Examples of such difficult to recover materials include reinforcing bars within concrete, wire products such as nails and fencing, buried piping and oil well casings. It is this quantity of steel that must ultimately be replaced by the mining and reduction of iron ore.

The use of scrap for steelmaking results in large reductions in air pollution, water use (40% saving) and pollution, mining wastes and total energy consumption (virgin steel requires 36 GJ per tonne while recycled steel requires only 18 GJ per tonne, SAEFL, 1998) while also conserving iron ore (1.5 tonnes per tonne scrap recycled), coal (0.5 tonnes per tonne scrap recycled) and limestone. Blast furnace slag is used by the construction industry as an aggregate and for road building. The savings in landfill space are also worth mentioning. Recycling operations do, however generate certain emissions and waste streams that are subject to increasingly strict legislation, thus increasing the cost of recycling in many cases. Furnaces that use large amounts of galvanised scrap generate dust from which zinc can be recovered and recycled. US legislation requires that the dust be processed for recovery if it contains $\geq 15\%$ zinc.

Ferrous metal within household and commercial waste is found in the form of iron and steel scrap, but the majority is in the form of tinplate in food and beverage cans. To reprocess steel from steel scrap merely involves a sort to remove contaminants, before the scrap is melted and recast. To produce high-grade steel from tinplate for further use, it must first be detinned. This process, shown in Figure 14.5, consists of shredding the incoming tinplate and removing contaminants, before electrolytic removal of the tin plating. The tin, a layer of 0.004 microns, represents between 0.25% and 0.36% of the input material (Habersatter, 1991; Boustead, 1993b), but the value of this metal makes it worthwhile to recover, melt and recast the scrap tin for further use. The detinned steel scrap needs to be washed thoroughly to remove process chemicals, and then baled for delivery to the steel converting plant.

Non-ferrous metal manufacture and recycling

The major non-ferrous metal to be recovered from household waste is aluminium, mainly in the form of used beverage cans, so discussion of reprocessing will be restricted to this metal.

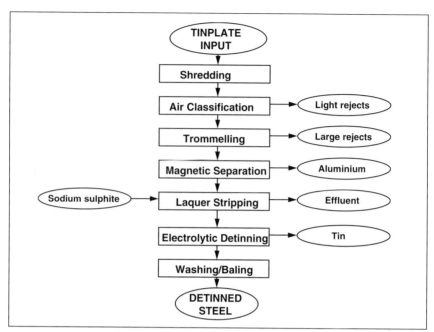

Figure 14.5 Stages in the reprocessing of tinplate steel. Source: AMG Resources (1992).

Aluminium is a silver-white metal obtained from bauxite, a residual rock composed almost entirely of aluminium hydroxides formed by weathering in tropical regions. Aluminium is the earth's third most abundant element (after oxygen and silicon) and the most abundant metal in the earth's crust (8% by mass).

Aluminium manufacturing is a two-stage process. In the first stage crushed bauxite is mixed with hot caustic soda, which dissolves the aluminium oxide. Impurities such as sand can be filtered out and the caustic solution is cooled to crystallise the dissolved aluminium oxide into a white sand-like powder. The second stage is the smelting process where the aluminium oxide is dissolved at 900°C. A powerful electric current is passed through the liquid that splits the aluminium oxide into aluminium and oxygen. The molten aluminium is drawn off and made into ingots.

The reprocessing of recovered aluminium is a much simpler and less energy intensive process than the production of virgin aluminium. The total energy required to produce aluminium from bauxite ore is around 183 GJ/tonne, whereas the total requirement from scrap is only 8 GJ/tonne (SAEFL, 1998). Reprocessing involves sorting of the recovered metal and then melting in a furnace (Figure 14.6). Since most aluminium is used in an alloy form with other metals or coatings, it is necessary to select an appropriate mix of recovered metal to give an ingot of the correct composition. Sometimes contaminants need to be diluted by the incorporation of high-grade virgin material to meet more strict specifications. Aluminium is very reactive and has a strong tendency to form a dross (an oxide containing other metals). Up to 5% of the molten metal forms a white dross during processing. To recover the aluminium from the white dross,

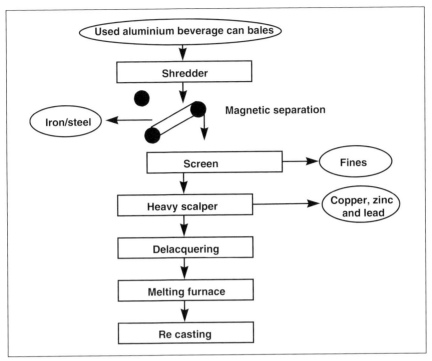

Figure 14.6 Used aluminium beverage can recycling process.

salts (usually chlorides) must be added. This treatment produces a secondary dross and disposal of this waste is becoming more carefully controlled and therefore more expensive. New technologies including centrifugation, flotation, plasma and electric arc melting are being investigated to recover white dross without the addition of chlorides.

Plastic manufacturing and recycling

Plastics are made from oil, natural gas, coal and salt. The major feedstock is oil; the petrochemicals industry supplies the monomers for plastics production and manufactures a wide range of additives to modify their behaviour. Plastics are produced by polymerisation, the chemical bonding of monomers into polymers. The size and structure of the polymer molecule determines the properties of the plastic material. In their basic form, plastics are produced as powders, granules, liquids and solutions. The application of heat and pressure to these raw materials produces the final plastic product.

Plastics are classified as thermoplastic or thermosetting resins. Thermoplastic resins, when heated, soften and flow as viscous liquids; when cooled they solidify. The heating and cooling cycle can be repeated many times without the loss of specific properties. Thermosetting resins liquefy when heated and solidify with continued heating. The polymer undergoes permanent cross-linking and retains its shape during subsequent cooling and heating cycles. Thermoset plastics cannot be reheated and remoulded; however, thermoplastics can be reprocessed by melting and hence readily recycled.

Elements of IWM

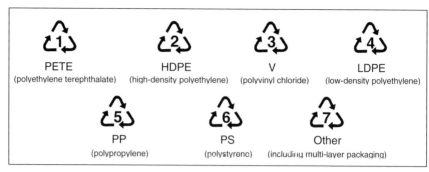

Figure 14.7 Categories of rigid plastic containers. Source: SPI (1988).

Almost 85% of all resins produced are thermoplastics and over 70% of the total volume of thermoplastics is accounted for by the resins: polyethylene, polypropylene, polystyrene and polyvinyl chloride (PVC) (Modern Plastics, 1994). They are made in a variety of grades and because of their low cost are the first choice for a large number of applications. A variety of processing and shaping methods are available to form the desired product, of these processes extrusion and injection moulding are the most common.

Plastics can be separated from Municipal Solid Waste by householders or at central sorting facilities. Householders are not easily able to identify different types of plastic and studies indicate that more people take part in source-separation schemes if they are not required to sort different plastic types (Warmer Bulletin, 1997d). To facilitate sorting ahead of reprocessing the American Society of the Plastics Industry developed a simple coding system in 1988 (Figure 14.7). The appropriate symbol is moulded into the base of rigid plastic containers. This system has now been widely adopted.

After separation of the resin types into individual, or at least compatible fractions, plastics can be either mechanically or chemically recycled. In mechanical recycling the plastic is shredded or crumbed to a flake form, and contaminants such as paper labels are removed using cyclone separators. The flake is then generally washed (this stage may also be used to separate different resins on the basis of density), dried and then extruded as pellets for sale to the plastics market.

Chemical recycling involves a more complex process whereby the plastic polymer is broken down into the monomer form, and then re-polymerised. In this case, as with glass and steel, the recycled product is indistinguishable from the virgin material. This recycling method has been developed for certain resins, notably polyethyleneterepthalate (PET), where chemical recycling via a methanolysis process is commonly used. However, the cost of the monomers obtained from chemical recycling is frequently higher than the cost of monomers derived from the traditional chemistry (La Manita and Pilati, 1996). This is often because the polymers obtained by traditional methods are produced in large plants benefiting from economies of scale whereas recycling plants of similar capacity would require collection over a vast area incurring high transportation costs. Also contaminants present in polymer wastes require costly purification technology to obtain polymerisation grade monomers.

Feedstock recycling can treat large volumes of mixed polymer wastes, as the recycled products (oil or gas) can usually be joined to the crude oil stream to undergo traditional treatments.

However, only the basic organic structure is preserved by this technology and the oil or gas obtained is currently more expensive than equivalent products from primary fossil resources.

Textiles

Textiles are manufactured from fibres by a variety of processes to form knitted, woven or non-woven fabrics. Textile fibres can be classified according to their origin, either from naturally occurring fibres (animal, vegetable or mineral sources) or from manufactured fibres (based on natural organic polymers, synthetic organic polymers or inorganic substances). The scale of the industry can be judged from the figures presented in Table 14.2.

Similarly to pulp and paper manufacturing, textile manufacturing requires large amounts of process water for preparing the fibres, large amounts of energy for processing the fibres into materials and large amounts of chemicals for dyeing and finishing the materials. From a Life Cycle point of view, the energy required to plant and harvest cotton and all pesticides and fertilisers associated with the growing phase must also be taken into account (as it should be for wood also), as should the consumption of oil for the production of synthetic fibres.

Natural dyes (colours derived from plant or animal material without chemical processing) have been used for thousands of years without showing any harmful effects. Synthetic dyes on the other hand are often resistant to degradation and a few exhibit toxic effects to microbial populations and can be toxic or carcinogenic to animals. The textile industry, again like the pulp and paper industry, uses physical, chemical and biological methods for the treatment of the effluents from manufacturing mills.

Region	Synthetic fibre*		Natural fibre	
	Synthetic polymers	Cellulosics	Cotton	Wool
Asia and Australasia	5.86	0.63	5.26	1.15
Canada and other Americas	1.08	0.11	1.26	0.20
China	2.03	0.34	4.51	0.25
Europe (including former Soviet Union)	3.88	0.79	4.09	0.04
Japan	1.40	0.22		
Middle East and Africa	0.28	0.02	1.40	0.64
USA	31.90	0.23	4.09	0.04
Total	17.70	2.32	19.00	2.80

Table 14.2 World fibre production by region, 1994 (in millions of tonnes). *Not including olefin fibre and glass fibre. Source: Fibre Organon (1995)

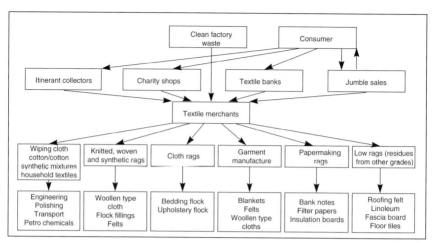

Figure 14.8 Schematic representation of textile recycling. Source: Textile Recycling Association (1999).

Textile recovery has a long history, and has mainly been practised for economic rather than environmental reasons. Unlike most recycling, a high proportion of textile reclamation takes the form of reuse and therefore does not involve resource intensive reprocessing. Where re-manufacturing is involved, such as the re-spinning of fibres to make new products (Figure 14.8), the procedures are generally less energy intensive and generally have fewer negative environmental impacts than the production of virgin textiles, especially if re-dying can be avoided.

Of the textiles currently recovered, the majority are re-used as cloth, rather than recycled as fibres. In the UK, for example, some 26% of recovered textiles are re-used as second-hand clothing, 40% used as wiping cloths and 22% used as filling material. Only 7% is actually reprocessed to produce recycled fibres for cloth production (UK Textile Reclamation Association, cited in Ogilvie, 1992). Within Europe the total amount of textile waste was estimated to be around 4.2 million tonnes in 1995; only 20% of this was recycled (Europe Environment, 1998). The recycling process uses toothed drums, combs and pulling machines to tear the textiles apart and extract the fibres. As with paper recycling, the process leads to a shortening of the fibres. Consequently, textiles cannot be recycled indefinitely since at some stage the fibres become too short to be re-used.

CHAPTER 15

IWM-2: A Life Cycle Inventory Model for Integrated Waste Management

Summary

The following chapters are a guide to the IWM-2 computer model. Each chapter describes the details of the system boundaries and any assumptions made for each section of the model. Where necessary, data tables and appropriate references are provided that support assumptions or the use of either default or generic data. Once this background has been established, each screen of the model is described in detail. The model does contain a help system and glossary.

The following text contains a detailed description of the program's potential applications and limitations.

Introduction

Who are the potential users of the model?

Decision makers who want to obtain data on the overall environmental burdens and economic costs of waste management systems, existing or proposed, are potential users. This group will include individuals with such diverse interests as policy makers, waste management officials, consultants, academics, students and environmental groups.

What are the potential uses of the model?

The model's most obvious uses are waste management scenario optimisation and comparisons. For example, investigating the environmental burdens associated with the implementation of a kerbside recycling system collecting either paper, or paper and glass, or paper, glass and metals.

What data are needed to run the model?

The actual data needs of the model are small, as wherever possible default data are provided. Although it must be emphasised that the more data that can be supplied by the user to describe the waste management system under study the more accurate the results will be. The model requires data on the number of inhabitants and households in the area under study, the amount of waste generated per person per year, a waste characterisation of the area under study (although default data from a number of countries are provided). Data (such as energy requirements, operating costs and operating efficiency) are also required to describe each of the waste management unit processes from collection through sorting, biological treatment, thermal treatment and landfilling.

An important data requirement of the model is the description of the System Area electricity generating grid (Hard coal, Brown coal, Oil, Natural gas, Nuclear and Hydro electric). The different methods of electricity generation each produce a significantly different range of environmental burdens. It is therefore essential to accurately represent the generating grid of the System Area to ensure that the environmental burdens associated with electricity use are correctly modelled. The default values in the Advanced screen of the IWM-2 model are based on the UCPTE model (1994). Users can select generic country data from the list in the Select Country button, but it is strongly recommended that more specific data are used if possible. To account for the savings of environmental burdens associated with electricity generation that is not required, due to the generation of electricity from thermal treatment, biogasification and landfill, it is necessary for the user to specify the makeup of the electricity generation grid that is being displaced. This would typically be the type of electricity-generating facility in, or closest to the waste management System Area. The model defaults to the current breakdown of the generating grid, but this can be edited by the user to displace a single energy type or a mixture of energy types.

The IWM-2 model has no built-in sensitivity analysis function. It is recommended that the user carries out their own limited sensitivity analysis, by changing individual parameters and re-running the model and examining the results. The objective of this exercise is to identify the most critical parameter with respect to the whole waste management scenario under investigation and examine the validity of this data.

What is the goal of the model?

The goal of this model is to be able to, as accurately as possible, predict the environmental burdens and economic costs of a specific waste management system.

What is the scope of the model?

The scope of this model is to enable a Life Cycle Inventory of a specific waste management system to be carried out. The unit processes included within the model are waste generation, waste collection, sorting processes, biological treatment, thermal treatment, landfill and energy generation. Second-level burdens (those associated with building and decommissioning of waste management facilities and equipment) are not included in the model, although they should be included in the economic analysis. Where assumptions have been made, they are highlighted (**bold italics**) and explained in the following text. Data quality is a major issue throughout the model, as it is with all waste management models. Although default data are provided where possible, it must be understood that the more high-quality data that can be collected and input to the model the more accurate the output of the model will be.

What is the functional unit of the model?

The functional unit of this model is the appropriate management of the total Municipal Solid Waste arisings of a defined geographical region during a defined period of time (e.g. a year).

What are the system boundaries (cradle and grave) of the model?

The system boundaries of this model are as follows:

1. Inputs (Waste): the point where the waste leaves the household.
2. Inputs (Energy): the extraction of fuel resources.

3. Outputs (Energy): the electric power leaving an Energy from Waste facility, or from the combustion of landfill gas (the electrical energy generated is subtracted from the energy consumed, so is effectively used within the system, and not exported).
4. Outputs (Recovered Materials): material collection bank or exit of Material Recovery Facility, Refuse-Derived Fuel plant or biological treatment plant.
5. Outputs (Compost): exit of biological treatment plant.
6. Outputs (Air Emissions): exhaust of transport vehicles, stack of thermal treatment plant, i.e. after emission controls, stack of power station (for electricity generation) or landfill lining/cap.
7. Outputs (Water Emissions): outlet of biological treatment plant thermal treatment plant or power station (electricity).
8. Outputs (Residual Solid Waste): content of landfill at end of biologically active period.

Allocation procedure

Allocation is defined as 'the partition of the input or output flows (or costs) of a unit process to the product system under study' in ISO 14040 (1997). In the IWM-2 model the inputs and outputs of each of the unit processes (waste collection, central sorting, biological treatment, thermal treatment and landfill) are all done on a mass basis, except for thermal treatment which is done on a mass and stoichiometric basis. Landfill gas and leachate are allocated on a component specific basis (i.e. based on the composition of the material landfilled); this approach takes into account the underlying physical relationships between landfilled waste and gas and leachate production as recommended by ISO 14041 (1998) and ISO TR 14049 (1999).

Conventions used in this chapter

Click   Point the mouse pointer at the object you want to select, and then quickly press and release the left mouse button.
Double-click   Point to the item and press and release the left mouse button twice in rapid succession.
On-screen text Text that appears on-screen in the computer model is shown in bold type in the guide document.
Scroll bars   Scroll bars may appear along the bottom and right edges of some of the windows in the program (depending on your computer's setup), where text takes up more space than the area shown. Using the scroll bars allows you to navigate the full window.

The IWM-2 computer model

Both the LCI and the economic assessment model are included in one computer model, which operates in Windows 95 or higher. Based on Figure 15.1, the model follows the solid waste stream through its Life Cycle. Each of the stages in the Life Cycle of waste is represented in the model by a window containing input questions. The answers to these questions

IWM2 Model Guide

define the waste management system considered. The first window of the model defines the input, the amount and composition of waste as it enters the waste management system from both household and commercial sources. Since the effectiveness of any treatment process, e.g. composting, thermal treatment, will depend on what is in the waste stream entering the process, it is necessary to keep the different materials separate in the model, even though they may be physically mixed together. By doing this, it is possible to characterise the material composition of the waste, and hence also its calorific value, at any point in the Life Cycle.

The windows in the model represent waste collection, sorting and materials recycling, biological treatment, thermal treatment and landfilling, respectively. The structure mirrors Chapters 8–14 of this book. A description of the input data required for each screen, their relevance to the model and the assumptions made by the model are presented in the following text.

Throughout the model the flow of the collected waste can be followed using the **Streams** button in the bottom left hand corner of each of the six input windows. Within each window, as materials are recovered, they are subtracted from the **Residue** stream and enter into the **Materials** stream. Other outputs from processes are entered into the relevant streams where they accumulate. Total costs for the system accumulate through the Life Cycle to produce the economic assessment.

By the end of the Life Cycle, all of the materials will have left the **Residual** stream and will have been entered into either the **Landfill** or the **Materials** streams. This emulates the definition of the 'cradle' and 'grave' of solid waste discussed in Chapter 5 – the cradle is the point at which the material is thrown away (i.e. ceases to have value), the grave is the point at which the material regains value (i.e. as secondary products) or is released as an emission to land, air or water. The model then totals the energy consumption, energy production, products, emissions to air, emissions to water and residual solid waste to produce the Life Cycle Inventory for the solid waste of the chosen region.

The source of each piece of default data provided within the model can be accessed (to conform with ISO 14041) by double-clicking the box in which the data appears. A window called **Variable Information** opens, containing the following information.

1. **Calculation** – if the variable is calculated, the full calculation is shown here. Any other calculated variable within this displayed calculation may also be viewed by double-clicking the text description of that variable. This drill-down function ensures that all of the calculations in the model are completely transparent (another ISO requirement). The drop-down list box in this window allows tracking of the variables and their calculations viewed using the drill down method.
2. **ISO reference** – the term ISO reference has been used to describe a data point or data set used in the model that requires a full source reference to maintain the transparency of the model as recommended by the ISO 14040 Standards (i.e. default data contained in the model rather than user input data).

New users are strongly recommended to initially navigate through the model screens in the order **Waste Inputs**, **Waste Collection**, **MRF/RDF Sorting**, **Biological Treatments**, **Thermal Treatments**, **Landfilling** and finally **Recycling**, until they are familiar with the layout and data requirements of each section of the model. New users are also advised to care-

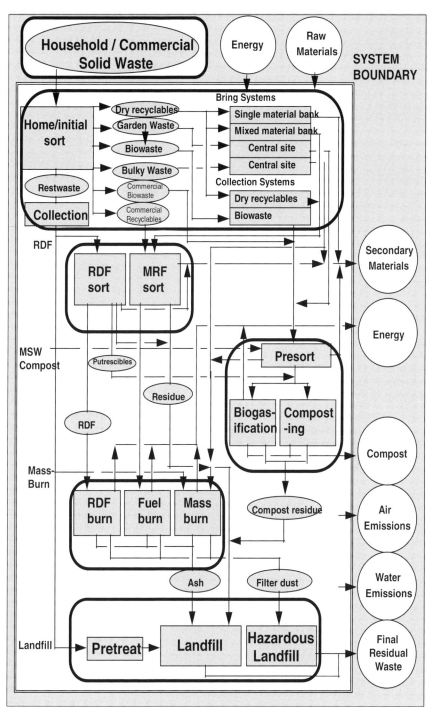

Figure 15.1 System boundaries and inputs/outputs for the IWM-2 model.

fully work through each screen tab by tab in order not to miss any details that may result in incorrect output in the results section of the model.

User note: use the Waste System Flow diagram to check that your waste management system has been accurately represented by the model as you planned.

The user guide

To install the IWM-2 computer model, put the CD-ROM in the computer's CD-ROM drive. The model will self-install, if it does not, use Windows Explorer to select the CD-ROM drive and double-click the setup.exe file. The software will self-extract and the automatic installation procedure will take place.

To run the IWM-2 software, double-click the red IWM-2 icon that appears on your computer's desktop.

Welcome to IWM-2

The IWM-2 model has been designed to work similarly to any Windows program, with the term 'scenario' replacing the term 'file'. A scenario is a description of a waste management system stored within the IWM-2 program. The model allows the development of new scenarios and the modification of existing scenarios. All scenarios are saved in a directory within the IWM-2 program.

The LCI model begins by offering the user five options (Screen 1).

- **Select a scenario to open** – opens previously saved scenarios.
- **Create a new scenario** – allows a new scenario to be created, using default data.
- **Setup configuration** – where the currency symbol and input screen position can be altered.
- **Go to menus** – gives access to the drop-down menus for **Scenario**, **Utilities** and **Help**.
- **Introduction to IWM-2** – offers a brief introduction to the LCI model, the concept and principles of IWM and a summary of waste management technology.

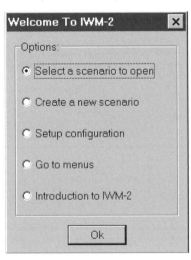

Screen 1 Welcome to IWM-2.

Click the option you require and then click **OK**. When selecting a scenario to open, highlight the desired scenario in the **Existing Scenario** list and click **OK**. The scenario is loaded and its name is displayed in the bottom left corner of the IWM-2 home screen.

When creating a new scenario the program asks **Create a blank scenario?** or **Use a template?** If a blank scenario is selected the program opens a new version of the model that only contains default data for certain of the waste management options. All of the default data can be replaced by user data, if the user has more accurate local figures than the generic data provided in the program. To use a template, the program allows the user to select an existing scenario and use it as a basis for the development of a new scenario. Upon completion, the program prompts the user to save this new scenario under a new name.

By selecting the **Go to menus** button, the user moves to the **IWM-2 Main Screen** (Screen 2) but can only can view and use the drop-down menus under **Scenario**, **Utilities** and **Help**. The **Scenarios** menu offers the user the following choices **New**, **Open**, **Revert**, **Delete** and **Exit**. These menu options allow the user to create a new scenario, open an existing scenario, move from a changed scenario back to the last saved version of that scenario, delete an existing scenario or exit IWM-2, respectively. The **Utilities** menu offers three choices. **Setup Configuration** allows the user to set the Window placement on screen and select an appropriate currency symbol. **Compare Scenarios** is a key function that loads the results of up to eight different scenarios, compares them and displays the results graphically. **What's Changed?** compares any differences between the mass flows of an existing scenario with the previous saved version of that scenario. The **Help** menu provides access to the **Help contents**, a **Glossary of terms** and a list of **Conversion factors**.

IWM-2 Main Screen

Once a scenario has been opened or a new scenario has been created the model moves to a window called **IWM-2 Main Screen** (Screen 2).

This screen contains the following control buttons arranged horizontally.

- **Save** – the save option (safety first).
- **Streams** – a full breakdown of waste arisings through each stage of the model.
- **Waste Flow** – a simple schematic diagram describing the flow of materials in the current scenario.
- **Results** – the overall environmental burdens and economic costs of the whole scenario.
- **Advanced** – where the advanced variables in the model can be altered by the user.
- **Notes** – where the user can add text, describing the scenario, assumptions, etc.

The buttons arranged vertically are the sub-components (sections) of the overall model and initially should be opened and data entered sequentially.

- **Waste Inputs** – defines the amount and composition of the waste material entering the waste management system.
- **Waste Collection** – describes the waste collection system, allowing its environmental burdens and costs to be calculated.
- **MRF/RDF Sorting** – describes Materials Recovery Facility and Refuse-Derived Fuel sorting processes.

IWM2 Model Guide

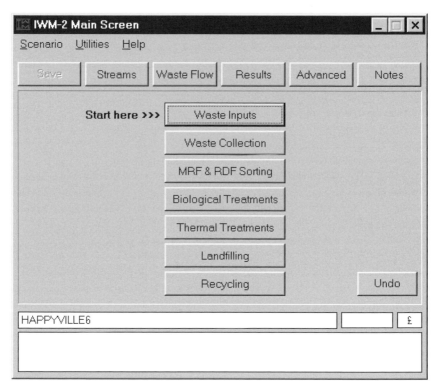

Screen 2 IWM-2 Main Screen.

* **Biological Treatments** – describes both composting and biogasification.
* **Thermal Treatments** – describes conventional incineration, Refuse-Derived Fuel burn and Paper and Plastic Fuel burn.
* **Landfilling** – describes the landfill process.
* **Recycling** – describes the recycling of the materials recovered throughout the whole waste management operation.

Once a scenario has been completed or if an existing scenario is to be edited then the user can go directly to the appropriate section.

The **Undo** button opens a window that lists the last 50 changes made to a Scenario. The user can select the number of changes they want the model to undo.

The default currency symbol to be used in the model can be changed by clicking the small button in the bottom right-hand corner of the main screen (a **£** sign in Screen 2).

The modelling procedure begins by the user clicking **Waste Inputs** and typing appropriate values into the white data boxes. The white boxes of each screen can be easily navigated by using the mouse or the **Tab** button (forwards), or using the **Tab** and **Shift** buttons (⇧) (backwards); the Enter/Return button (⏎) has no effect on the model.

IWM2 Model Guide

CHAPTER 16

Waste Inputs

Defining the waste input for the LCI computer model – data sources

The first step in carrying out an LCI for household and commercial waste is to define the amount and composition of such waste generated by the area being investigated. As discussed above, local weighbridge data on amounts and waste analysis results are needed for an accurate estimate. The quantity and quality of household waste will depend on factors such as population density, levels of affluence, housing types and efforts at source reduction. Commercial waste will reflect both the types and level of local commercial activity.

If locally sourced data are not available, generic data contained in the model can be used, but with caution. It is important to know what the generic data used reflects. Does household waste include garden waste and bulky waste, or does it only include 'collected household waste', i.e. the dustbin contents? Most national waste composition figures are based on collected household waste; data on delivered household waste often do not exist. Similarly, little reliable data on commercial waste arisings are available. Thus while generic information can be used in the absence of local data, it is not an entirely satisfactory substitute.

Classification of solid waste used in the Life Cycle Inventory

The categories of solid waste used in this analysis, and their definitions, are presented in Table 16.1. It will be seen that only the most basic level of the proposed European Recovery and Recycling Association (ERRA) classification (Figure 8.11) has been used. We still consider that this level of detail remains sufficient, due to the overall (relative) simplicity it affords. Using more detailed categories will lead to a more accurate prediction of the overall environmental burdens and economic costs of a waste management system, but will considerably increase the complexity of the model. Here we believe that we have achieved an acceptable balance between simplicity and accuracy. More sophisticated models are available from the UK Environment Agency and the US Environmental Protection Agency.

Category	Description
MSW fractions	
Paper (PA)	Paper, board and corrugated board, paper products.
Glass (GL)	Glass bottles and jars (all colours), sheet glass.
Metal (ME)	All metals including cans.
	Further subdivided into: ferrous (ME-Fe) and non-ferrous (ME-nFe).
Plastic (PL)	All plastic resin types, including bottles, films, laminates.
	Further subdivided into rigid plastics (PL-R) and plastic film (PL-F).
Textiles (TE)	All cloth, rag, etc., whether synthetic or natural fibres.
Organic (OR)*	Putrescible kitchen and garden waste, food-processing waste.
Other (OT)	All other materials, including fines material, leather, rubber, wood.
Waste treatment residues	
Compost residues (CO)	Residues from biological treatment (composting or anaerobic digestion), that cannot be marketed as products due to contaminant levels (stones, plastics, metals or textiles) or lack of suitable markets.
Ash (AS)	Bottom ash, clinker or slag from incinerators, RDF or alternate fuel boilers.

Table 16.1 Classification of solid waste used in the IWM-2 Life Cycle Inventory model.
*Paper and plastic fractions are also strictly of organic origin, but to maintain alignment with the ERRA classification system, the term 'organic' is used here to describe putrescible kitchen and garden waste only

IWM2 Model Guide

The Waste Input screen

This window in the model gathers data on the amount and composition of solid waste entering the system, from all sources. The instructions given in the window are listed below, with additional comments on the type of data needed, and how the model handles the input.

The four sections (tabs) within this screen define the system and all of the household and commercial waste managed within the system.

Tab 1 System area (Screen 3)

The user first defines the system area.

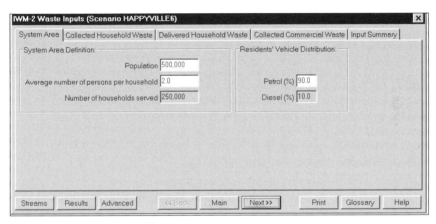

Screen 3 Waste Inputs – System Area.

System Area Definition:
Population
Average number of persons per household

From these figures the model calculates the number of households served, which is displayed below. User data are entered in the white boxes.

Number of households served

This is a calculated value and is therefore displayed in a grey box. This convention is followed throughout the model. Both the population and the average number of persons per household need to be entered, as along with the number of households in the specified area, this allows both per capita and per household data to be used.

Residents' Vehicle Distribution: Petrol (%)   Diesel (%)

The split between petrol and diesel cars is required for the calculation of the emissions from fuel used during journeys to and from bring systems modelled in the Waste Collection window. *Default values of 90% petrol and 10% diesel are provided in the model*; these values can be changed by simply typing over them. If region-specific data are available, they should be entered here.

IWM2 Model Guide

Tab 2 Collected Household Waste (Screen 4)

This section defines the amount of household waste generated in the system area and its composition.

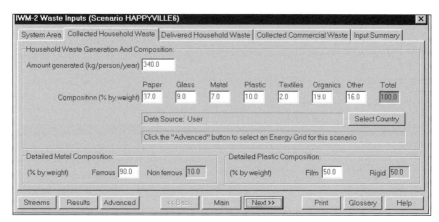

Screen 4 Waste inputs – Collected Household Waste.

Household Waste Generation and Composition:
Amount generated in system area (kg/person/year)
Composition (% by weight) Paper, Glass, Metal, Plastic, Textiles, Organic, Other, Total

The **Total** is calculated by the model and can only result in a total composition of 100% (shown in a green box); if the composition does not equal 100% the **Total** is shown in a red box and the user must amend the input data. The model will not allow the user to progress through the model if this **Total** does not equal 100%. If region-specific data are available on the amounts of household waste generated per capita, and its composition, these should be inserted. In the absence of region-specific data the '**Select Country**' button takes the user to a Country data table where generic national data on the average amounts and composition of waste generated is stored. This national data can be selected (by double-clicking any data value in the row of the desired country) for use in the model. The full reference for each country can be accessed by double-clicking any one of the data values.

Detailed Metal Composition: (% by weight)   **Ferrous   Non-Fe**
Detailed Plastic Composition: (% by weight)   **Film   Rigid**

The model supplies default data for the detailed composition of both metals and plastic in household waste but again if region-specific data are available they should be entered here.

Tab 3 Delivered Household Waste (Screen 5)

This section details the amount of household waste delivered to Central Collection sites. For simplicity in the model this material has been separated into Bulky Waste (such as furniture and household appliances) and Garden Waste.

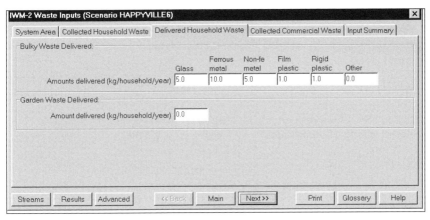

Screen 5 Waste Inputs – Delivered Household Waste.

Bulky Waste Delivered:
  Amounts delivered from system area (kg/household/year)
  Glass, Ferrous Metal, Non-Fe Metal, Film Plastic, Rigid Plastic, Other

Garden Waste Delivered:
  Amount delivered from system area (kg/household/year)

The amount of garden waste (kg/household/year) inserted is converted into the total amount of organic material collected for the area per year, which is added to the biological treatment stream.

Tab 4 Collected Commercial Waste (Screen 6)

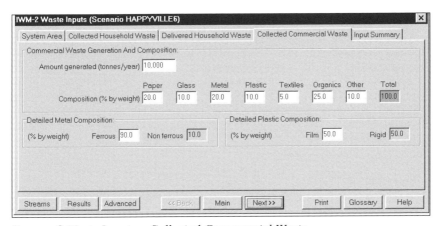

Screen 6 Waste Inputs – Collected Commercial Waste.

Commercial Waste Generation and Collection:
   Amount generated in system area (tonnes/year)
   Composition (% by weight) Paper, Glass, Metal, Plastic, Textiles, Organic, Other, Total

As in the Household waste collection screen the **Total** is calculated by the model and can only result in a total composition of 100%.

Detailed Metal Composition: (% by weight)	**Ferrous**	**Non-Fe**
Detailed Plastic Composition: (% by weight)	**Film**	**Rigid**

The model supplies default data for the detailed composition of both metals and plastic in commercial waste but again if region-specific data are available they should be entered here. Estimates of the total amount generated and composition of commercial waste in the region need to be entered in the white boxes. No reliable generic data on commercial waste generation are available for use as default values.

The model calculates the amounts of each waste material entering the system, using the categories in Table 16.1, and then adds these totals to the appropriate waste-stream columns.

Tab 5 Input Summary (Screen 7)

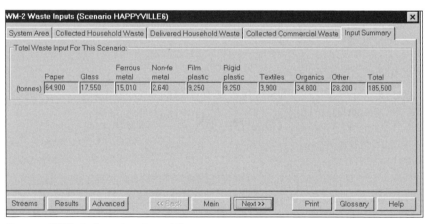

Screen 7 Waste Inputs – Input Summary.

This screen calculates and displays the total waste input, in tonnes per fraction of the waste stream, for the scenario being described.

CHAPTER 17

Waste Collection

Summary

The major environmental burdens associated with waste collection systems will be due to the transport required, which consumes energy and results in significant air emissions. The function of the collection system, after all, is to transport the waste from the household or commercial property to the sorting or treatment site. There may be other burdens, however, such as the production of plastic bags used in the collection, or the cleaning of bins. This chapter describes the assumptions made and the data required to complete the Waste Collection section of the model.

Defining the system boundaries

The system boundaries of the waste collection section of the IWM-2 model are presented in Figure 17.1. The inputs to the system are energy, raw materials and the waste at the point it

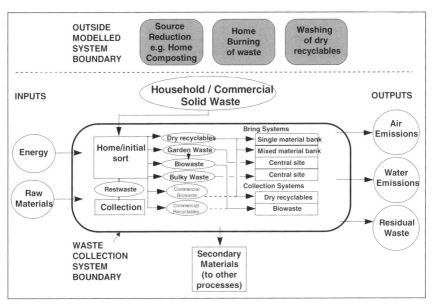

Figure 17.1 System boundaries of the waste collection section of the model.

leaves the household or commercial premises. The collection of household and commercial waste by kerbside or bring systems, either as sorted fractions or as co-mingled restwaste, are within the system boundary. The outputs of the system are air and water emissions, residual waste and secondary materials that are the inputs for other processes (central sorting, biological treatment, thermal treatment and recycling).

Source reduction (e.g. home composting) is considered to be an activity that takes place outside of the system boundary. The effects of source reduction on a waste management system can be modelled by reducing the **Amount Generated (kg/person/year)** and appropriately adjusting the **Composition (% by weight)** in the **Waste Input** screen. Home or backyard burning of both household and garden waste is not included in the model as this activity (which is illegal in many countries) results in uncontrolled emissions to air and should not be encouraged. The burdens associated with the washing of dry recyclables (such as glass bottles and jars, plastic containers and metal cans) are also considered to be outside of the system boundary. This is mainly due to the difficulty of gathering suitable data to model this activity. If dry recyclables are washed and/or rinsed as part of the ordinary household dish-washing process, then little in the way of extra burdens could be associated with the waste management system. However, if householders wash each dry recyclable item individually then the overall increase in burdens from energy use (water heating) and water emissions (detergent use) may become significant. This model assumes that if householders have been encouraged to separate dry recyclables correctly, then they have also been encouraged to wash this separated material (only if necessary) as part of their regular dish-washing process.

Environmental burdens due to transport

The transport involved in collection systems involves a mixture of householders' cars and municipal waste collection vehicles, the exact combination varying with the collection method used. At one extreme, in the use of central collection sites, most if not all of the transport will involve the use of householders' vehicles. Low-density materials banks will involve householders driving to the bank sites, and then special collection vehicles emptying the banks and transporting the materials to bulking depots, prior to sale and transport on to the materials processors. At the other end of the spectrum, close-to-home bring schemes, within walking distance of each property, and kerbside collections involve only municipal or contractors' collection vehicles.

Calculation of the energy consumption and emissions resulting from each type of transport require data on the distances driven and the average fuel consumption of the vehicles used.

Whilst the latter information is available (Table 17.1), details of distances driven will vary widely between different areas, so cannot easily be generalised. As a result, these are included as variable data in this analysis, and have to be inserted by the user for the geographical area under study.

For bring systems (both central collection sites and low-density materials banks) the number of special journeys made by car, each year, to the collection sites needs to be estimated, as well as the average distance. Note that depositing recyclables in a materials bank in a supermarket car park during a shopping trip would not count, as the transport there could be allocated to the shopping trip. Theoretically the burdens should be divided between the two functions, but

	Fuel consumption l/100 km
Private car	
Petrol – average	7.9
Diesel – average	5.3
Heavy goods vehicle	
Diesel average (20-tonne load)	32.8

Table 17.1 Fuel consumption data for different road vehicles. Source: ETSU (1996)

this is beyond the level of detail of the present study. Data of this type are not readily available, but could be acquired for any given region via a consumer survey. Fuel consumption and emission levels for private cars, (petrol and diesel), are presented in Table 17.1.

Calculation of the energy consumption and emissions from kerbside collection is less straightforward. Clearly the stop–start nature of most kerbside collection makes the use of standard heavy goods vehicle (HGV) data inappropriate. Data for such stop–start kerbside collections of dry recyclables from schemes in Adur and Milton Keynes (UK) report fuel consumption of 44 and 20 litres per 100 km, respectively (Adur District Council, personal communication; Porteous, 1992). What are needed for this LCI study, however, are the burdens per household serviced, or per tonne of material collected. Calculating this from fuel consumption data requires the average distance travelled by the collection vehicle per household visited. Again this will vary with housing density, and with the distance that the collection vehicle travels from the collection area to the sorting or treatment plant for emptying. This problem can be avoided by using data that should be available to every collection system operator – the average fuel consumption per collection round. Knowledge of the average number of households served then allows calculation of the average fuel used per household visit. Data of waste collected allow conversion into fuel used per tonne collected. The above reported results from Adur give values of 32 litres per 1000 households visited or 14.3 litres/tonne collected (calculated from data in IGD, 1992), and 7.2 litres/tonne for Milton Keynes (Porteous, 1992). Kerbside Dublin, a similar scheme collecting dry recyclables, has reported a fuel consumption of 17 litres per 1000 households visited (5.8 litres per tonne collected) over the first 6 months of 1993 (ERRA, personal communication). Fuel consumption per household visited is likely to be the more reliable measure, however, since the same distance must be covered by the vehicle, no matter how much waste is picked up from each property. Consumption of diesel fuel can then be converted into primary (thermal) energy consumption, emissions to air and water, and solid waste, using the generic data presented earlier in Chapter 5 (Table 5.7).

Other burdens

The input boundary for this LCI study has been defined as waste at the point that it leaves the waste generator, i.e. the household or commercial property (Chapter 5). Any materials needed to get the waste from this point to a collection site or vehicle, (e.g. refuse bags, refuse bins and recycling bags or bins) will therefore fall within the defined boundaries, and the relevant life cycle inputs and outputs, from cradle to grave, should be included. It is important to consider these additional burdens, since the range of collection systems discussed above differ in the number of different waste fractions collected, and hence in the relative need for different bags or bins for the collection process.

Collection bags

Although they are normally included in analyses of household waste, collection sacks are not strictly 'waste' during the collection stage as they are performing a useful function, i.e. containing the waste; they do become waste when the waste is subsequently delivered to a treatment site (e.g. a Materials Recovery Facility or composting plant).

Collection bags in common use are either paper or plastic, so the Life Cycle inputs and outputs in terms of energy, emissions and solid waste can be calculated from generic production data for these materials, given the average weight per bag and the average number of bags used per household per year. The number of bags used will vary with the quantity of waste generated, size of bags and the degree of waste sorting and separation required in the home. Sorting the waste into many fractions could lead to the collection of numerous half-filled collection bags. The generic data for production of both paper and plastic bags used in this study are presented in Table 17.2.

Note that burdens due to small refuse bags that are used to convey waste from the house to a dustbin or large refuse sack will not be included, since these are part of the waste before it leaves the property. Similarly, some schemes (e.g. Chudleigh, Devon, UK) use ordinary carrier bags (plastic grocery bags) for collecting materials such as dry recyclables. Since these would have been included in the waste stream in any case, the upstream burdens of producing these bags should not be included.

Collection bins

Inclusion of the production burdens for collection bins in this LCI study is debatable, since they can be considered to represent capital equipment, rather than operating consumables. Under the goal definition section in Chapter 5, capital equipment was excluded from the study. However, this would artificially bias any comparisons between collection systems using different collection container types. Bins consume a large amount of material (most often plastic) initially, rather than on a weekly basis, but like bags, bins will also enter the solid waste stream eventually at the end of their useful life. The burdens of bin use is therefore included; once calculated, it will be possible to determine whether the burden of this 'capital equipment' is insignificant, as originally predicted.

It is possible to calculate the relevant burdens from the use of bins given generic data on the materials used, the weight of material per bin/container, and the expected useful life-span. For example, the Blue Boxes used for dry recyclables in the Adur scheme in the UK are made from 1.6 kg of injection moulded polypropylene. After 3 years of operation, the boxes start to be replaced (D. Gaskell, ERRA, personal communication). This gives a requirement of around 0.5 kg of material, per household per year for this part of the collection. Wheeled bins (capacity 240 litres) have a weight of around 15 kg; a 10-year life expectancy would give a material requirement of 1.5 kg of polypropylene per household per year.

Table 17.2 Inventory Data for the production of paper and low-density polyethylene (LDPE) bags and polypropylene (PP) bins used in waste collection systems

Source	Virgin LDPE (per tonne) BUWAL (1998)	Bag Production (per tonne) Habersatter (1991)	Total for LDPE bags (per tonne) Calculated	Virgin PP (per tonne) BUWAL (1998)	Injection moulding of bins (per tonne) Habersatter (1991)	Total for PP bins (per tonne) Calculated	Paper bags (unbleached sulphate pulp) (per tonne) BUWAL (1998)
Energy consumption (GJ)	88.55	9.5	98.05	80.03	9.5	89.53	54.46
Air emissions (g):							
Particulates	3000	197	3197	2,000	197	2197	316
CO	900	349	1249	700	349	1049	359
CO_2	2,320,000	441,657	2,761,657	1,800,000	441,657	2,241,657	360,000
CH_4	4,400		4400	3,400		3400	469
NOx	12,000	1,236	13,236	10,000	1,236	11,236	2,870
N_2O	6.7	70	76.7	5.7	70	75.7	11.6
SOx	9000	2,502	11,502	11,000	2,502	13,502	4,600
HCl	70		70	40		40	8.26
HF	5	0.01	5.01	1	0.01	1.01	0.838
H_2S			0				
Total HC	16,600	2,112	18,712	9,600	2,112	11,712	1,030
Chlorinated HC							
Dioxins/Furans (TEQ)							
Ammonia	1.1	0.49	1.59	0.83	0.49	1.32	16

(continued)

Source	Virgin LDPE (per tonne) BUWAL (1998)	Bag Production (per tonne) Habersatter (1991)	Total for LDPE bags (per tonne) Calculated	Virgin PP (per tonne) BUWAL (1998)	Injection moulding of bins (per tonne) Habersatter (1991)	Total for PP bins (per tonne) Calculated	Paper bags (unbleached sulphate pulp) (per tonne) BUWAL (1998)
Air emissions (g):							
Arsenic							
Cadmium	0.018		0.018	0.018		0.018	0.0479
Chromium							
Copper							
Lead	0.1		0.1	0.088		0.088	0.12
Manganese	0.035		0.035	0.027		0.027	0.0139
Mercury	0.036		0.036	0.023		0.023	0.00173
Nickel	1.0		1	0.96		0.96	1.29
Zinc	0.4		0.4	0.42		0.42	0.367
Water emissions (g):							
BOD	200	0.15	200.15	60	0.15	60.15	6,760
COD	1500	0.44	1500.44	400	0.44	400.44	23,400
Suspended solids	500	0.15	500.15	200	0.15	200.15	5,240
TOC	100	4.7	104.7	300	4.7	300.7	44.9
AOX				0.075		0.075	1.01
Chlorinated HCs				0.021		0.021	0.0044
Dioxins/Furans(TEQ)							

(continued)

Table 17.2(continued) Inventory Data for the production of paper and low-density polyethylene (LDPE) bags and polypropylene (PP) bins used in waste collection systems

Table 17.2 (continued) Inventory Data for the production of paper and low-density polyethylene (LDPE) bags and polypropylene (PP) bins used in waste collection systems

Source	Virgin LDPE (per tonne) BUWAL (1998)	Bag Production (per tonne) Habersatter (1991)	Total for LDPE bags (per tonne) Calculated	Virgin PP (per tonne) BUWAL (1998)	Injection moulding of bins (per tonne) Habersatter (1991)	Total for PP bins (per tonne) Calculated	Paper bags (unbleached sulphate pulp) (per tonne) BUWAL (1998)
Water emissions (g):							
Phenol	2.4	0	2.4	2.6	0		0.697
Ammonium	5	0.62	5.62	10	0.62		9.3
Arsenic	0.34		0.34	0.26		0.26	0.0387
Cadmium	0.028		0.028	0.028		0.028	0.0073
Chromium	1.8		1.8	1.4		1.4	0.213
Copper	0.83		0.083	0.65		0.65	0.0904
Iron	220	0.003	220.003	170	0.003	170.003	12.4
Lead	1		1	58		58	0.337
Mercury	0.002		0.002	0.0013		0.0013	0.000082
Nickel	0.85		0.85	0.67		0.67	0.0994
Zinc	1.8		1.8	1.4		1.4	0.225
Chloride	130	0.02	130.02	800	0.02	800.02	2,900
Fluoride		1.335	1.335		1.335	1.335	
Nitrate	5	1.32	6.32	20	1.32	21.32	163
Phosphate	5		5	13.5		13.5	0.999
Sulphate	2100		2100	1,700		1,700	1,200
Sulphide	0.54		0.54	0.61		0.61	0.151
Solid Waste (kg)	39.1	49.1	88.2	31.2	49.1	80.3	129.1

Use of bins can lead to a further source of burdens, due to the need to wash the bins. This is likely to be relevant where unlined bins are used to collect biowaste, since this can cause severe odour and/or fly nuisance in hot weather. This source of burdens should also be included, especially in comparison of bin versus bag collections for biowaste, but data on the level of bin cleaning, and on the typical amounts of water, etc. used are not readily available. **Assuming that around 25 litres of warm water (heated 20°C above ambient) are used per bin, 2.14 MJ would be needed. For simplicity it will be assumed that water is heated by electricity with 100% efficiency, so around 0.6 kWh would be consumed per wash.** Any burdens due to the use of cleaning agents should also be included. A Life Cycle study by Procter & Gamble on hard surface cleaners has shown, however, that the heating of the water represents the major source of both energy consumption and solid waste generation from such cleaning operations (P&G internal report), so the burdens of bin cleaning included here will be restricted to this element.

Pre-treatment of waste

There will also be some environmental burdens due to waste sorting and treatment within the household. Some collection schemes for dry recyclables request that food cans, for example, are rinsed out prior to collection. These burdens occur prior to waste leaving the property, so are not included in this study, but could be included in other LCI studies with more widely defined boundaries.

Economic costs

Care is needed when extracting data on the actual costs of individual collection systems. Although costs are often quoted for various collection and sorting schemes, it is important to understand exactly what is included in these costs, and equally importantly, what is excluded. Often, quoted figures for materials recovery schemes include not only the collection system, but also the sorting system and the revenues from the subsequent sale of material. Some costs also include disposal savings for any material that is diverted from landfill. If the collection systems are operated by a municipality the costs of waste collection are often not separated from other areas of expenditure, so the actual collection cost is not known.

As with materials recovery systems, comparisons between the costs of different collection systems require that standard accounting methods are used. A study by ERRA (1998) observed that 'the accounting systems used by 11 different European waste management systems were as different as the systems themselves'.

To calculate the economic costs of different collection systems it is necessary to use local data. Salaries, a major component of collection costs, will vary geographically, so no general figures are applicable. This section presents collected data to demonstrate the typical ranges of figures for different countries.

Material bank systems

Data for the collection costs of bring systems are often presented inclusive of the sale of the collected material. This revenue is likely to be significant compared to the collection cost. This inclusive figure will therefore vary with market price fluctuations of the materials. For a modelling purpose, it is more useful to have the collection cost separated from any subsequent revenue.

Kerbside collection systems

Collection costs for kerbside collection systems are also quoted in a variety of ways: collection only, collection plus sorting, collection plus sorting plus sale of recovered materials, collection plus disposal, etc. Where full and transparent accounts have been published (e.g. IGD, 1992), it is possible to calculate the contributions of these various components. In the Adur Blue Box Scheme (W. Sussex, UK) for example, collection of dry recyclables costs 23.16 euro per household per year (equivalent to 166.51 euro per tonne collected and sold on). Including subsequent sorting of the material raises this figure slightly, to 26.45 euro per household per year (190.17 euro per tonne). When the revenue from sale of materials is included, this falls again to 22.37 euro (160.83 euro per tonne) (IGD, 1992). Therefore, in contrast to bring systems, income from materials is relatively insignificant in the inclusive cost; the key factor is the actual cost of the collection.

The Waste Collection screen

This screen contains five tabs: **System Area**, **Collected Household Waste**, **Delivered Household Waste, Collected Commercial Waste** and **Summary**.

At the outset, it is necessary to stress the need for compatibility between the waste input data and the waste collection data to be inserted here. The collection section attempts to cover the majority of possible ways that waste can be collected, for all possible wastes, including bulky household wastes, garden waste and commercial waste. Clearly all of these must have been included in the waste generation data used. If, for example, the waste generation data inserted does not include garden waste or commercial waste, the collection system should not include their collection.

The program allows up to four Kerbside Collection Systems (KCS) and four Material Bank Collection Systems (MBCS) to be modelled together, as often the complete collection system of a large city or region is made up of a combination of more than one type of collection system.

Tab 1 System Area (Screen 8)

The input data here defines the number and type (up to four of each) of collection systems used for household waste within the system area. The user must specify the percentage of collected household waste managed by Kerbside Collection Systems and/or Material Bank Collection Systems. The model displays the number of households served by each system. Individually neither the Kerbside Collection Systems nor the Material Bank Collection Systems can serve more than the total number of households within the system area, but together they can serve the number of households in the system area twice. For example, a region may have a Kerbside Collection System for every household but also provides a system of material banks for the collection of recyclable material. In this case each household is being served by two systems. In the model as in real life, it is not possible to collect more waste than is generated within the system area as defined in the Waste Input screen.

IWM2 Model Guide

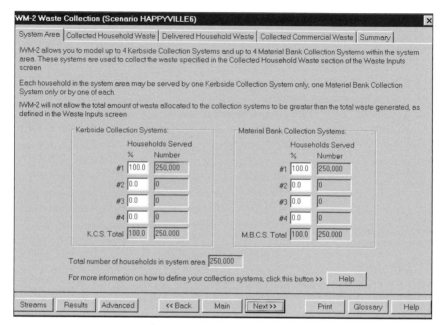

Screen 8 Waste Collection – System Area.

Kerbside Collection Systems:			Material Bank Collection Systems:		
	Households served			Households served	
	%	Number		%	Number
#1			#1		
#2			#2		
#3			#3		
#4			#4		
K.C.S. Total			M.B.C.S. Total		

The sum of the percentages of the Kerbside Collection Systems and the Material Bank Collection Systems must both equal 100%, as the user has already defined the total amount of household waste collected. In the example shown in Screen 8, the user has specified that a single Kerbside Collection Systems services 100% of the population in the system area, a single Material Bank Collection System also services 100% of the population in the system area.

To illustrate the flexibility of this approach to modelling collection systems, another example is presented below. Here, three different Kerbside Collection Systems operate within the system area, servicing 75%, 15% and 10% of the population, respectively. A single Material Bank Collection System also serves the whole population; this is represented schematically in Figure 17.2.

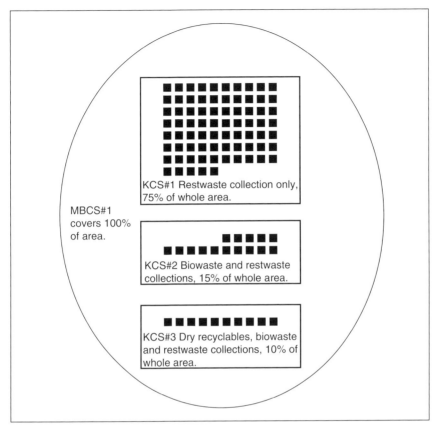

Figure 17.2 Schematic representation of a system area comprising three kerbside collection systems (servicing 75%, 15% and 10% of households) and one material bank collection system (servicing all households in the system area).

Tab 2 Collected Household Waste (Screen 9)

The tab calculates and displays the total amount of each material that is available in kg/household/year. The number of Kerbside Collection Systems (KCS) and Material Bank Collection Systems (MBCS) that were specified in the **System Area** Tab will be available within the **Collected Household Waste** tab (up to a maximum of four sub-tabs for each system).

Tab KCS#1

This tab allows the description of a Kerbside Collection System, where collection vehicles pick up biowaste, recyclable material and restwaste. If no figures are entered on this tab, the model assumes that all household waste is collected as co-mingled restwaste either in plastic bags or bins.

IWM2 Model Guide

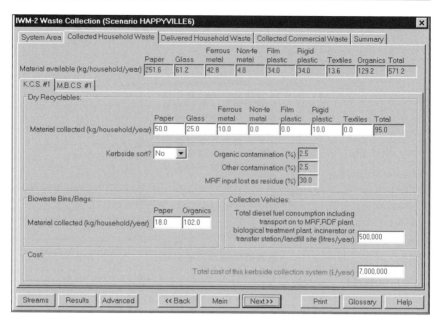

Screen 9 Waste Collection – Collection of Household Waste – KCS#1.

Dry Recyclables:
Material Collected (kg/household/year) Paper, Glass, Ferrous Metal, Non-Fe Metal, Film Plastic, Rigid Plastic, Textiles, Total

If separated, the amount of each material collected as dry recyclables is inserted here in the form of amount per household per year. The model adds the amount of each material collected and displays the **Total** in the grey box.

Kerbside sort?

The model asks whether kerbside sorting occurs, e.g. as in Blue Box schemes. 'Yes' is selected if it occurs, 'No' if it does not. Kerbside sorting helps prevent contaminants in the form of non-requested materials (e.g. organic or other waste) or non-targeted items of requested materials (e.g. films, foils) from entering the recyclables collection system. In the absence of a kerbside sort, a default figure of 5% contamination (2.5% organic material and 2.5% other material) by other waste materials is added at this stage (this can be changed in the **Advanced** window, **Waste Collection** Tab, **KCS#** Tab if the user has better data). A lower sorting efficiency resulting in 30% residue (which includes contaminants collected and sorting inefficiency) is assumed in the subsequent MRF sorting stage (this default value can also be changed if required in the **Advanced** window). If kerbside sorting does occur the default value for contamination is 0% and the sorting efficiency improves; only 8% residue is assumed in the MRF sorting stage (again this default value can be changed in the **Advanced** window).

Biowaste Bins/Bags:

Material collected (kg/household/year)   Paper   Organic

If biowaste is separated, the amount of organic material and paper (if paper is included in the biowaste definition) collected in this stream needs to be entered. The model will remove 5% of the total collected of both paper and organic material and add an equivalent weight of plastic as contamination (plastic is assumed to be the most common contaminant found in source-separated biowaste). As discussed in Chapter 9, even where biowaste is collected in bins rather than bags, there is typically this level of plastic contamination. This default value can be altered if necessary in the **Advanced** window, **Waste Collection** Tab, or **Bins & Bags** Tab.

Collection Vehicles:

Total diesel fuel consumption including transport to MRF, RDF plant, biological treatment plant, incinerator or transfer station/landfill site (litres/ year)

This includes collections for all of the different fractions (biowaste, dry recyclables and rest-waste). This figure is simply added to the fuel consumption column.

Cost:

Total cost of this Kerbside Collection System (£/year)

This figure should include all of the collections made, but not include any element for subsequent sorting or other treatments.

Tab MBCS #1 (Screen 10)

This sub-tab (one of up to four) describes Material Bank Collection Systems. These are bring systems where householders deliver their (sorted or unsorted) waste arisings to collection banks or containers placed on the street and at street corners.

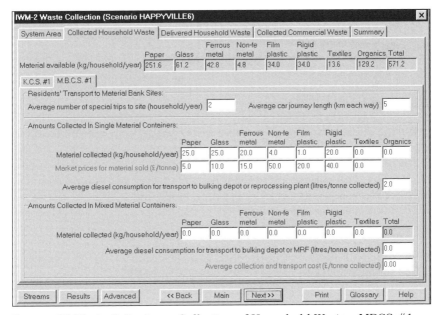

Screen 10 Waste Collection – Collection of Household Waste – MBCS #1.

Residents' Transport to Material Bank Sites:
   Average number of special trips to site (household/year)
   Average car journey length (km each way)

The numbers of journeys and average distances allow total fuel consumption of petrol and diesel for the area to be calculated. Only special journeys to materials banks should be included; visits to materials banks in supermarket car parks as part of regular shopping trips should not be included. Similarly, where close to home materials banks are used and transport to the bank is on foot, zero special car journeys should be inserted.

Amounts Collected In Single Material Containers:
   Materials collected (kg/household/year) Paper, Glass, Ferrous Metal, Non-Fe Metal, Film Plastic, Rigid Plastic, Textiles, Organics

This accounts for materials such as colour-separated glass or aluminium cans, which can be transported to material processors without further sorting. This material is added directly to the secondary materials stream. Organic material is added to the biological stream.

   Market price for material sold (£/tonne) Paper, Glass, Ferrous Metal, Non-Fe Metal, Film Plastic, Rigid Plastic, Textiles

The market prices that need to be inserted are per tonne of this material, ex-collection bank. No price is input for organic material as it must be treated prior to it having any market value.

   Average diesel consumption for transport to bulking depot or reprocessing plant (litres per tonne collected)

The average fuel consumption per tonne of material collected must be entered here. The model adds this directly to the fuel stream.

Amounts Collected in Mixed Material Containers:
   Material collected (kg/household/year) Paper, Glass, Ferrous Metal, Non-Fe Metal, Film Plastic, Rigid Plastic, Textiles

This allows for collection of material that needs a subsequent sort, e.g. at an MRF, prior to sale. This could apply to the collection of mixed plastics, for example, or to a system where mixed recyclables from high-rise housing are collected in communal containers. This material is added to the sorting stream, which forms the input to the MRF.

   Average diesel consumption for transport to bulking depot or MRF (litres per tonne collected)

This accounts for the transport from materials banks to either a central bulking site or transfer station (for separated materials prior to sale and onward shipment to materials reprocessors) or to a regional MRF. Since a variety of transport types is likely to be used, the simplest form of data that can be used here is the average diesel consumption per tonne of material collected. This value should be available to system operators.

Cost:
   Average collection and transport cost (£/tonne collected)

This needs to be exclusive of revenue from the sale of the recovered materials, which has already been accounted for above. If only net data inclusive of revenues are available, these can be used if zero sales values are inserted for the collected materials.

Tab 3 Delivered Household Waste (Screen 11)

This tab defines the materials that are delivered to central collection sites for garden and bulky wastes. This is another type of bring system.

```
IWM-2 Waste Collection (Scenario HAPPYVILLE6)                                    [x]

 System Area | Collected Household Waste | Delivered Household Waste | Collected Commercial Waste | Summary |

 ┌─ Residents' Transport To Central Site: ──────────────────────────────────────────────┐
 │  Average number of special trips to site (household/year) [0]    Average car journey length (km each way) [0] │
 └──────────────────────────────────────────────────────────────────────────────────────┘

 ┌─ Bulky Waste Delivered: ─────────────────────────────────────────────────────────────┐
 │                                        Ferrous   Non-fe    Film     Rigid                │
 │                                Glass    metal     metal    plastic  plastic   Other*     │
 │   Recovery of materials (as % of delivered) [0.0]  [0.0]  [0.0]  [0.0]  [0.0]  [0.0]     │
 │             Market prices (£/tonne) [0.0]  [0.0]  [0.0]  [0.0]  [0.0]                     │
 │                              * If materials cannot be identified, they cannot be recovered │
 │                                       Incineration  Landfill                              │
 │         Bulky waste residue treatment (%) [0.0]    [100.0]                                │
 │         Transport distance (km each way) [0]       [0]                                    │
 │     Average diesel consumption for transport to bulking depot or reprocessing plant (litres/tonne collected) [0.0] │
 └──────────────────────────────────────────────────────────────────────────────────────┘

 ┌─ Garden Waste Delivered: ────────────────────────────────────────────────────────────┐
 │             Transport distance to biological treatment plant (km each way) [0]           │
 └──────────────────────────────────────────────────────────────────────────────────────┘

 ┌─ Cost ───────────────────────────────────────────────────────────────────────────────┐
 │  Cost of central collection site and transport to treatment plants excluding material revenue (£/tonne handled) [0.00] │
 └──────────────────────────────────────────────────────────────────────────────────────┘

 [Streams] [Results] [Advanced]    [<< Back] [Main] [Next >>]    [Print] [Glossary] [Help]
```

Screen 11 Waste Collection – Delivered Household Waste.

Residents' Transport to Central Site:
Average number of special trips to site (household/year)
Average car journey length (km each way)

The model calculates the total consumption of petrol and diesel for the whole area on a round-trip distance basis and adds this to the fuel stream. When the total fuel consumption over the whole life cycle has been calculated, this is multiplied by the burdens of petrol or diesel use, respectively, to give the overall primary energy consumption and emissions.

Bulky Waste Delivered:
Recovery of materials (as % of delivered) Glass, Ferrous Metal, Non-Fe Metal, Film Plastic, Rigid Plastic, Other

For the amount of bulky waste delivered, the percentage recovered is subtracted from the total (as defined in the **Waste Input** window) and added to the products stream. This allows for the recovery, mainly of metals, which occurs at such sites due to removal of used domestic appliances from the waste stream.

Market prices (£/tonne) Glass, Ferrous Metal, Non-Fe Metal, Film Plastic, Rigid Plastic

The market prices obtained from the sale of materials recovered from bulky waste are entered here.

Average diesel consumption for transport to bulking depot or reprocessing plant (litres/tonne collected)

This adds the burdens of transporting the materials recovered from the bulky waste to a bulking depot or a reprocessing plant.

Bulky waste residue treatment (%)   **Incineration**   **Landfill** (calculated)

The remaining bulky waste will enter the streams for landfilling or thermal treatment by mass-burn incineration as defined here by the user.

Transport Distance (km each way)   **Incineration**   **Landfill**

This adds the burdens of transporting the bulky waste residue for thermal treatment or disposal to landfill.

Garden Waste Delivered:
Transport distance to biological treatment plant (km each way)

Transport distances to the biological treatment plant, thermal treatment plant and landfill are used to calculate overall fuel (diesel) consumption, using the data in Table 17.1 above for a 20-tonne truck load, based on a round-trip distance (i.e. assuming no return load is carried). Again this default value can be changed in the **Advanced** window, **General** Tab. The amount of Garden Waste delivered has already been defined by the user on the **Waste Input** Screen, **Delivered Household Waste** tab.

Cost:
Cost of central collection site and transport to treatment plants, excluding material revenue (£ per tonne handled)

The cost of the central collection site, excluding any revenues from the sale of recovered materials (mainly metals), needs to be inserted in terms of cost per tonne of material handled per year.

Tab 4 Collected Commercial Waste (Screen 12)

This tab defines any additional collection systems that deal with commercial waste that is not managed by the household waste collection vehicles.

Dry Recyclable Fractions Collected:
Material available (tonnes/year) Paper, Glass, Ferrous Metal, Non-Fe Metal, Film Plastic, Rigid Plastic, Textiles, Total
Material collected (tonnes/year) Paper, Glass, Ferrous Metal, Non-Fe Metal, Film Plastic, Rigid Plastic, Textiles, Total

The model calculates and displays the total amount of material available in tonnes per year. The amounts of each material collected in these separate fractions needs to be inserted here. It is assumed that dry recyclables are transported to a MRF for sorting prior to sale, though this need not always be the case. The model adds the specified amounts of material to the sorting stream (dry recyclables). Since materials from commercial sources are likely to be less mixed than from household sources, no contamination is added to the streams at this point, although contamination rates can be added if desired in the **Advanced** section, **Waste Collection** tab, under **Collected Commercial Waste**.

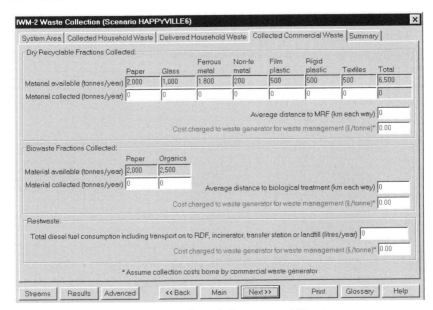

Screen 12 Waste Collection – Collected Commercial Waste.

Average distance to MRF (km each way)
Costs charged to waste generator for waste management (£/tonne)*
Again the model uses these inputs to calculate the respective fuel consumption, assuming a 20-tonne load and a round-trip distance (no return load). This fuel consumption is then added to the fuel consumption column.

Biowaste Fractions Collected:

	Paper	**Organic**
Material available (tonnes/year)		
Material collected (tonnes/year)		
Average distance to Biological treatment (km each way)		
Costs charged to waste generator for waste management (£/tonne)*		

*Assume collection costs borne by commercial waste generator.

Again the model calculates and displays the total amount of material available in tonnes per year. The model adds the specified amounts collected to the biological stream and uses these inputs to calculate the respective fuel consumption, assuming a 20-tonne load and a round-trip distance (no return load). This fuel consumption is then added to the fuel consumption column.

Restwaste:
Average diesel consumption for transport to RDF, incinerator, transfer station or landfill (litres/year)
Costs charged to waste generator for waste management (£/tonne)*
*Assume collection costs borne by commercial waste generator.

The model assumes that transport costs to all treatment sites are paid for by the waste generator. The treatment/disposal charges levied are inserted here. These represent an income for the waste management system, so will be subtracted from overall costs.

IWM2 Model Guide

Tab 5 Summary (Screen 13)

This summary screen has been included in the model to allow the user to view the flow of material within the waste input and waste collection sections of the model. This acts as a check, to ensure that all material specified in the waste input section is accounted for by the waste collection system that has been described. Collection systems are often a major burden on the rest of the waste management system so it is important that the user has defined the system as accurately as possible.

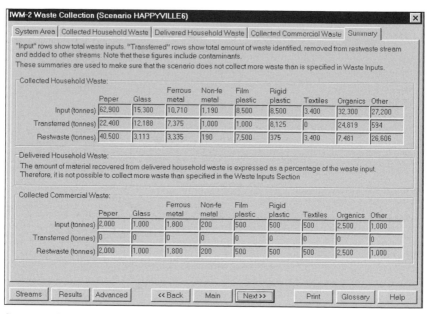

Screen 13 Waste Collection – Summary.

The **Input** rows should be equal to the sum of the **Transferred** and **Restwaste** rows in the **Collected Household Waste** and **Collected Commercial Waste** summary tables. The term transferred refers to materials that have been separated (and therefore subtracted) from the restwaste stream during the collection process and have been added to either the products or biological stream. The calculations used to obtain the figures shown in the tables can be examined by double-clicking any of the values; this opens a **Variable Information** window where the calculation is fully described.

CHAPTER 18

MRF and RDF Sorting

Summary

The major environmental burdens of both types of central sorting considered in this model are associated with their usage of energy in whatever form: electrical, gas or diesel. If consumption data can be obtained for the relevant processes, these can be converted into primary energy consumption and emission, using the generic data for fuel and energy usage presented in Table 5.6 and Table 5.7. It is also necessary to consider the input and output material streams, which determine where the products and residues of the process are destined to go, as these will eventually cause environmental burdens in subsequent processes.

Defining the system boundaries

The system boundaries of the MRF and RDF sorting section of the IWM-2 model are presented in Figure 18.1. The inputs to the system are energy, raw materials and the waste from the collection system. The outputs of the system are air and water emissions, residual waste and secondary materials (RDF, paper, glass, metals, plastics, organic material for biological treatment and residues to be incinerated) to be sent to other processes.

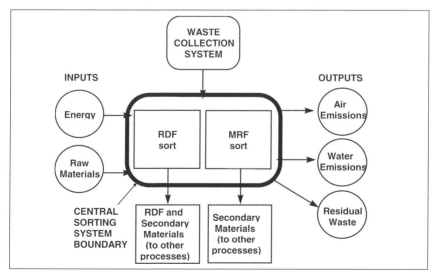

Figure 18.1 System boundary for central sorting.

Scheme	Electricity (kWh/tonne)	Diesel (litres/tonne)	Natural gas (m^3/tonne)	Source
Adur, UK	24	0.87	–	R. Moore, Community Recycling, Sompting MRF, personal communication (1993)
Dublin, Eire	22.1	35	2.3	Kerbside Dublin (1993)
Prato, Italy	27	n/a	–	ERRA (1993)

Table 18.1 Energy and fuel consumption data (per input tonne) for materials recovery facilities

MRF sorting

Inputs

Average energy consumption figures for Material Recovery Facility (MRF) sorting are likely to vary significantly between schemes, since there is no standard MRF process. Energy consumption is likely to be higher where more materials are separated in the MRF, as opposed to at the kerbside, and to increase with the level of mechanised sorting, in replacement of hand picking.

Recyclables streams that are collected co-mingled, such as mixed plastics and metals, need more sorting, and hence more energy, than streams collected in a pre-sorted fraction, such as glass or paper. To fully predict the likely energy consumption of any particular MRF process, it is necessary to have data on the sorting energy needed for each of these individual streams. Table 18.1 provides information that has been collected for MRF processing. Electrical energy is used to power conveyor belts, ferrous/eddy current separators and other equipment. Diesel is consumed mainly by auxiliary vehicles such as fork-lift trucks, mechanical shovels, etc., and gas where used is normally for heating. Energy and fuel consumption have been averaged over the total input to the MRF, since the individual fuel consumption cannot be allocated to individual materials.

Outputs

Residues from MRFs arise from two sources:

1. Material collected but not requested (i.e. the collection contaminants discussed in Chapter 9). The amount of such contamination can only be determined by a waste analysis of what is collected in the recyclables fraction. This material will not be selected in the MRF, so will become part of the residue.
2. Requested (i.e. targeted) material that is not separated out in the MRF (i.e. a sorting efficiency below 100%).

The total amount of residue from a MRF will represent the sum of these two contributions, and can vary from around 5% of the collected material, for Blue Box schemes with a kerbside sort,

Material	High throughput (%)	Low throughput (%)
Paper/board	31.2	35.4
Plastic	45.8	40.9
Glass	2.6	1.1
Metal	4.0	3.9
Food/garden	4.6	0.5
Textiles	4.2	11.8
Other	7.4	6.4
Total	99.8%	100%
Level of targeted material in residue	36.7%	13%
Level of targeted material in residue that could be recovered	25.3%	n/a

Table 18.2 MRF residue composition, sorting efficiency versus throughput rate, Prato, Italy. Source: ERRA (1993c)
n/a = data not available

to over 50% for co-mingled recyclables collected in a communal 'bring' container. Manual sorting of plastic bottles can give a sorting efficiency of around 97% whereas automated sorting results in between 88% and 95% efficiency (Resource Recycling, 1999). Analysis of the residue is necessary to determine the exact contribution of the two factors. Table 18.2 shows residue analysis results from the MRF at Prato, Italy, where on average the residue accounted for 35% of the input material (this level of contamination has since fallen). During trials, the level of targeted material in the residue varied from 13% to 37%, depending on the rate of throughput, leaving 63–87% of the residue made up of non-requested contaminating materials. This represents a loss of 7–17% of the targeted material entering the MRF, or a sorting efficiency of 83–93%. Of the targeted material in the residue, some was damaged or contaminated, so not all would actually be recoverable. This would mean that the true sorting efficiency would be actually higher still. Sorting efficiency will clearly vary with the level of contaminants present and the throughput of the MRF.

RDF sorting

Inputs

The input to the Refuse-Derived Fuel RDF process is normally mixed or residual waste, which has been collected co-mingled and unsorted. Data are available on the inputs and outputs of the RDF process for wastes with typical MSW compositions (Figure 18.1). The introduction of separate collection systems for individual waste fractions such as the dry recyclables, whether

IWM2 Model Guide

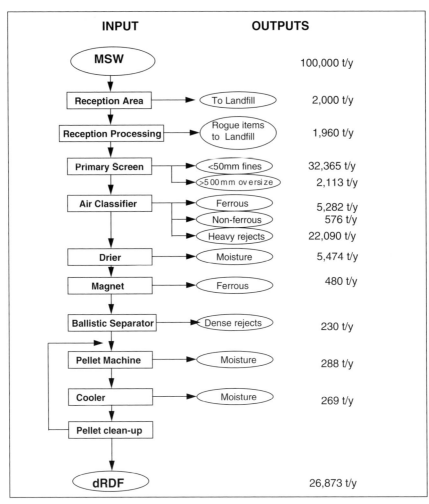

Figure 18.2 Detailed input and output analysis for production of densified Refuse-Derived Fuel (dRDF). Source: ETSU (1993).

using a bring or kerbside system, however, is likely to significantly alter the composition of the residual waste. Therefore it is important to be able to predict the inputs and outputs of the RDF process for any input waste composition.

Energy consumption

Several operations in the RDF process have significant electrical energy consumptions, in particular primary shredding (12.5 kWh per tonne of rated capacity), secondary shredding (8.5 kWh/t) and pelletising (9.5 kWh/t). The overall electrical energy consumption for the dRDF process has been estimated as **55.5 kWh** per tonne of rated capacity, i.e. per tonne of annual plant input (this figure is used as a default value in the model). In addition, the drying process prior to pelletising requires around 400 MJ of heat energy per tonne of rated capacity (ETSU, 1993b).

In plants where on-site combustion of RDF occurs, this drying heat requirement can be met by burning some of the RDF produced, or by using waste heat from the power generation system. Where no on-site burning of RDF occurs, heating by gas or other fuels will be needed.

The coarse Refuse-Derived Fuel (cRDF) process does not involve so many energy intensive stages, nor the drying process. As a result, the energy consumption is lower and has been estimated at 6 kWh per tonne of plant input for the crude Type A cRDF and **21.5kWh** per tonne of input for the more refined Type B cRDF (ETSU, 1992). (This figure of 21.5 kWh is used as a default value in the model.)

Outputs

The individual outputs from the processes in a typical dRDF flow line are shown in Figure 18.2. When aggregated, these outputs give the total amounts of RDF, recovered materials, putrescible fines, residue and air emissions from the dRDF process, as shown in Figure 18.3. This mass

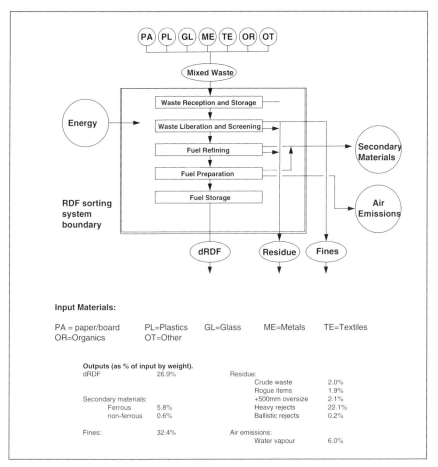

Figure 18.3 Mass balance for dRDF process with typical mixed waste input. Source: ETSU (1993b).

				Composition of output streams (%)							
Stream	Output (%)	Paper	Plastic film	Plastic rigid	Textile	Glass	Metal Fe	Metal non-Fe	Organic	Other*	Total
Fines	33.7	8.0	0.6	0.9	0.4	16.5	0.9	0.6	37.5	34.6	100.0
Gross oversize	2.2	23.9	24.5	2.4	21.6	1.0	5.6	0.7	4.7	15.6	100.0
Fe fraction	6.7	1.1	0.0	0.1	0.4	0.2	96.3	0.0	0.7	1.2	100.0
Heavy rejects	15.4	14.8	0.5	10.0	4.1	4.9	3.5	5.2	15.2	41.8	100.0
Scalp magnet fraction	0.1	16.6	9.3	0.2	14.2	0.0	19.8	0.5	0.6	38.8	100.0
Fuel fraction	41.9†	56.0	6.3	1.7	4.8	0.0	0.0	0.4	2.1	28.7	100.0
Total	100.0										

Table 18.3 RDF processing; typical composition (% by weight) of the process output streams. *Includes miscellaneous combustibles, miscellaneous non-combustibles and <10mm fines categories. Source: ETSU (1993b)
†Fuel fraction includes moisture content, which will be removed by drying before final fuel is produced.

Material	Process outputs				
	Fuel fraction	Fe fraction	Fines	Residue	Total
Paper/Board	81.3	0	8.8	9.9	100%
Plastic (film)	80.8	0	4.5	14.7	100%
Plastic (rigid)	28.3	0.3	8.8	62.6	100%
Glass	0	0.5	70.5	29.0	100%
Metal (Fe)	0	86.7	4.0	9.3	100%
Metal (non-Fe)	14.1	0	17.1	68.8*	100%
Textiles	61.0	0.8	4.1	34.1	100%
Organic	11.6	0.6	56.2	31.6	100%
Other†	14.2	0.2	61.5	24.1	100%

Table 18.4 dRDF processing: distribution of incoming waste materials (% by weight) between process outputs streams. Table gives distribution of materials between fractions, not composition of fractions. Thus for any individual category of material entering the dRDF process, this table will show the distribution of this material between the different process outputs. Source: calculated from data in ETSU (1993)
*Assuming no recovery of non-Fe material from heavy reject stream. If recovery occurs, residue contains 18.8% of non-Fe material; 50% of non-Fe material is recovered. †Calculated from original data assuming that 'other' category contains 28% miscellaneous combustibles, 13% miscellaneous non-combustibles and 59% –10 mm fines (taken from core waste analysis used in ETSU, 1993b). Source: Calculated from data in ETSU (1993b)

balance applies to a 'typical' input waste composition, and shows that on average around 27% of the input, by weight, is converted into dried, pelletised dRDF.

To estimate the mass balance for any input waste composition it is necessary to have information on the composition of the different output streams (Table 18.3) and the distribution of each material in the **Input waste** stream that enters the plant between the various RDF sorting process outputs (Table 18.4). Using this information, it is possible to determine both the amount and composition of the various process outputs (RDF, recovered materials, residue, etc.) for any input waste amount or composition.

Data used in RDF sorting section of the model

The data presented in Tables 18.5 and 18.6 are also used in the RDF sorting section of the model.

	cRDF	dRDF
Fuel consumption		
Electricity	21.5 kWh/input tonne	55.5 kWh/input tonne
Natural Gas	10.3 m^3/input tonne	
Screening of input material:		
Waste rejected due to unavailability of plant	2% of waste input	2% of plant input
Unsuitable items rejected	2% of plant input (1.96% of waste input)	2% of plant input (1.96% of waste input)
Process input	96.04% of waste input	96.04% of waste input

Table 18.5 RDF data 1. Fuel consumption and input screening losses. Source: ETSU (1992, 1993b)

Economic Costs

MRF sorting

As with environmental burdens, it is impossible to predict an average MRF processing cost as there is no standard MRF process. Schemes that collect co-mingled recyclables, so simplifying and perhaps saving costs in collection, are likely to have higher MRF processing costs than schemes with kerbside sorting, where the MRF processing will be simpler. Thus there is a trade-off of costs between the collection and sorting parts of the Life Cycle.

In most cases it is not possible to allocate costs to the processing of individual materials. One study in the USA has tackled this problem, and has published the MRF processing costs for different materials (NSWMA, 1992). Ten privately owned MRFs were examined, and the results show two distinct features: firstly there are clear differences between individual materials, as might be expected due to the level of sorting needed. The second finding was that the costs vary widely between MRFs, generally by a factor of three, but sometimes by up to a factor of five for the same material. Data are also now becoming available for the sorting of packaging materials from the Dual System in Germany, see for example Berndt and Thiele (1993).

The other economic factor that needs to be considered at this point is the revenue obtained from the sale of materials recovered at an MRF. Like any commodity, these are affected by market forces and their prices will fluctuate with supply and demand over time and geography.

RDF sorting

Two studies by ETSU (1992, 1993b) have attempted to predict the economic costs of processing Refuse-Derived Fuel. The resulting costs depend on the capacity of the plant used,

Input Materials to RDF process

Process Outputs	Paper/board	Plastic film	Plastic rigid	Glass	Metal – Fe	Metal – non-Fe	Textile	Organic	Other
cRDF									
Fuel	89.5	83.6	89.3	28.7	7.6	31.5	80.7	42.7	36.5
Fe	0	0	0.3	0.5	86.7	0	0.8	0.6	0.2
non-Fe	0	0	0	0	0	50.0	0	0	0
Fines	8.8	4.5	8.8	70.5	4.0	17.1	4.1	56.2	61.5
Residue	1.7	11.9	1.6	0.3	1.7	1.4	14.4	0.5	1.8
dRDF									
Fuel	81.3	80.8	28.3	0	0	14.1	61.0	11.6	14.2
Fe	0	0	0.3	0.5	86.7	0	0.8	0.6	0.2
non-Fe	0	0	0	0	0	50.0	0	0	0
Fines	8.8	4.5	8.8	70.5	4.0	17.1	4.1	56.2	61.5
Residue	9.9	14.7	62.6	29.0	9.3	18.8	34.1	31.6	24.1

Loss of moisture due to drying and pelletising of dRDF = 18.3% of fuel fraction; final amount of dried dRDF produced = 81.7% of fuel fraction

Table 18.6 RDF data 2. Distribution (%) of process input between outputs. Non-ferrous separation is assumed in both cRDF and dRDF production. cRDF data calculated for a refined cRDF product, from dRDF data. Sources: ETSU (1992; 1993b)

due to economies of scale. The predicted break-even gate fee for a plant producing dRDF pellets for sale in the UK was 33.88 euro per input tonne for a 100,000 tonnes/year plant, falling to 31.66 euro for a plant with double this capacity. This gate fee is inclusive of revenues from the sale of the fuel pellets and recovered ferrous and non-ferrous metals, and the costs of transport and disposal of the residues. When these are split away, the processing cost for dRDF can be estimated at 24.39 euro per input tonne, for a plant of 200,000 tonnes/year capacity. For comparison, a plant of 100,00 tonnes/year capacity including a composting operation for further treatment of the putrescible fines fraction, is estimated to have a break-even gate fee of 36.28 euro (ETSU, 1993b). Similar estimates have been made for the production and, in this case, on-site combustion of cRDF (ETSU, 1992), giving a break-even gate fee of 47.79 euro for a 200,000 tonnes/year capacity plant.

MRF/RDF Sorting screen

The central sorting operations of both Material Recovery Facilities (MRF) and Refuse-Derived Fuel (RDF) facilities are modelled in this window.

Tab 1 MRF Sorting (Screen 14)

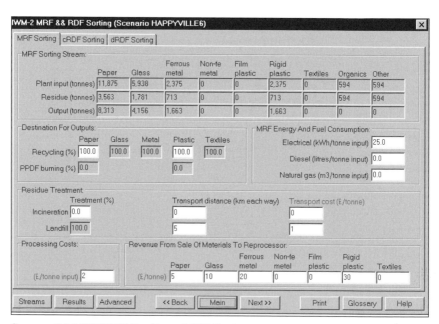

Screen 14 MRF/RDF Sorting – MRF Sorting.

MRF Sorting Stream:

Plant input	**(tonnes)**
Residue	**(tonnes)**
Output	**(tonnes)**

These grey boxes display the materials in the sorting stream available to MRF sorting processes. This convention of displaying the input to each process will continue throughout the model. This is included for both transparency and to allow the user to design a waste management system based on the actual materials in each waste stream at each stage in the waste management system.

Destination for outputs:

Recycling (%)		**Paper**	**Glass**	**Metal**	**Plastic** **Textiles**
PPDF Burning (%)		**Paper**			**Plastic**

As it is possible to use paper and plastic as a fuel for energy recovery, as well as for materials recycling, it is necessary to insert the relative usages of paper and plastic at this point. It is assumed that glass, metal and textiles (if collected and sorted) are destined for materials recycling. The model calculates the amount of each material left after removal of the residue, and then adds the relative amounts to either the **Materials** stream (materials recycling) or to the PPDF (Paper and Plastic Derived Fuel) burning inputs as appropriate.

MRF Energy and Fuel Consumption:

Electrical (kWh/tonne input)	**25**
Diesel (litres/tonne input)	
Natural Gas (m^3/tonne input)	

Although there is no standard MRF process, an energy consumption of **25 kWh** per input tonne is provided in the model as a default value, based upon the data in Table 18.1. If the user has access to more representative data, it must inserted here. The model adds the total energy/fuel requirement to the Life Cycle energy/fuel consumption totals.

Residue treatment:

	Treatment (%)	**Transport distance (km each way)** **(£/tonne)**	**Transport cost**
Incineration			
Landfill	(calculated)		

The user defines the treatment method used for the residue and the average distance (one way) to each type of treatment plant. The model adds the appropriate amounts of each material to either the **Thermal** or **Landfill** streams, and calculates the fuel (diesel) consumed by transport (assuming a 20-tonne load on a round trip basis, i.e. no return load), which is added to the fuel consumption totals. The user also needs to insert the transport cost per tonne of residue transported to thermal treatment or landfill.

User Note: The percentage of MRF input lost as residue is defined in the **Waste Collection** window, **Collected Household Waste** tab, **KCS** (Kerbside Collection Systems) tab, where the model asks **Kerbside sort ?**- Yes/No. The model automatically inserts a value of 30% where co-mingled recyclables are collected with no kerbside sort (if **No** is selected) and 8% where a kerbside sort is used (if **Yes** is selected). This loss includes both contaminants collected, and sorting inefficiency. The model removes this amount of each material collected, and adds it to the incineration or landfill streams, as appropriate. These default values can be changed by the user in the **Advanced** window, **Waste Collection** tab.

IWM2 Model Guide

Processing Costs:
(£/tonne input)*
*excluding sale of material and cost of residue disposal

This value should exclude revenues from sale of recovered materials and costs of residue disposal. If data that include sales revenue are used, then zero revenues should be inserted in the next boxes.

Revenue from Sale of Materials to Reprocessor:
(£/tonne)

Recovered materials destined for recycling leave the waste management system defined in this study at this point. The revenue received, per tonne of material ex-MRF, needs to be inserted in the relevant boxes.

Tab 2 cRDF Sorting (Screen 15)

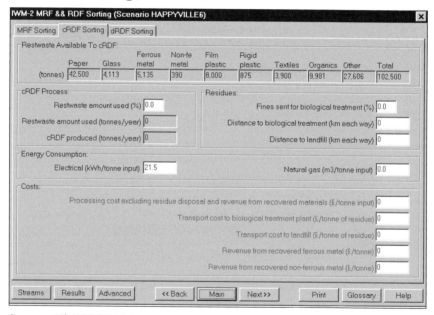

IWM-2 MRF && RDF Sorting (Scenario HAPPYVILLE6)

MRF Sorting | cRDF Sorting | dRDF Sorting

Restwaste Available To cRDF:

	Paper	Glass	Ferrous metal	Non-fe metal	Film plastic	Rigid plastic	Textiles	Organics	Other	Total
(tonnes)	42,500	4,113	5,135	390	8,000	875	3,900	9,981	27,606	102,500

cRDF Process:
- Restwaste amount used (%) 0.0
- Restwaste amount used (tonnes/year) 0
- cRDF produced (tonnes/year) 0

Residues:
- Fines sent for biological treatment (%) 0.0
- Distance to biological treatment (km each way) 0
- Distance to landfill (km each way) 0

Energy Consumption:
- Electrical (kWh/tonne input) 21.5
- Natural gas (m3/tonne input) 0.0

Costs:
- Processing cost excluding residue disposal and revenue from recovered materials (£/tonne input) 0
- Transport cost to biological treatment plant (£/tonne of residue) 0
- Transport cost to landfill (£/tonne of residue) 0
- Revenue from recovered ferrous metal (£/tonne) 0
- Revenue from recovered non-ferrous metal (£/tonne) 0

Streams | Results | Advanced | << Back | Main | Next >> | Print | Glossary | Help

Screen 15 MRF/RDF Sorting – cRDF Sorting.

Restwaste Available to cRDF:
(tonnes)

The grey boxes display the materials in the **Restwaste** stream available to coarse RDF sorting processes.

cRDF Process:
Restwaste amount used (%)
Restwaste amount used (tonnes/year)
cRDF produced (tonnes/year)

This defines whether coarse Refuse-Derived Fuel (cRDF) is included in the waste management system or not. If the user inputs a value, then that percentage of restwaste is used as the input to the cRDF process. The model also automatically inserts both the amount of restwaste that will be processed for cRDF, and the final amount of cRDF that this will produce.

If the cRDF process is not part of the waste management system being modelled the default value can be left as zero. Select the **dRDF Sorting** tab if this process is part of the waste management system to be modelled, or select the **Next** button.

The cRDF process does not include drying, so the weight is on an 'as received' basis. The cRDF tab in the **Advanced** window allows the user to change the default values for the screening process (prior to the RDF process) and those relating to the actual process outputs (fuel, metals, residue and fines). The amount of fuel produced is inserted into the thermal stream, which forms the input for the **Thermal Treatment** section of the model.

Residues:
Fines sent for biological treatment (%)
If a value is entered here then that amount of putrescible fines is added to the **Biological treatment** stream; the remainder is added to the **Landfill** stream. If no value is entered, all of the fines are added to the **Landfill** stream.

Distance to biological treatment (km each way)
Distance to landfill (km each way)
This allows calculation of fuel consumption burdens for the transport to the biological treatment plant and to landfill. If the fines are treated on site, no transport is needed and a zero should be inserted. This is calculated on a round-trip basis, assuming no return load.

Energy Consumption:
Electrical (kWh/tonne of input)
Natural gas (m³/tonne of input)
This is inserted automatically using the default data from the **Advanced** window in the cRDF tab. The fuel/electricity used per tonne is multiplied by the input tonnage to get totals for consumption, which are added to the **Fuel** stream.

Costs:
Processing cost excluding residue transport and disposal and revenue from recovered materials (£/tonne of input):
This should be exclusive of the costs of residue transport or disposal, and the revenues from sale of recovered materials or fuel.

Transport costs to biological treatment plant (£/tonne of residue)
Transport costs to landfill (£/tonne of residue)
Revenue from recovered ferrous metal (£/tonne)
Revenue from recovered non-ferrous metal (£/tonne)
These costs need to be accounted for here, but the revenue from the fuel should not be included as it does not leave the waste management system as defined in this model. The model assumes 89.1% ferrous and 50% non-ferrous metal recovery in the cRDF and dRDF processes, respectively. These default values and the sources of each of the metal fractions can be changed in the **Advanced** window in the cRDF tab, if the user has local data.

Tab 3 dRDF Sorting (Screen 16)

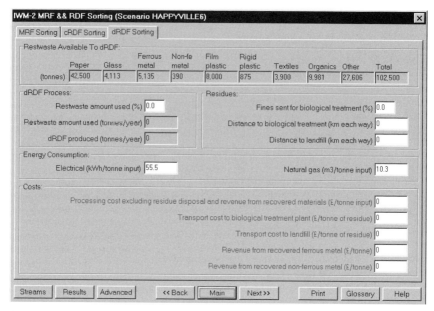

Screen 16 MRF/RDF Sorting – dRDF Sorting.

This tab defines densified Refuse-Derived Fuel (dRDF) production. The grey boxes here show the **Restwaste Available to dRDF**, by subtracting the amount of restwaste used by the cRDF process (if used) from the total amount of restwaste remaining in the system. The layout of the tab is exactly the same as the cRDF tab, but the model uses a different **Process Output** data table (which again can be edited in the **Advanced** window in the dRDF tab).

Note that the dRDF process includes drying, so there is a weight loss due to loss of moisture; this value can also be changed in the **Advanced** window in the dRDF tab.

CHAPTER 19

Biological Treatment

Summary

This chapter provides environmental and economic data on the burdens and range of possible costs associated with both composting and biogasification. The model allows the user the option of composting and/or biogasifying separately collected biowaste (see Chapter 11). The model also allows the user to treat restwaste biologically (although this is not recommended if the aim is to produce a marketable product). The model does not allow the user to model more than one composting (or biogasification) facility. Therefore, a separate but simultaneous composting of biowaste and composting of restwaste (perhaps as part of a mechanical–biological pre-treatment prior to landfilling) is not possible.

Defining the system boundaries

The system boundaries for biological treatment are defined here as the physical boundaries of the plant (Figure 19.1). Thus both the pre-sorting treatment and the biological process are

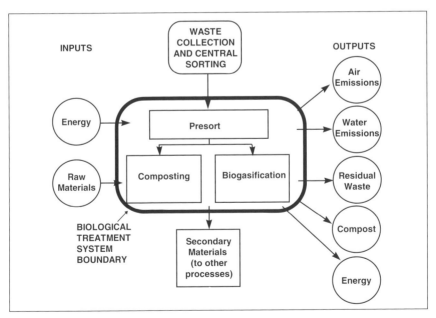

Figure 19.1 System boundary for biological treatment processes.

IWM2 Model Guide

included. Materials enter the system as waste inputs and leave as compost, recovered second-ary materials, residues (from sorting and composting), or as air or water emissions. Energy enters the system as either electrical energy from the national grid, or as fuels (e.g. diesel). In the case of biogasification, some of the energy recovered in the biogas is consumed on site for process heating, and it is assumed that the rest of the biogas is burned on site in a gas engine-powered generator to produce electricity. Again some of this is used on site, but the rest is exported from the site as electrical energy.

Typical flow diagrams for composting plants and biogasification plants are presented in Figures 19.2 and 19.3.

Waste Inputs

The feedstock for biological treatment can arise from at least three different sources: separate-ly collected organic/paper material, mechanically separated putrescibles from an RDF process or mixed and unsorted MSW.

The current trend in Europe is towards a separated collection of organic material from households. The exact composition of this feedstock will vary according to the definition of 'green waste' or 'biowaste' used by the collection scheme (see Chapter 9), but will generally consist of kitchen and garden waste, plus in many cases non-recyclable soiled paper and paper products. Even in narrowly defined feedstocks there will always be a level of nuisance materi-als, requiring a pre-sorting stage. Such nuisance materials arise from the inclusion of (1) bags (often plastic) used to contain organic material; (2) other materials, which form a small part of otherwise organic materials; (3) materials included in the biowaste by mistake.

Mixed waste inputs, such as MSW will need extensive pre-sorting to remove all of the non-organic material, which is not suitable for biological treatment. In contrast, finely sorted putrescible feedstock from an RDF type process will have already undergone a sorting stage, so will not require another pre-sort prior to the biological treatment process.

Energy consumption

The energy consumption of the pre-treatment process will depend on the feedstock used. Mixed feedstocks, such as MSW will need more extensive sorting per tonne of input, with associated energy requirements, than more narrowly defined feedstocks or those that have already been mechanically sorted as part of an RDF process, irrespective of the subsequent method of treatment. The energy consumption of the biological treatment process itself will depend on the technology employed.

Composting

Composting involves a net consumption of energy, consuming process energy and not producing any energy in a usable form. The German Government (1993) report a typical energy consumption of between 20 and 50 kWh (electrical energy) per input tonne for plants capable of processing 10,000 tonnes of biowaste per year. Available data on the overall energy consumption for various methods of composting are presented in Table 19.1, and suggest a range of 18–50 kWh (electrical) per tonne of input. This variability will reflect both the different feedstocks used, the different sizes of the composting plants, and the maturity of the

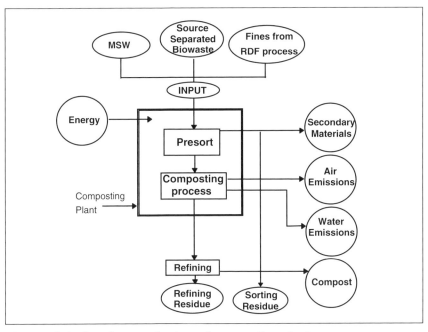

Figure 19.2 Flow diagram for typical composting plant.

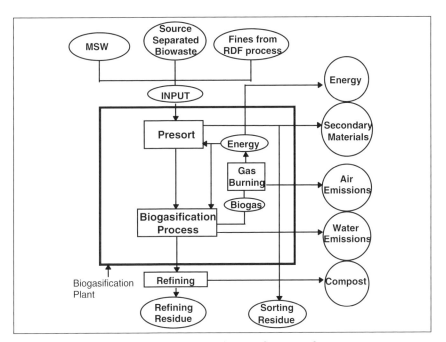

Figure 19.3 Flow diagram for typical biogasification plant.

Plant type	German Range	Windrow composting	Box composting	Tunnel reactor	Drum composting
Capacity t/year	10,000	13,700	6,800	18,000	35,000
Feedstock	BW	BW/GW	BW/GW	WW	BW/GW
Energy consumption (kWh/tonne)	20–50	18	18	50	40
Source	1	2	2	2	2

Table 19.1 Energy consumption of various composting plants. Sources: (1) German Government Report (1993); (2) Bergmann and Lentz (1992) Key to feedstocks: WW= 'wet waste'; BW= 'Biowaste'; GW= 'Greenwaste'.

compost produced. Kern (1992) looked at several different composting methods and calculated an average energy consumption of 21 kWh/tonne input: for plants producing less mature compost (rotte grades I–II) the average consumption was 18.3 kWh/tonne, whilst for plants producing mature compost (rotte grades III–IV) the average was 30.7 kWh/tonne. *For the purposes of the LCI model, a default energy consumption of 30 kWh of electrical energy per tonne of input to the composting plant is assumed.*

Biogasification

Biogasification involves both consumption of energy during processing, plus the production of useful energy as biogas. Since some of the biogas can be burned to produce steam to heat the digester, and more can be burned in a gas engine to produce electricity, the energy requirement for the process can be met from within the biogas produced. The remaining biogas can either be exported as biogas (i.e. as fuel) or burned on site to provide heat or to generate electricity (both for export). *In this LCI model it is assumed that the biogas is burned on-site for power generation, and that surplus energy is exported as electrical energy.*

The electrical energy requirement for biogasification has been reported as 50 and 54 kWh per input tonne for two different processes (Schneider, 1992; Schön, 1992). This represents around 32–35% of the gross electricity produced by the plant. In another example, a biogas plant operated using the dry process consumed between 30 and 50% of the electricity produced (De Baere, 1993). Thermal energy is also required for the process, but this can be obtained by using waste heat left after electricity generation, or by burning some of the biogas. Therefore no additional energy needs to be imported into the site for this. **In the LCI model, biogasification is assumed to consume 50kWh of the generated electricity, for every input tonne (including nuisance materials and recoverable materials).**

IWM2 Model Guide

Outputs

Examples of mass balances for both composting and biogasification plants are presented in Figures 19.4–19.6.

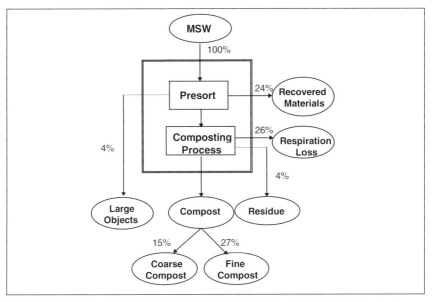

Figure 19.4 Typical mass balance (on the basis of dry weight) for lumbri-composting plant (La Voulte, France). Source: SOVADEC; Schauner (1994).

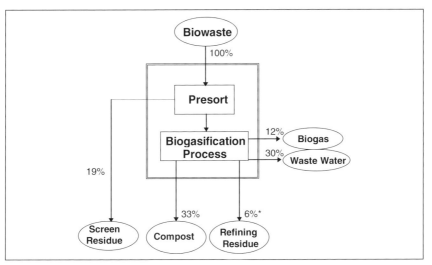

Figure 19.5 Typical mass balance (on the basis of wet weight) for dry process biogasification plant. Source: De Baere (1993).
*Estimated.

IWM2 Model Guide

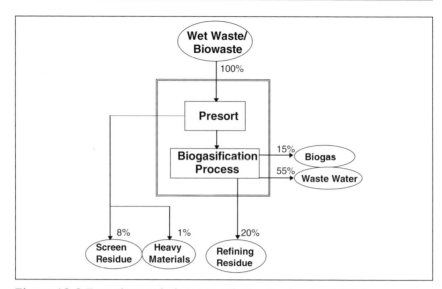

Figure 19.6 Typical mass balance (on the basis of wet weight) for a two-stage wet biogasification plant (Garching, Germany). Source: German Government Report (1993).

Secondary materials from pre-sorting

The amount of secondary materials that will be produced by a composting plant depends on the composition of the input stream, and on the pre-sorting equipment installed. A narrowly defined input, such as biowaste or Vegetable Fruit and Garden waste (VFG) will contain a certain level of contamination; this material is not suitable for recovery. A mixed waste stream input (MSW or household waste), will contain considerable amounts of glass, plastic, metal, etc. that could be recovered for use as secondary materials, but levels of contamination are likely to be high, and the quality of the material recovered is likely to be lower than that from source-separated collection of recyclables (Chapter 9). Recovery of recyclables from the input requires suitable sorting equipment or manual sorting; in most cases this is limited to magnetic separation, which can remove up to 90% of incoming ferrous material.

Biogas/energy

The amount of biogas produced during anaerobic digestion will depend on the nature of the organic material used as feedstock, as well as the process used. Biogasification of a range of industrial organic wastes from vegetables to dairy and brewing wastes in Switzerland produce from 200 to 600 Nm^3 per tonne of input (dry weight) (BfE, 1990). Biogas from household wastes are normally expressed per tonne as received (i.e. wet weight): grass clippings, for example, may be expected to produce 100 Nm^3 per input tonne (German Government report, 1993). Putrescible material mechanically sorted has been shown to produce 130–160 m^3 of biogas per tonne (Van der Vlugt and Rulkens, 1984).

Process type	Dry one-stage	Dry one-stage	Dry one-stage	Dry one-stage	Wet one-stage	Wet two-stage	Wet two-stage
Capacity (tonne/year)	10,000	100,000	Pilot	Pilot	30,000	20,000	1000
Feed stock*	BW	HW	BW	VFG	MSW + Sewage sludge	BW/WW	BW/MW
Energy requirement /tonne input (Th = Thermal) (El = Electrical)	Total 840 MJ Th	270 MJ El ? Th	76 MJ El 263 MJ Th	144 MJ El ? Th		288 MJ El 130 MJ Th	
Biogas production Nm³/tonne	90	100–140	100		50	115	90–150
Methane content (%)	55	60–65	60–65	54		70–80	60–75
Source	1	2	2,3	2	2	2,4	2

Table 19.2 Energy requirements and biogas generation for biogasification plants. Sources: (1) De Baere (1993); (2) Bergmann and Lentz (1992); (3) Korz and Frick (1992); Schön (1992)
*BW= Biowaste; HW= Household Waste; WW = Wet Waste; MSW = Municipal Solid Waste; VFG = Vegetable, Fruit and Garden waste.

IWM2 Model Guide

Production figures for the different processes reflect the amount of organic decomposition that is achieved. The more complex two-stage process converts more organic material to biogas (around 65–70% by dry weight) than single-stage processes (around 45% by dry weight), giving typical production rates of 115 and 75 m^3 per tonne of biowaste, respectively (Korz and Frick, 1993). Biogas composition, especially methane content, also varies with process type (Table 19.2), again being generally higher in the two-stage process. The methane content generally varies from 50–75%, the rest of the biogas comprising carbon dioxide and some trace components (Table 19.3).

If it is assumed that 100 Nm^3 of biogas are produced per tonne of input to the digestor (i.e. after nuisance materials have been removed in the pre-treatment stage), with a methane content of 55% (methane has a calorific value of 37.75MJ/Nm^3 (Perry and Green, 1984)), this will give a gross energy potential of 2076 MJ thermal energy per tonne digested. If this is burned in a gas engine to produce electricity with an efficiency of 33% (ETSU, 1995), this will give a gross electricity production of 190 kWh per tonne digested. Figures reported by De Baere (1999) for actual plant performances in Austria, Belgium and Germany agree with this; they report gross electricity generation rates of 220 kWh/tonne, 165 kWh/tonne and 245 kWh/tonne, respectively. *For the LCI model, a gross production of 190 kWh/tonne of digestor input will be used as a default value.*

Compost

The quantity and quality of the compost produced by biological treatment are clearly not independent. The more the product is refined to improve the quality, the less the final quantity, (and hence the greater the residue). In many cases, different grades of compost will be produced, the important factor being the existence of a market for the material. Put simply, if there is no market for the compost, regardless of its quality, it will be a residue rather than a valuable product. One possible alternative use for compost, where there is no other market, is as landfill cover material. This has the potential to save costs where soil has to be transported to a landfill site, only to be used as daily cover material.

Compost quantity

For composting, the final amount of compost produced (wet weight) is in the region of 50% of the input of organic material (organics plus paper) (ORCA, 1992a). The other 50% is lost due to evaporation and respiration. Where further refining of the compost occurs, the amount of compost actually marketed may be considerably less than this. For biogasification, the amount of final compost-like material will depend on the extent to which the organic material is broken down into biogas. Production can account for 33% by weight of a plant's input (equivalent to 41% of the input to the digester after the pre-sort) (De Baere, 1993). By contrast, the two-stage wet process produces more biogas with a higher methane content than the dry process, leaving around 20% of the plant input (22% of the digester input) as composted residue (see Figures 19.5 and 19.6).

For the purposes of the LCI model, it is assumed that in composting the mass loss due to evaporation and biodegradation of the organic fraction of final compost accounts for 50% of the input to the composting process (i.e. after any pre-sorting); for biogasification, an average figure of 70% is used.

Compost quality is the key factor that determines whether the output from biological treatment processes is a valuable product or a residue. A valuable material is one that has a market, hence the need to develop markets for different grades of compost. Producers will then either

	Biogas	Biogas	Biogas after combustion
Source	Lentz *et al.* (1992)	BTA (1992)	IFEU (1992)
% vol			
CO_2	26.8%	45%	
CH_4	71.4%	54%	
N_2	1.4%		
O_2	0.3%		
mg/m³			
NOx			100
SOx			25
sum Chlorine	0.6	0.9	11
sum Fluorine	0.1		0.021
HCl			
HF			
H_2S	700	420	0.33
Total HC		<1.5	0.023
Chlorinated HC		<1.5	7.3 E–03
Dioxins/furans			1.0 E–07
Ammonia			
Arsenic			
Cadmium			9.4 E–06
Chromium			1.1 E–06
Copper			
Lead			8.5 E–06
Mercury			6.9 E–08
Nickel			
Zinc			1.3 E–04

Table 19.3 Biogas composition

IWM2 Model Guide

	Biowaste compost	Biowaste with paper compost	Greenwaste compost	Wetwaste* compost	Total waste compost
H_2O% wet wt	37.7	45.0	34.8	44.2	35.6
pH-value	7.6	7.5	7.6	7.5	7.3
Salt g/l wet wt	3.9	3.6	2.3	5.8	7.3
OS % dry wt	33.3	42.0	32.5	55.4	39.7
C/N ratio	17.0	21.8	20.0	18.8	17.8
N total % dry wt	1.2	1.1	0.8	1.7	1.1
P_2O_5 % dry wt	0.6	0.6	0.4	0.9	0.9
K_2O % dry wt	1.0	0.9	0.8	1.2	0.6
MgO % dry wt	0.8	0.8	0.6	2.0	0.7
CaO % dry wt	4.0	4.1	3.0	10.0	4.9

Table 19.4 Physical characteristics and plant nutrient contents of different types of compost produced by aerobic process. Source: Fricke and Vogtmann (1992)
*Fraction remaining after separate collection of dry waste, e.g. recyclables such as glass, paper, metal, wood, etc. OS, organic substances.

be able to produce large amounts of lower grade composts, or smaller amounts of higher grade material.

Compost quality is determined by the feedstock type, type of technology used and level of process control. The physical characteristics, plant nutrient and heavy metal contents of a range of composts derived from different feedstocks are presented in Tables 19.4 and 19.5. These can be compared with the standards discussed in Chapter 11. It can be seen that the major variability occurs in the heavy metal content. Not surprisingly, the more mixed the feedstock, the higher the heavy metal content of the compost. The same pattern is shown for the compost-like residues from biogasification (Table 19.6). So, while it is possible to make compost from mixed waste streams, the high level of contamination may mean that no market for this material can be found. It is accepted in Germany, for example, that the composted residue from biogasification is not marketable as compost (German Government report, 1993) and that it needs to be disposed of. In 2005 the T.A. Siedlungsabfall ordinance will no longer permit this material to be landfilled directly in Germany, so incineration will need to be used.

IWM2 Model Guide

Table 19.5 Heavy metal content of different composts produced by aerobic process (mg/kg dry wt). *BGGK = Bundesgütegemeinschaft Kompost (1997) (German Federal Association for Quality Compost)

Element	Biowaste compost	Biowaste with paper compost	Greenwaste compost	Wetwaste compost	Total waste compost	BGGK* limits
Based on material as produced						
Pb	77.6	78.6	60.8	449	513	
Cd	0.8	0.7	0.7	2.6	5.5	
Cr	33.7	31.7	27.0	72	71.4	
Cu	43.2	58.2	32.7	228	274	
Ni	19.1	16.1	17.5	30	44.9	
Zn	232.8	273.8	167.8	850	1570	
Hg	0.3	0.4	0.3	1.0	2.4	
Based on standardised organic matter content of 30% (dry wt)						
Pb	83.1	116.2	63.1	705	596	150
Cd	0.8	1.0	0.7	4.1	6.4	1.5
Cr	35.8	39.8	28.4	113.0	82.9	100
Cu	46.8	76.2	34.5	357.8	318	100
Ni	20.5	21.4	18.6	47.1	52.1	50
Zn	249.1	350.3	176.9	1334	1823	400
Hg	0.4	0.5	0.3	1.6	2.8	1.0

Process	Composting			Biogasification					
	Windrow	Box composting	Tunnel reactor	Dry one-stage	Dry one-stage	Dry one-stage	Dry one-stage	Wet two-stage	Wet two-stage
Feed stock*	BW	BW	WW	VFG	VFG	BW	GM	BW	BW
Nutrients: (% by wt)									
N-total		1.1	n.a.	1.8–2.1		1.2	0.8–0.9	1.92	1.24
P_2O_5		0.73	n.a.	0.15–20		0.7	0.6	0.9	0.6
K_2O		1.4	n.a.	1.0–1.2		1.1	2.3	0.49	0.5
$CaCO_3$		1.9	10.0	5.2–6.4		2.7	2.7	6.6	3.82
C/N ratio		17–18	16	12–15		n.a	15	20	11–15
pH		7.8	7.5	8.0–8.6		n.a	8	7.6	n.a
Heavy metals (mg/kg TS)									
Zn	324	247	850	138	173	253	122	135	491
Pb	139	84	449	67	75	100	43	85	155
Cu	61	36	228	20	27	54	27	52	27
Cr	32	55	72	n.a.	30	36	15	44	34
Ni	17	41	35	25	9	20	7	27	16
Cd	1.8	0.7	2.6	1.8	0.8	1.3	0.4	1.0	1.1
Hg	0.8	0.2	1	n.a.	n.a.	0.7	0.3	<0.25	0.2

Table 19.6 Quality of compost produced by a variety of biological treatment processes. Source: Bergmann and Lentz (1992) *BW= Biowaste; HW= Household Waste; WW= Wet Waste; MSW= Municipal Solid Waste; VFG= Vegetable, Fruit and Garden waste.

Environmental benefits of using compost

The compost produced by both the aerobic composting process and the anaerobic biogasifi-cation processes (after de-watering and aerobic composting) can be used as low-quality fertilis-er (or soil improver). Based on the amount of marketable compost produced, the model calculates and credits the recycling column of the inventory results with the savings in environ-mental burdens associated with the production of fertiliser containing the equivalent amount of N, P_2O_5 and K_2O as is in an average compost. *The model assumes that the average mois-ture content of compost is 40% and that the average N, P_2O_5 and K_2O content of compost is 1.18, 0.68 and 0.9 % dry weight, respectively* (see Table 19.4). Based on these figures, the amounts of N, P_2O_5 and K_2O available in 1 tonne of wet compost are 0.71, 0.41 and 0.54%, respectively. The emissions to air associated with the production of fertilisers are presented in Table 19.7.

The data presented in Table 19.7 includes the emissions from the fertiliser production processes and the emissions from energy utilisation of each of the production processes.

Sorting residue

This will consist of two types of materials: (1) non-biodegradable materials arriving as nuisance materials in biowaste, or materials in mixed waste that have not been recovered as secondary materials; and (2) degradable material (organic or paper) that is either unsuitable for biological processing (e.g. too large) or is removed adhered to nuisance materials. There is little data available that distinguish between these types, however. Where the feedstock is source-sepa-rated biowaste, a nuisance level of around 5% is typical (Chapter 9). Where the feedstock is mixed waste such as MSW then the level of sorting residue is likely to be much higher, although where recovery of other materials occurs (e.g. La Voulte, Figure 19.4), residue rates as low as 4–5% may be found.

In the LCI model it is assumed that all of the categories other than paper and organics are removed as residue during the pre-sort. In addition, 2.5% of the organ-ic and 2.5% of the paper fractions are added to the residue to account for material that is not readily biodegradable, or that adheres to the nuisance materials as they are removed. These figures can be edited by the user if better data are available.

Compost-refining residue

This represents the composted/digested output that is not marketed. The amount will range from zero, if a use can be found for all of the compost, to 100% of the output if no market can be found.

Air emissions

The major air emission by volume from biological processing will be carbon dioxide, which is a contributor to the greenhouse effect. In aerobic processing the organic material is broken down directly to carbon dioxide and water. In anaerobic processing, biogas containing methane and carbon dioxide is produced, of which the methane also forms carbon dioxide when burned.

The amount of emissions per tonne of process input will depend on the moisture content of the incoming material. *In the following calculations an average moisture content of 50% is assumed.* The actual level will depend on the ratio of paper to wet organic material present, but ORCA (1992a) suggest that 50% is the optimum moisture content level for com-posting feedstocks.

IWM2 Model Guide

Air emissions	N fertiliser (1 kg)	7.1 kg N (equivalent to 1 tonne compost)	P_2O_5 (1 kg) fertiliser	4.1 kg P_2O_5 (equivalent to 1 tonne compost)	K_2O (1 kg) fertiliser	5.4 kg K_2O (equivalent to 1 tonne compost)	Total emissions (equivalent to 1 tonne compost)
	(1)	(2)	(3)	(4)	(5)	(6)	Sum of (2)+(4)+(6)
CO_2	2404	17068.4	448	1836.8	443	2392.2	21297.4
CH_4	0.45	3.195	0.018	0.0738	0.02	0.108	3.3768
N_2O	9.63	68.373	0.031	0.1271	0.0089	0.04806	68.5482
SO_2	3.3	23.43	8.25	33.825	0.012	0.0648	57.3198
CO	2.15	15.265	0.42	1.722	0.2	1.08	18.067
NOx	9.64	68.444	3.42	14.022	0.54	2.916	85.382
Particles	0	0	0.041	0.1681	0.028	0.1512	0.3193
HCl	0.11	0.781	0.016	0.0656	0.048	0.2592	1.1058
NH_3	4.93	35.003	0.0016	0.00656	0.0011	0.00594	35.0155
Dioxins	1.19E-09	8.45E-09	1.70E-10	6.97E-10	2.10E-10	1.13E-09	1.03E-08

Table 19.7 Air emissions from chemical fertiliser production. Source: Patyk (1996)

For composting, the dry weight loss during composting is approximately 40%, giving a dry weight loss of 200 kg per wet input tonne. Assuming that most of the organic material decomposed is cellulose, with a carbon content of 44% (from formula), composting will evolve approximately 323 kg of CO_2 (164 Nm^3) per tonne of wet organic feedstock.

For biogasification, the dry weight loss varies with process type, and reports vary from 45 to 70%. Assuming a mid-range dry weight loss of 55% means that 275 kg of organic matter is converted into gas. If all this was converted to carbon dioxide the total emitted would be 444 kg (226 Nm^3) per tonne of digester input. This does not agree with published biogas production figures, however. Given the composition of biogas in Table 19.3, combustion will convert the methane to carbon dioxide and water in the reaction:

$$CH_4 + 2O_2 = CO_2 + 2H_2O$$

Complete combustion of the biogas will therefore produce 0.982 Nm^3 of CO_2 per Nm^3 of biogas burned, equivalent to 1.93 kg of CO_2. Given a production of around 100 Nm^3 of biogas per tonne of organic material feedstock, this produces a CO_2 emission of 193kg per input tonne, considerably less than that predicted from the dry weight loss during the process. This discrepancy probably reflects some process losses, and more importantly the aerobic maturation stage that follows the anaerobic stage. During this stage the material needs to be aerated and heats up, demonstrating considerable aerobic microbial activity, during which further carbon dioxide is likely to be released.

For the purposes of the LCI model, overall carbon dioxide emissions are assumed to be 320 kg and 440 kg per tonne of wet organic material, for composting and biogasification, respectively.

Few complete data sets are available at the time of writing with respect to air emissions from composting processes, although the odour problems that can occur around compost plants demonstrate that air emissions are significant. Air emissions resulting from the combustion of biogas are presented in Table 19.3.

Feedstock	Biowaste (narrow definition)	Biowaste + 10% (by wt) paper	Biowaste + 20% paper	Biowaste + 30% paper
Leachate production (litres/tonne)	13.5	1.6	0	0
Leachate composition:				
COD (mg O_2/l)	33,100	30,200	–	–
BOD_5 (mg O_2/l)	19,000	19,000	–	–

Table 19.8 Effect of including paper in biowaste on leachate production during windrow composting. Source: Verstraete *et al.* (1993)

Process	Composting				Biogasification			
	Worm Compsting	Box composting	Drum composting	Tunnel reactor	Dry one-stage	Dry one-stage	Dry one-stage	Wet two-stage
Amount (litres/tonne)	0	300	n.a.	n.a.	290	490	540	500
Composition (mg/l)								
BOD_5		270–485	50–600	3300–7050	<65	n.a.	740	60
COD		458–808	150–7000	6200–15.100	<250	n.a.	1400	200
NH_4		48–117	n.a.	n.a.	<100	n.a.	250	100
N total		0–1	6–36	0–3	<100	n.a.	6	n.a.
pH		7.9	7.1–7.8	7.1–8.1	n.a.	n.a.	8.0	n.a.
Source*	1	2	2	2	2	2	2	2

Table 19.9 Water emissions from biological treatment processes. Sources: *1, Schauner (1995); 2, Bergmann and Lentz. (1992)

Water emissions

The aqueous effluents reported for biological treatment vary widely in both amounts and composition, depending on both the process used and the feedstock. In composting, considerable evaporation will take place during the process. Any run-off collected is often sprayed back onto the composting material to maintain sufficiently high moisture contents. If waste paper is included in the feedstock, this will absorb much of the water, and so little or no leachate is actually produced (see Table 19.8).

In biogasification, water is produced when the digested material is pressed or filtered. Large amounts will be produced, especially in the wet (low solids) process type. Some of this water will be recirculated to adjust the water content of the incoming feedstock, but the rest needs to be treated prior to discharge. Typical amounts and compositions of the leachates produced by both composting and biogasification are presented in Table 19.9.

Economic costs

Data on the economic costs of biological treatment are not always reported on a consistent basis, so comparisons are difficult to make. In many cases biological treatment is considered as a final disposal option, and consequently costs given are as an all-inclusive 'gate fee' or 'tip fee'. This cost will include allowance for any revenues collected from the sale of recovered materials, compost, and energy from biogas utilisation, and include disposal costs for any residues requiring thermal treatment or landfilling. The problem with this level of accounting is that the cost of biological treatment will vary with the market prices of energy, compost and recovered materials and the cost of landfill. Alternatively, cost data for biological treatment can refer to the biological processing itself (e.g. as in ORCA, 1991b). This is more useful when modelling the economics of the overall waste management system, since it is independent of the cost of other parts of the system, but this type of data is not widely available.

Biological Treatments

The Biological treatments section of the model begins with a window that provides a breakdown of the input materials available to the composting and biogasification processes. Again this has been included in the model to ensure complete transparency for the user.

Tab 1 Process Input (Screen 17)

Biological Stream Input:
  (tonnes)
The values in the grey boxes show the total material available to the biological treatment processes at this stage in the Life Cycle. The glass, metal, plastic, textile and other fractions in the biological stream are due to contamination from the collection and sorting processes.

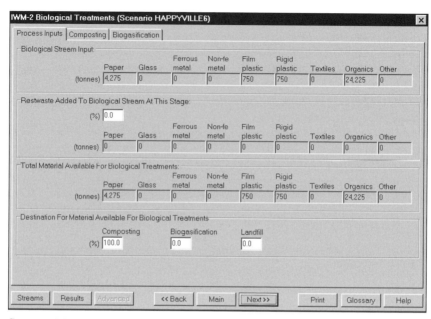

Screen 17 Biological Treatments – Process Input.

Restwaste Added To Biological Stream At This Stage:
  (%)
  (tonnes)
If restwaste is to be added to the **Biological** stream, the user must enter the percentage of the total **Restwaste** stream here. The model calculates this amount, and displays it in the second row of grey boxes.

Total Material Available For Biological Treatments:
  (tonnes)
The model adds the restwaste fraction to the material available in the biological stream and displays the new total amount of material available for biological treatment here.

Destination For Material Available To Biological Treatments:
  (%)   Composting   Biogasification   Landfill (calculated)
This allows the user to identify what fraction of the total material available for biological treatment is sent to either composting or biogasification. If the sum of composting and biogasification does not equal 100%, then by default the remaining material is sent to landfill. This allows users to identify collection systems that may result in the collection of more biological material than their existing facilities can treat, and alter the waste management system accordingly.

IWM2 Model Guide

Tab 2 Composting (Screen 18)

IWM-2 Biological Treatments (Scenario HAPPYVILLE6)   ⊠

Process Inputs | Composting | Biogasification

Composting Input And Presort:

	Paper	Glass	Ferrous metal	Non-fe metal	Film plastic	Rigid plastic	Textiles	Organics	Other
Plant input (tonnes)	4,275	0	0	0	750	750	0	24,225	0
Presort recovery (%)	0.0	0.0	0.0	0.0	0.0	0.0	0.0	0.0	0.0
Presort residue (%)	2.5	100.0	100.0	100.0	100.0	100.0	100.0	2.5	100.0
Process input (tonnes)	4,168							23,619	

Composting Process:

Mass loss (%)* 50.0

Compost produced (tonnes) 13,894

Compost marketable (%) 100.0

Energy consumption (kWh/tonne of plant input) 30

* due to moisture loss and degradation. Range: 30% to 60%

Residue:

	Incineration	Landfill
Sorting residue treatment (%)	0.0	100.0
Compost residue treatment (%)	0.0	100.0
Transport distance (km each way)	0	5
Transport cost (£/tonne)	0.0	1.0

Revenue From Recovered Materials:

	Glass	Ferrous metal	Non-fe metal	Film plastic	Rigid plastic	Textiles
(£/tonne)	0.0	0.0	0.0	0.0	0.0	0.0

Costs:

Processing (£/tonne plant input) 2.0

Market price for compost (£/tonne) 8.0

Streams | Results | Advanced | << Back | Main | Next >> | Print | Glossary | Help

Screen 18 Biological Treatments – Composting.

Composting Input and Presort:
Plant input (tonnes)
Presort recovery (%)
Presort residue (%)

The model shows the composting plant input in the first row of grey boxes. The user inserts the level of recovery for glass, metals and plastics during any **Pre-sort recovery** process. If no recovery occurs, these boxes are left blank. The model assumes that all of the material except for the paper and organics fractions are removed as **Pre-sort residue**, along with 2.5% of the organic material and 2.5% of the paper (to account for material that is not biodegradable or adhering to nuisance materials).

Process input (tonnes)   Paper   Organics

The model calculates the remaining amounts of both paper and organic material and displays these as **Process inputs** (in grey boxes).

Composting Process:
Mass loss (%)
Compost produced (tonnes)
Compost marketable (%)

Due to the wide range of different composting processes available, the model requires the user to input the **Mass loss** (due to both moisture loss and degradation of the organic fraction) of

IWM2 Model Guide

the particular process being modelled. If this figure is not known, the default value is a generic figure of 50%. The model calculates the total amount of compost produced. The user specifies the percentage of final compost that is sold as opposed to treated as a residue; the model's default value is set to zero.

User Note: the amount of compost that is sold is given a recycling credit, based on the production of chemical fertiliser containing an equivalent amount of N, P and K. This means that the emissions that are avoided by using compost instead of chemical fertiliser are credited to the waste management system. For consistency within the model, if compost is sold then a recycling credit is added to the recycling section of the model, as is done for the sale of other recycled materials (such as paper, plastic and metals, etc.) see Chapter 22 for details.

Energy consumption (kWh/tonne of plant input)

The energy consumption per tonne of the composting plants input is added to the **Fuel** stream. Again, a very generic figure (30 kWh/input tonne) is used by the model as a default value.

Residue: **Incineration**   **Landfill**
   Sorting residue treatment (%)
   Compost residue treatment (%)
   Transport distance (km each way)
   Transport cost (£/tonne)

The user specifies how much of each residue is treated by incineration or landfilled directly. The user must also insert the one-way distance in km from the biological treatment plant to the incinerator or landfill. The model calculates the fuel consumption assuming a 20-tonne load and that no return load is carried, and adds the fuel used to the **Fuel** stream. The transport cost is added to the costs stream.

Revenue From Recovered Materials:
   (£/tonne)

If any income is generated by the sale of material recovered prior to the composting process it must be entered here.

Costs:
   Processing costs (£/tonne plant input)
   Market price for compost (£/tonne)

The **Processing cost** is the overall cost for pre-sorting and composting, not including revenues from sale of products, or costs of transporting residues. (If only an inclusive cost is available, this can be inserted if the revenue and other cost boxes on this tab are left blank). These costs and revenues are inserted by the user to calculate the overall cost of composting.

Tab 3 Biogasification (Screen 19)

As can be seen from Screen 18, the **Biogasification** tab is similar to the **Composting** tab. The **Plant input**, which was defined in the **Process Inputs** tab is calculated by the model and displayed in the first row of grey boxes. The only differences in this tab are that the default value for **Mass loss** is 60% (higher than the aerobic composting process as the organic material is digested, pressed to remove excess moisture then composted); the default value for

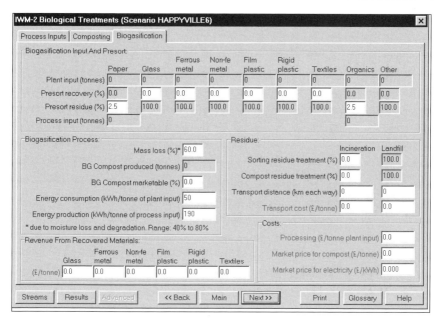

Screen 19 Biological Treatments – Biogasification.

energy consumption by the process is 50 kWh/tonne (higher than for aerobic composting) and the model provides a default value of 190 kWh/tonne energy production from the process. The user is also required to enter the market price, £/kWh, for any surplus electricity that is exported from the biogasification process. These costs and revenues are entered by the user to enable the model to calculate the overall cost of the biogasification process described.

CHAPTER 20

Thermal Treatment

Summary

The major environmental burdens of mass burn incineration, Refuse-Derived Fuel burn incineration and Paper- and Plastic-Derived Fuel burn incineration are considered in this model. The model does not include approaches such as pyrolysis, gasification and the incineration of waste in cement kilns due to the relative lack of suitable operational data available at the time of writing. It is recognised that as these approaches become more common they should be included in the model. The system boundaries, inputs and outputs and energy balance of Mass Burn incineration, Refuse-Derived Fuel burn incineration and Paper- and Plastic-Derived Fuel burn incineration are discussed.

Defining the system boundaries

Before the inputs and outputs of thermal treatment can be assessed, it is necessary to define the system boundaries for the process. These are shown in Figure 20.1, and include the

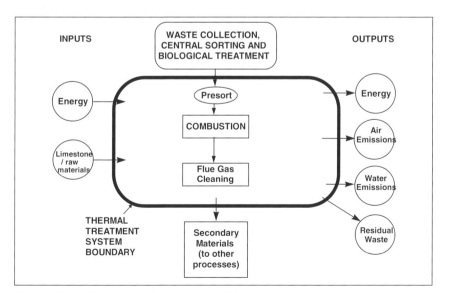

Figure 20.1 The system boundaries of the thermal treatment processes.

emission abatement processes within the system. The actual emissions to air will therefore depend on the type of emission abatement equipment installed, and the efficiency of its operation. For this study, emission data from both old incinerators and modern, state-of-the-art equipment are available to the user, but as the overall objective is to assess possible effects of future IWM schemes, rather than what is currently being achieved it is expected that the modern incinerator data will be used most often. As with all aspects of this LCI approach, if local data on existing or planned incinerators are available, these should be used in preference to generic figures.

The amounts of solid and aqueous wastes generated will also depend on the type of gas-cleaning process used: a dry-scrubbing system will generate only solid waste whereas a wet gas-scrubbing system will produce both solid and aqueous wastes.

For the purpose of this LCI, it is assumed that recovered energy is used to generate electricity only, and that this electricity is exported from the system.

Data availability

The outputs of the incineration process clearly also depend on the inputs, i.e. what is burned. Most incinerators are mass-burn facilities, with an input of mixed MSW, and data are generally available on the associated outputs. However, as fractions of the waste stream are removed for recycling or composting, or when RDF or source-separated materials are burned as fuel, the input to the incineration process may be markedly different (see Chapter 12).

To predict the outputs from such different combustion scenarios it is necessary to be able to attribute energy production and emissions to individual materials or fractions within the incinerator's input. Whilst this is possible for energy production, emissions cannot be reliably attributed for two reasons. Firstly, there is a serious lack of data in this area, since most experience has been with the mass-burn approach. Secondly, there are interactions between different materials in the combustion process, such as in the *de novo* production of dioxins, and these mechanisms are not yet fully understood.

Available data are presented for incineration of mixed MSW, dRDF and for combustion of individual materials. It needs to be borne in mind, however, that not all emission data sets are complete, and that incineration of mixtures may not give rise to the sum of their parts, since interactions may occur.

Waste Inputs

The input to the thermal treatment system will be either MSW, Refuse-Derived Fuel (RDF) or source-separated paper and plastic depending on the process involved. Burning RDF or paper and plastic as fuels involves a relatively consistent and well-defined feedstock. The input to MSW incinerators, however, comprising of restwaste and residues from other waste treatment processes, is much more variable. This feedstock will vary with the waste composition, which has been shown to vary both geographically and seasonally (Chapter 8).

Further variability in the input to MSW incinerators will also occur when materials are recovered for recycling or composting, so affecting the composition of the remaining restwaste. This will affect the overall calorific value of what is burned, and has led to concern over the compatibility of recycling with Energy from Waste schemes (e.g. ENDS, 1992). Whilst any materials recovery and recycling scheme will reduce the overall throughput of a mass-burn incinerator, the average calorific content of the throughput may rise or fall, depending on which fractions are removed. Several studies have looked at this effect and in general, recovery of glass and steel for recycling will increase the average calorific content, as will separate collection of putrescible material for composting.

The calorific value of the feedstock, and its variability, needs to be considered when planning an MSW incinerator. Problems with the calorific value of the waste input rising above the heat capacity of the incinerator plant have been encountered in both Switzerland and Japan. In both these countries, plastics now need to be excluded from the feedstock to prevent damage to some incinerators. In Greece, by contrast, the high level of putrescible waste has resulted in a calorific value too low to give effective burning.

Energy consumption

The thermal treatment process, as well as liberating energy from the feedstock, also consumes energy. This is required for operating the cranes, moving grates or fluidised beds, fans for air injectors, and emission control equipment, as well as for general heating and lighting. The on-site energy consumption will particularly reflect the level of flue-gas treatment. For MSW incineration, ETSU (1993a) suggest that around 14% of the electrical power generated is consumed on-site, with a specific consumption of around 70 kWh per tonne incinerated. Energy is also required in the form of natural gas, to heat up the incinerator during start-up. For this ETSU (1993a) give an average figure of 0.23 m^3/per tonne. *For the purposes of the LCI model, it will be assumed that there is an electric energy consumption of 70 kWh per tonne incinerated and a natural gas consumption of 0.23 m^3/per tonne incinerated.*

For RDF and alternate fuel (paper/plastic) burning, there will be some energy consumed by the flue-gas cleaning equipment, although no appropriate data are available for this. It is likely to be lower than for MSW incineration, since there is less plant to operate. *For the purposes of the LCI model, it will be assumed that there is an energy consumption of 20 kWh per tonne of RDF or paper/plastic fuel burned.*

Outputs

Energy

Energy released during incineration may be used for several purposes (heat and steam production, electricity generation), each with its own conversion rate for the amount of useful energy produced. *For the purposes of this LCI, the user can select whether the recovered heat is used to generate electricity or both electricity and steam and alter the gross energy recovery rate accordingly.*

IWM2 Model Guide

Mass burn

Energy production will depend on the composition of MSW, and will be higher in countries with high levels of paper and especially plastic (generally Northern Europe), and lower where there is a high level of wet organic/putrescible waste (Southern Europe). For Germany values in the range 8.3–9.2 MJ/kg are given (Bechtel and Lentz, unpublished), Sweden 7–11 MJ/kg (Svedberg, 1992), and for the UK a gross calorific value of 8.4–10 MJ/kg has been reported (Barton, 1986; Porteous, 1991). It is necessary to take account of the water content of MSW as received at the incinerator, since this will reduce the overall calorific value per kilogram, due to both its weight (dilution effect) and latent heat of evaporation. In the case of the UK, this reduces a typical gross calorific value of MSW from 8.4 MJ/kg to 7.06 MJ/kg. For comparison, the calorific value for fuel oil is 40–42 MJ/kg, for hard coal 25–30 MJ/kg, for peat 9–13 MJ/kg and for wood chips 7–13 MJ/kg (Svedberg, 1992).

In the LCI model, to allow for variability in the feedstock, the calorific value of the input to MSW incinerators will be calculated from the composition, using the material-specific calorific values given in the Advanced Variables window – Thermal Treatment tab of the model.

RDF

Ranges given for the calorific value of RDF vary from 11-18 MJ/kg (Swedish data; Svedberg, 1992) to 18–20 MJ/kg (UK data; Ogilvie, 1992). The exact value will vary with the composition of the original waste, and the process used to produce the RDF. *For the purpose of the LCI model, a calorific value of 18 GJ/tonne will be assumed for RDF.*

Source-separated fuel

The calorific value of this alternative fuel will depend on the ratio of paper to plastic burned. *In the LCI model the net calorific values of each fraction given in the Advanced Variables window – Thermal Treatment tab of the model are used to calculate the calorific values of any specified mixture.*

Energy recovery

Clearly not all of the primary heat energy released in an incinerator can be recovered in a useful form, and the level of recovery depends on the use to which the energy is put. Since boilers attached to municipal waste incinerators must operate at lower steam temperatures to reduce corrosion, incinerators producing electricity only have a conversion efficiency of around 20% (RCEP, 1993). ETSU (1993) predict a gross power production of 520 kWh per tonne of waste with a net calorific value of 8.01 GJ/tonne, giving a conversion efficiency of 23% for a plant of 200,000 tonnes per year capacity. Once the on-site power consumption of 70 kWh/tonne has been subtracted, this gives a net export of 450 kWh/tonne from the site.

Energy recovery for district heating schemes, such as used extensively in Japan, recover around 70% of the energy released, whereas combined heat and power (CHP) schemes, which utilise the residual heat after generation of electricity, achieve an overall conversion efficiency of around 70–90%. This means that a plant such as in Frankfurt, burning 420,000 tonnes of waste per year can provide 30,000 people with both electricity and district heating, replacing the need for some 64,000 tonnes of fuel oil per year (Bechtel and Lentz, unpublished).

For the purposes of the LCI model, a Gross electricity production efficiency of up to 30% is recommended, depending on the age of the facility, etc. for MSW incinerators recovering electricity, whereas a Gross electricity production efficiency of up to 90% is recommended, again depending on the age of the facility, etc. for MSW incinerators recovering both electricity and steam. For plants burning RDF or a paper/plastic fuel, an efficiency of 30% will be assumed due to the more homogenous and controlled nature of the feedstock.

Energy recovered during incineration is credited as an energy gain in the LCI model and is assumed to displace electricity production that depends on conventional fuel. The exact mix of the displaced energy can be specified by the user in the **Advanced window – General tab**. Therefore net emissions from the incineration process are the emissions from the process minus the emissions that would have been produced by the mixture of electricity selected by the user.

Air emissions

It is important to remember that due to the conservation of mass, whatever materials enter the incineration process will leave it in one state or another. Although organic pollutants may be broken down into harmless molecules, heavy metals within the waste stream will either be left in the ash (clinker), removed as filter dust or will escape to the air. The total amount produced will remain the same, so as flue gas cleaning becomes more effective and less inorganic pollutants are emitted to the atmosphere, the amounts collected from the flue gas cleaning process will correspondingly increase.

Mass burn

The most significant pollutants emitted from unsorted MSW incineration are the acid gases (hydrogen chloride, sulphur dioxide, nitrogen oxides), carbon dioxide, particulate matter (PM), dioxins/dibenzofurans (PCDD/PCDFs) and heavy metals (mercury, cadmium, lead) (Clayton, 1991).

The approach used in the LCI model to estimate emissions from the incineration process is based upon the model developed by the US EPA and described in 'Application of Life Cycle Management To Evaluate Integrated Municipal Solid Waste Strategies' (Research Triangle Institute, 1997). This approach splits the emissions into non-metal emissions (CO, SO_2, NO_X, HCl, PM, dioxins/furans) and metal emissions (As, Cd, Cr, Cu, Hg, Ni, Pb, Zn). Allowing a different modelling approach to be taken for each group of emissions.

The non-metal emissions CO_2, CO, SO_2, NO_X, HCl, PM and dioxins/furans are calculated using a stoichiometric approach involving a combustion equation and ultimate analysis of the components of solid waste (RTI, 1997, Appendix F, Combustion Process Model, pp B2–3). **With the exception of CO_2 for which there are no air pollution controls, the non-metal pollutants are assumed to be controlled to specific concentration levels that are independent of the waste stream composition.**

Carbon dioxide emissions are based upon a stoichiometric equation for waste combustion for each waste component (RTI, 1997, Appendix F, Combustion Process Model, pp. B10–11). Flue gas production is also based upon this stoichiometric approach. The 'emission factors' for CO_2 and flue gas production are presented in Table 20.1.

The tonnages of each material in the incinerator process input are multiplied by the emission factor for CO_2 and flue gas to calculate the emission of CO_2 and flue gas.

	Paper	Glass	Fe metal	Non-Fe metal	Film plastic	Rigid plastic	Textiles	Organics	Other
CO$_2$ emission factors (kg/tonne)	1279	59	0	0	2740	2652	1280	586	1280
Flue gas (dry std m^3/tonne)	5016	228	0	0	12991	12532	5206	2409	5206

Table 20.1 Emission factors for CO$_2$ and flue gas. Note: figures based on aggregated data. Source: RTI (1997)

The non-metal pollutants are assumed to be controlled to specific concentration levels (as facilities are operated to comply with target emission levels) that are independent of the waste stream composition. For these pollutants the user may select default values based upon either the US Federal Standards (1995) for Municipal Waste Combustors (Subpart Eb) or the average performance of a new facility. These values are presented in Table 20.2.

The 'emission factors' for each of these non-metal pollutants is calculated by multiplying the standard (for either US Federal or New Facility) by the volume of flue gas generated per tonne of material combusted. These 'emission factors' are shown for each material component in Table 20.3 (US Federal Regulations) and 20.4 (emissions from a new facility).

Metal emissions are based on the metals composition of individual waste components. **The model assumes that the emissions attributed to a component are in proportion to its metals content. This also assumes that the tendency of a metal to volatise and escape through the stack is the same regardless of how the metal is bound to the waste.**

The metal emissions to air are calculated by multiplying the emissions of metal per tonne of waste component before gas cleaning (uncontrolled emission factors) by the removal efficiency for that metal of the air pollution control devices (RTI, 1997, Appendix F, Combustion Process Model, pp. C1–4). The uncontrolled emission factors for metals are presented in Table 20.5.

Emission	Units	US Federal Regulations Total	Av. Emissions from a new facility Total
SO$_2$	mg/Nm^3	88	23
HCl	mg/Nm^3	41	15
NOx (as NO)	mg/Nm^3	308	279
Dioxins/furans	ng/Nm^3	13	4.5
CO	mg/Nm^3	125	33
PM	mg/Nm^3	24	4

Table 20.2 Default target values for non-metal emissions. Source: RTI (1997)

	Paper	Glass	Fe metal	Non-Fe metal	Film plastic	Rigid plastic	Textiles	Organics	Other
SO_2	0.429	0.019	0	0	1.109	1.07	0.445	0.206	0.445
HCl	0.204	0.009	0	0	0.528	0.509	0.212	0.098	0.212
NOx (as NO)	1.005	0.046	0	0	2.604	2.512	1.043	0.483	1.043
Dioxins/furans	6.50E–08	2.96E–09	0	0	1.68E–07	1.62E–07	6.75E–08	3.12E–08	6.75E–08
CO	0.625	0.028	0	0	1.62	1.563	0.649	0.3	0.649
PM	0.109	0.005	0	0	0.311	0.299	0.124	0.057	0.124

Table 20.3 Emission factors based upon US Federal Regulations (kg/tonne waste component) Note: figures based upon aggregated data. Source: RTI (1997)

	Paper	Glass	Fe metal	Non-Fe metal	Film plastic	Rigid plastic	Textiles	Organics	Other
SO_2	0.114	0.005	0	0	0.296	0.286	0.118	0.055	0.118
HCl	0.072	0.003	0	0	0.188	0.181	0.075	0.035	0.075
NOx (as NO)	0.948	0.041	0	0	2.361	2.277	0.946	0.437	0.946
Dioxins/furans	2.25E–08	1.02E–09	0	0	3.50E–07	1.69E–07	2.33E–08	1.08E–08	2.33E–08
CO	0.168	0.007	0	0	0.168	0.406	0.169	0.078	0.169
PM	0.02	0.001	0	0	0.051	0.15	0.021	0.009	0.021

Table 20.4 Emission factors based on performance of new facility (kg/tonne waste component) Note: figures based upon aggregated data. Source: RTI (1997)

Metals (kg/tonne)	Paper	Glass	Fe metal	Non-Fe metal	Film plastic	Rigid plastic	Textiles	Organics	Other	Removal efficiency (%) New incinerator
As	1.65E–06	9.00E–06	6.55E–03	1.32E–03	1.07E–06	1.06E–06	1.29E–03	7.35E–06	1.29E–03	99.9
Cd	4.08E–05	3.27E–04	3.59E–03	2.51E–03	1.01E–03	4.51E–04	3.12E–03	4.60E–04	3.12E–03	99.7
Cr	9.40E–05	1.56E–03	1.32E–03	2.21E–03	2.11E–04	8.40E–05	1.29E–03	3.38E–04	1.29E–03	99.3
Cu	4.93E–06	7.90E–06	6.25E–02	1.22E–04	5.90E–06	5.65E–06	1.19E–02	7.85E–05	1.19E–02	99.6
Hg	2.55E–04	1.39E–04	3.21E–03	2.26E–04	9.99E–05	9.85E–05	7.60E–04	4.31E–04	7.60E–04	92.7
Ni	1.60E–04	4.80E–04	1.84E–03	4.31E–04	1.52E–04	1.26E–04	4.99E–04	2.66E–04	4.99E–04	96.6
Pb	2.65E–03	4.34E–03	2.38E–02	2.43E–03	5.50E–03	3.20E–03	1.90E–02	5.80E–03	1.90E–02	99.8
Zn	1.50E–03	2.55E–03	1.24E–01	2.15	4.31E–03	2.94E–03	3.33E–02	5.99E–03	3.33E–02	99.7

Table 20.5 Uncontrolled emission factors (kg metal/tonne waste component) and air pollution control removal efficiency.
Note: figures based upon aggregated data. Source: RTI (1997)

It is possible that the use of this methodology will result in an estimate of controlled emissions that breach an emissions limit. The model does not warn the user of any such violation. Care must be taken when modelling the incineration process that the process input, derived from previous stages of an Life Cycle Inventory strategy, does not result in breach of emissions standards.

RDF and source-separated fuels

Data for burning RDF and paper and plastic are presented (as well as emissions data for MSW incineration as a comparison) in Table 20.6. Note that the data for RDF burning are from UK trials with only dust precipitation equipment. Use of gas-scrubbing equipment would reduce airborne emissions, but would increase solid waste generation from the filters.

Water emissions

This arises only from use of wet gas-scrubbing equipment. Figures given for the amount of sewage produced range from 200 to 770 litres per tonne of waste input (SPMP, 1991; ETSU, 1993; Bechtel and Lentz, unpublished). *In the LCI model it is assumed that all water emissions are treated on-site, with only resultant sludges leaving the site.*

Solid waste

Solid residues from incineration arise from two main sources: combustion residues (bottom ash and fly ash) and solid residues from the gas-cleaning system. The latter will include both filter dusts and sludge residues resulting from water treatment (where wet gas-scrubbing systems are installed).

Mass burn

Mass burn of typical MSW results in 250–300 kg of bottom ash per tonne of waste (SPMP, 1991; IFEU, 1992; ETSU, 1993, IAWG, 1997). To allow for variability in the mass burn incinerator feedstock in an IWM system, however, *the amount of bottom ash in the LCI model is calculated from the ash content of each fraction of the waste in the Advanced window – Thermal Treatment tab of the model.*

Dry gas cleaning systems produce approximately 45–52 kg of dust and residues (calcium chloride and surplus lime) per tonne of waste. The semi-dry or semi-wet processes are similar, producing 40 kg of loose ash mixed with calcium chloride and surplus lime per tonne of waste. Wet gas-scrubbing results in 20–30 kg of dust and 2.5–12 kg of sludge residue per tonne of waste (SPMP, 1991; IFEU, 1992; IAWG, 1997; Bechtel and Lentz, unpublished). *In the LCI model it is assumed that for each tonne of MSW incinerated, 20 kg of filter dust and 12 kg of sludge residues from the gas-scrubbing system are produced.*

RDF

RDF combustion will typically leave a residue of approximately 86 kg ash/carbon per tonne input, with a further 1.8 kg of ash filtered from the flue gas (Ogilvie, 1992). Data are not available for residues from further emission control, but are likely to be similar to those above for mass burn. *It is therefore assumed that a further 12 kg of sludge residues result from gas scrubbing, giving a total of 13.8 kg from the gas-cleaning process.*

IWM2 Model Guide

Source	Municipal Solid Waste (g/tonne burned) IFEU (1992)*	Municipal Solid Waste (g/tonne burned) EC Directive 89/369/EEC (1989)†	Refuse-Derived Fuel (RDF) (g/tonne burned) Hickey and Rampling, (1989)‡	Paper/board (g/tonne burned) Habersatter (1991)§	Plastic (no PVC) (g/tonne burned) Habersatter (1991)§
Particulates		150	1,710	20	50
CO		500	347	400	1,250
CO_2			1,421,000		
CH_4					
NO_x	500		2,019	1600	5,000
N_2O					
SO_x	125	1500	5,197	300	360
HCl	25	250	55,70	100	0
HF	2.5	100			
H_2S					

Table 20.6 Air emissions from combustion of MSW, Refuse-Derived Fuel (RDF) and source-separated paper and plastic.
*based on plant meeting or exceeding German 17th Emission Regulation, and assuming a production of 5000 m³ of stack gas per tonne of MSW incinerated; †calculated for plant meeting EC Directive for new MSW incinerator plants, assuming 5000 m³ of stack gas per tonne incinerated; ‡data are averages of three trials of burning dRDF, on as received basis with average moisture content of 11%. Note gas clean-up was limited to cyclone for particulates only; §data assumes flue gas cleaning capable of removing 90% of SOx and 95% of HCl; ¶Limits are for the sum of heavy metals, so the limit has been divided evenly between the metals combined.

Source	Municipal Solid Waste (g/tonne burned) IFEV (1992)*	Municipal Solid Waste (g/tonne burned) EC Directive	Refuse Derived Fuel (RDF) (g/tonne burned) Hickey & Rampling (1989)	Paper/board (g/tonne burned) Habersatter (1991)	Plastic (no PVC) (g/tonne burned) Habersatter (1991)
Total HC	0.0384	100	46		
Chlorinated HC	0.0836				
Dioxins/furans	0.0000005		0.0000678		
Ammonia					
Arsenic	0.008	2.5¶			
Cadmium	0.0135	0.5¶	2.8		
Chromium	0.013	6.25¶			
Copper	0.55	6.25¶			
Lead	0.245	6.25¶	63.2		
Mercury	0.0085	0.5¶			
Nickel	0.002	2.5¶			
Zinc	0.85				

Table 20.6 (continued) Air emissions from combustion of MSW, Refuse-Derived Fuel (RDF) and source-separated paper and plastic.

*based on plant meeting or exceeding German 17th Emission Regulation, and assuming a production of 5000 m^3 of stack gas per tonne of MSW incinerated; †calculated for plant meeting EC Directive for new MSW incinerator plants, assuming 5000 m^3 of stack gas per tonne incinerated; ‡data are averages of three trials of burning dRDF, on as received basis with average moisture content of 11%. Note gas clean-up was limited to cyclone for particulates only; §data assumes flue gas cleaning capable of removing 90% of SOx and 95% of HCl; ¶limits are for the sum of heavy metals, so the limit has been divided evenly between metals combined.

Source-separated fuel

Ash contents of different fractions of the waste stream are given in the Advanced window – Thermal Treatment tab of the model. To these must be added the solid waste generated from flue gas cleaning, to give total solid waste produced. Figures given by Habersatter (1991) for total solid waste from combustion in Switzerland are typically 35 kg/tonne input for plastics, and 87 kg/tonne for paper/cardboard, but these are more conservative than those given for the ash content alone by Barton (1986) for UK waste. *In the LCI model, therefore, the bottom ash production is calculated from the ash content levels in the Advanced window – Thermal Treatment tab of the model; the gas-cleaning residues are assumed to be the same as for RDF combustion above.*

Economic costs of thermal treatment

Mass burn

The economic costs of incineration are generally considered to be high, because of the capital investment required to set up a plant. When calculated per tonne of waste input, economic costs of mass-burn incineration vary widely across Europe (Table 20.7). Variability within countries results from four key factors: incinerator capacity, level of gas-cleaning equipment installed, whether energy is recovered or not, and whether economic instruments exist to encourage the generation of power from waste. With the implementation of the EC directives on new and existing MSW incinerators, all facilities now need to have extensive gas-scrubbing equipment, so making incineration more expensive, and less variable in cost. The majority of new incinerators will have energy recovery facilities. The economics of incineration with energy recovery are further improved when there are fiscal instruments designed to encourage the use of waste as a 'renewable' energy resource. For example, in the UK the Non-Fossil Fuel Obligation (NFFO) of the 1989 Electricity Act required electricity supply companies to purchase electricity generated from non-fossil fuel (including MSW). This guaranteed a market at a price premium for the Energy-from-Waste projects accepted under NFFO, which was worth 6–13 euro per tonne of

Country	Average cost (euro/tonne)
Sweden	30
Denmark	42
France	45
Belgium	70
The Netherlands	78
Germany	125
Switzerland	130

Table 20.7 Average cost for mass-burn incineration. Note: euro is at 1995 value. Source: Juniper (1995)

waste input (DoE, 1993). A similar scheme operates in Germany, whereby energy generated from waste can be supplied to the grid at any time for a guaranteed premium price.

RDF and source-separated materials

The original aim of RDF was to produce a readily transportable fuel that could be sold as a commodity alongside other fuels such as coal, hence the need for pelletising. This requires markets, however, which have proved difficult to develop for RDF. One reason is that in several countries when burned, it is still legally considered to be waste, and therefore subject to the same emission controls as mass-burn incineration, so differs from the combustion of coal and other fuels. As a result, surviving RDF plants burn the pellets on site, and export electricity into the national grid, and/or district heating. The additional cost involved here is the cost of operating the dedicated boiler, less the revenue from the sale of electricity. Since many RDF plants integrate RDF production and combustion on one site, these costs may not be separated out.

The concept of burning source-separated materials collected for recycling has not been fully explored, and here too markets for the paper and plastic fuel must be developed. If this can be done, this energy producing outlet would fix a lower economic value for this material, i.e. its calorific value. If the recycling markets for these materials are weak due to oversupply/lack of demand, the material could be sold for its calorific content (given that the calorific content of a 50:50 mix of paper and plastic would have about the same calorific content per kg as industrial coal).

The additional costs that need to be considered here are those for operating the boiler and associated emission control equipment, net of any revenues from the sale of exported electricity. Since this process has yet to be fully developed, no appropriate data on such costs are available.

Thermal treatments

Tab 1 Process Inputs (Screen 20)

Thermal Stream Input:

(tonnes)

The values in the grey boxes show the total material available to the thermal treatment processes at this stage in the Life Cycle. The compost fraction in the thermal stream input is the compost residue that cannot be sold.

Restwaste Added To Thermal Stream At This Stage:

(%)

(tonnes)

If restwaste is to be added to the **Thermal** stream, the user must enter the percentage of the total remaining **Restwaste** stream here. The model displays the composition of this amount of restwaste in the second row of grey boxes.

Total Material Available For Thermal Treatments:

(tonnes)

The model adds the restwaste fraction to the material available in the **Thermal** stream and displays the new total amount of material available for thermal treatment here.

IWM2 Model Guide

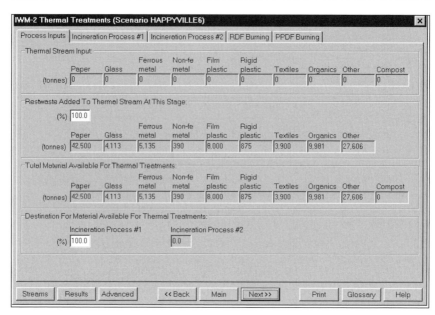

Screen 20 Thermal Treatments – Process Inputs.

Destination For Material Available To Thermal Treatments:
(%)   Incineration (1)   Incineration (2)
This allows the user to identify what fraction of the total material available for thermal treatment is sent to new or old incineration facilities. If the sum of the two incineration processes does not equal 100%, then by default the remaining material is sent to landfill. This allows users to identify systems that may result in the availability of more material suitable for thermal treatment than their existing facilities can treat, and enables them to alter the waste management system accordingly.

Tab 2 Incineration #1 (Screen 21)
Incineration Input and Presort:
Plant input (tonnes)
Presort residue (%)
Process input (tonnes)
The model calculates and displays the composition of the input to the incineration plant as defined in the **Process Inputs** tab. If a pre-sort of the material entering the incinerator occurs the user can enter the percentage of each material that is discarded as residue. This allows the user to account for unsuitable material such as large metal objects or potentially hazardous material that may damage the process equipment or pose a risk to the operators. The model calculates and displays the composition of the final process input.

IWM2 Model Guide

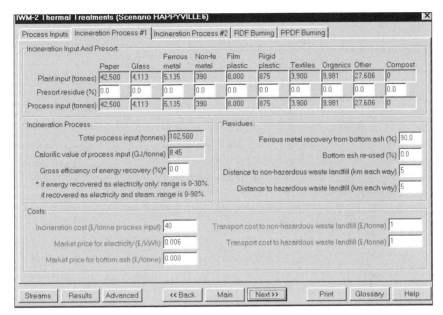

Screen 21 Thermal Treatments – Incineration.

Incineration Process:
  Total process input (tonnes)
  Calorific value of process input (GJ/tonne)
  Gross efficiency of energy recovery (%)*
*if energy recovered as electricity only: range is 0–30%; if recovered as electricity and steam: range is 0–90%

The model displays the total process input and calculates from this the total calorific value of this material. No default value for the gross efficiency of energy recovery is given as this depends very much on the technology used. The model does provide ranges for energy recovered as electricity only and energy recovered as electricity and steam. The model uses the figure entered here by the user to calculate the total energy balance of the incinerator and adds this to the fuel stream.

Residues:
  Ferrous metal recovery from ash (%)
  Ash/clinker re-used (%)

The model calculates the amount of ferrous metal recovered and adds this to the **Products** stream. The amount of ash or clinker re-used is subtracted from the residue needing disposal.

IWM2 Model Guide

Distance to non-hazardous waste landfill (km each way)
Distance to hazardous waste landfill (km each way)

The model calculates the fuel consumption for residue disposal assuming that a 20-tonne load is used and that no return load is carried.

Costs:
Incineration cost (£/tonne process input)

The cost entered should include any revenue from sales of electricity and ferrous metal, but not the cost of residue transport or disposal. Where revenue includes both electricity and steam generation, it would be appropriate to replace the price of electricity with a weighted price to take into account the larger price paid for steam (heat) energy.

Transport cost to non-hazardous waste landfill (£/tonne)
Transport cost to hazardous waste landfill (£/tonne)

The model calculates the transport costs assuming that fly ash, filter dust and gas-cleaning residues are sent to a hazardous waste landfill, while bottom ash is sent to a non-hazardous waste landfill.

Tab 3 Incineration #2

This screen is exactly the same as the **Incineration #1** tab, but allows the user to input different default values for the operation of the incineration process. The default data sets included in the model are for: (1) an incinerator that meets the US EPA regulations on air emissions; and (2) a new incinerator with high-specification gas-cleaning technology that performs considerable better than the minimum regulations (Table 20.2). This gives users the ability to model two different incinerators within a single scenario.

Tab 4 RDF Burning (Screen 22)
RDF Burning Process:

Amount of fuel burned (tonnes/year)	**cRDF**	**dRDF**
Calorific value of process input (GJ/tonne)	**cRDF**	**dRDF** (calculated)
Gross efficiency of energy recovery (%)*	**cRDF**	**dRDF**

*if energy recovered as electricity only: range is 0–30%; if recovered as electricity and steam: range is 0–90%

The amount of cRDF and dRDF produced earlier in the model is displayed here by the model. As in the **Incineration** tabs above, no default value for the gross efficiency of energy recovery is given as this again depends very much on the technology used. The model does provide ranges for energy recovered as electricity only and energy recovered as electricity and steam. The model uses these data to calculate the amount of electricity that will be produced from the burning of RDF.

Residues:
Distance landfill for non-hazardous waste (km each way)
Distance to landfill for hazardous waste (km one way)

The model calculates the fuel consumption for residue disposal assuming that a 20-tonne load is used and that no return load is carried.

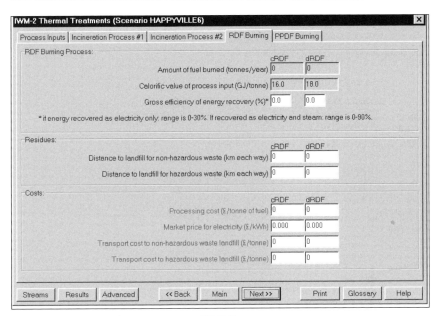

Screen 22 Thermal Treatments – RDF Burning.

Costs:
Processing cost (£/tonne fuel)
Market price for electricity (£/kWh)
If the revenue from energy sales exceeds the cost of operating the RDF-fired boiler and power generation plant, this profit should be inserted as a negative cost. Costs for residue disposal should not be included as they are entered below.

Transport cost to landfill for non-hazardous residues(£/tonne)
Transport cost to landfill for hazardous residues (£/tonne)
The model calculates the transport costs assuming that bottom ash is sent to a non-hazardous waste landfill, while fly ash, filter dust and gas-cleaning residues are sent to a hazardous waste landfill.

Tab 5 PPDF Burning (Screen 23)
This tab describes the information necessary to model the burning of source-separated paper and plastic as fuel.

PPDF Burning Process:
Amount of fuel burned (tonnes/year)   Paper   Plastic   Total
Calorific value of process input (GJ/tonne)
Gross efficiency of energy recovery (%)*
*if energy recovered as electricity only: range is 0–30%; if recovered as electricity and steam: range is 0–90%

IWM-2 Thermal Treatments (Scenario HAPPYVILLE6)   [×]

Process Inputs | Incineration Process #1 | Incineration Process #2 | RDF Burning | PPDF Burning

PPDF Burning process:

	Paper	Plastic	Total
Amount of fuel burned (tonnes/year)	0	0	0
Calorific value of process input (GJ/tonne)	0.00		
Gross efficiency of energy recovery (%)*	0.0		

* if energy recovered as electricity only: range is 0-30%. If recovered as electricity and steam: range is 0-90%.

Residues:

Distance to landfill for non-hazardous waste (km each way) 0

Distance to landfill for hazardous waste (km each way) 0

Costs:

Processing cost (£/tonne of fuel) 0

Market price for electricity (£/kWh) 0.000

Transport cost to non-hazardous waste landfill (£/tonne) 0

Transport cost to hazardous waste landfill (£/tonne) 0

Streams | Results | Advanced | << Back | Main | Next >> | Print | Glossary | Help

Screen 23 Thermal Treatments – PPDF Burning.

The model displays the amounts of paper and plastic that were not sent for recycling but were recovered in the MRF sorting section. As in the **Incineration** tabs above, no default value for the gross efficiency of energy recovery is given as this depends very much on the technology used. The model does provide ranges for energy recovered as electricity only and energy recovered as electricity and steam. The model uses this data to calculate the amount of electricity that will be produced from the burning of the paper and plastic-derived fuel.

The layout of the **Residues** and **Costs** sections are the same as in the **RDF Burning** tab.

CHAPTER 21

Landfilling

Summary

The environmental burdens associated with the landfill process, including leachate generation and treatment and biogas generation and treatment, are considered. The system boundaries, inputs and outputs and energy balance are discussed.

Defining the system boundaries

This chapter describe the inputs and outputs of the process, in both environmental and economic terms, so that the Life Cycle Inventory and economic assessment of solid waste systems can be completed. Considering landfilling as a waste treatment process, rather than simply a sink for the final disposal of solid waste, the inputs and outputs for the process are shown in Figure 21.1.

The environmental burdens of landfilling waste will depend both on the landfill design and method of operation, and the nature of the waste deposited (see Table 21.1). While there has

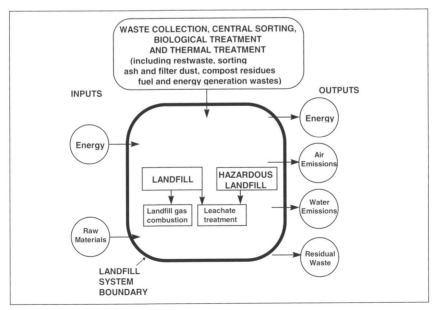

Figure 21.1 System boundaries of the landfilling process.

Inputs

Restwaste (can be total MSW)

Sorting residues

Biologically treated material

Thermal treatment residues (ash, filter dust and other residues from gas cleaning)

Outputs

Landfill gas from:

– biodegradable fraction in MSW	250 Nm^3/tonne
– biologically treated material	100 Nm^3/tonne
Leachate	150 litres/tonne

Inert solid waste from landfilled waste

Solid waste residue from leachate treatment

Table 21.1 Landfill inputs and outputs

been a lot of attention paid to the details of landfill site location and the design of various systems for gas and leachate containment and/or treatment, there has been less emphasis on the effect of future changes in the composition of the waste materials that will be destined for landfills. As more materials are recovered from waste streams for recycling, both the amount and the composition of the residual waste, from both household and commercial sources, may alter significantly.

Although some data are available for the inputs and outputs of the landfilling process, these generally refer to the landfilling of mixed waste streams, such as MSW. Data relating to the individual material fractions within such waste streams are not normally available. Thus it is difficult to extrapolate from the effects of present waste streams to the environmental effects of future waste streams with different compositions. This will be attempted, however, in this section: by collating data from landfills containing mixed waste streams and the limited amount of data on the behaviour of individual materials within landfills, it should be possible to estimate the outputs from landfills for a range of different waste inputs.

Waste inputs

There are four main waste streams from municipal solid waste management systems that are landfilled: Restwaste, Sorting residues, Biological treatment residues and Ash. Landfills are also likely to receive some industrial wastes, sewage sludges, etc., but these are not included within the boundaries of the system under study here, except those resulting from energy or raw material consumption by the waste management system itself.

Restwaste

Restwaste or residual waste is collected from both household and commercial sources and landfilled directly. This will vary in composition with geography and time of year and according to what fractions of the waste stream have been recovered or removed for biological or thermal treatment. Where the landfill is located near to the collection area the restwaste will be delivered directly by the collection vehicles; in the case of a distant landfill, the restwaste may be bulked up at a transfer station (where bulky wastes may also be crushed) ready for transport to the landfill in larger capacity trucks, rail cars or barges.

Sorting residues

Sorting residues from waste sorting processes at Materials Recovery Facilities (MRFs) or Refuse-Derived Fuel (RDF) plants (Chapter 10), or from pre-sorting processes at biological treatment plants (Chapter 11).

Biological treatment residues

Residual material from composting or biogasification processes. This is the stabilised organic material produced that does not become valorised as compost, due to contamination or lack of a suitable market.

Ash

Ash from thermal processing, whether it has been burned as mixed MSW, as RDF or as source-separated material (Chapter 13). This will include the bottom ash (or clinker/slag), which is disposed of in non-hazardous landfills and the fly ash and residues from gas cleaning systems, which are disposed of with the bottom ash in some countries but in others are deposited separately in hazardous waste landfills.

Solid waste from energy production or raw material manufacture

Solid waste generated during the production of energy, fuel and other raw materials (such as the plastic for refuse sacks), which are consumed within the system is another source of solid waste resulting from the waste management system itself, which is included in Figure 21.1. In a full 'dustbin to grave' analysis these also need to be taken into consideration. The data given for the amounts of these wastes (Chapter 9) often do not specify the composition of these wastes, though it is likely that, in most countries, a large part will be comprised of ash from energy generation plants. While some of these materials may have possible further uses, or alternative methods for treatment, it is assumed for the purposes of this study that they are all landfilled.

Energy consumption

The landfilling process will consume energy both in the form of vehicle fuel and electricity. For household waste that is landfilled directly, where the distance from the collection area to the landfill site is large, a transfer station may be used to bulk up the waste for more efficient transport by larger truck or rail. No data are available on the energy consumption of transfer stations, although generic fuel consumption data for road transport (Chapter 9) can be used to estimate fuel consumption for onward transport to the landfill site. For all waste types landfilled, fuel and electricity will also be consumed in the operation of the site itself. Data suggest that the fuel consumption for the landfilling process is around 0.6 litres of diesel per cubic metre of void space filled (Biffa Waste Services, 1994).

Outputs

The inputs to a landfill system occur over a limited time period – essentially the working life of the site. The outputs from the system occur over a much longer time span, which may involve

IWM2 Model Guide

at least tens and maybe even hundreds of years. The outputs calculated in the following sections are therefore integrated over time, since the gas and leachate produced by each tonne of waste landfilled will eventually be released.

Landfill gas production

Gas production

This must be considered separately for the types of waste that are landfilled. As well as posing a local health and safety concern, landfill gas has a more global environmental effect. Consisting mainly of methane and carbon dioxide, both 'greenhouse gases' (especially the former), landfill gas has become significant in the debate over global warming and climate change. Methane has been reported to be responsible for about 20% of recent increases in global warming (Lashof and Ahuja, 1990) and landfills are thought to be a major source of methane. In the UK, landfills are the single largest source of methane, contributing an estimated 46% of total production in 1996 (AEA, 1999), with over 1000 sites reported to be actively producing gas (Brown, 1991). Globally, it has been estimated that methane from decomposition of municipal solid waste, whether in crude dumps or organised landfills, could account for 7–20% of all anthropogenic methane emissions (Thorneloe, and referencess therein, 1991).

Landfill gas from Municipal Solid Waste, Restwaste and Sorting residues

In the literature, the amount of landfill gas generated per tonne of municipal solid waste deposited has been estimated by three different methods:

1. by theoretical calculations using the amount of organic carbon present in the waste
2. from laboratory-scale lysimeter studies
3. from gas production rates at existing landfills.

Not surprisingly, therefore, there is considerable variability in the estimates of landfill gas production (Table 21.2). Theoretical yields tend to be high (e.g. Gendebien et al., 1991), since they often assume that all of the degradable material does break down, but there may well be pockets within a landfill where little decomposition occurs due to insufficient moisture content. Lab-scale lysimeter studies use actual refuse, but are not likely to reflect fully the conditions existing in a real landfill. Data from existing landfills should be the most appropriate, but these too are difficult to interpret.

Gas production rates vary over the active life of a landfill (Figure 21.2), so it is necessary to extrapolate from measured current gas production rates to the total gas production integrated over this active period. Gas yields from landfill sites also do not fully reflect the amount of gas generated within the landfill, since the yield will also depend on the gas collection efficiency. Estimates of collection efficiencies vary (20–25%, De Baere et al., 1987; 40–70%, Carra and Cossu, 1990: 40–90%, Augenstein and Pacey, 1991; 40%, RCEP 1993), and will depend on size, shape and engineering design of the landfill site. For the purposes of the LCI model the default value for landfill gas collection is set to zero; this represents a landfill with no gas collection system. Users modelling landfills with gas collection systems can either input their own data or enter a figure of between 40 and 90 %, as supported by the references above.

Given the different methods used to reach the estimates of gas production, Table 21.2 shows some consensus on the amounts of gas produced by landfilling MSW, at around

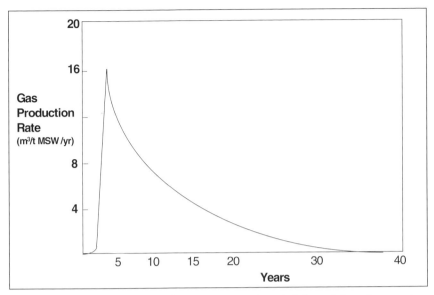

Figure 21.2 Landfill gas production rates estimated from laboratory scale experiments. Source: Ehrig (1991).

150 Nm^3 per tonne of waste as received (i.e. including moisture content). Some laboratory studies produced only 120 Nm^3 per tonne (Ehrig, 1991), but equally, the average gas yield from landfill sites in the UK has been estimated to be 222 Nm^3 per tonne (Richards and Aitchison, 1991), so the amount of gas actually produced may well be in excess of 150 Nm^3 per tonne of Municipal Solid Waste.

Landfill gas is only produced from the biodegradable fractions of MSW, however, which are essentially the putrescible organic fraction, the paper and board fraction and any non-synthetic textiles. Together these typically constitute around 60% by weight of MSW in Europe (Chapter 8). Therefore, gas production for these fractions might be expected to average around 250 Nm^3 per tonne. Other fractions such as glass, plastics and metals will probably affect the rate of decomposition, since their presence is likely to facilitate water percolation of the waste and diffusion of gases, but they will not markedly affect the total level of decomposition over time. Lysimeter studies of these putrescible fractions have shown similar levels of gas production from food waste (191–344 Nm^3/tonne) and cardboard (317 Nm^3/tonne), but lower values for magazines (100–225 Nm^3/tonne) and newspaper (120 Nm^3) (Ehrig, 1991). Comparisons with the biogasification process (Chapter 11) are also useful. In a much shorter time span (≤ 20 days) up to 150 Nm^3 of biogas can be produced per tonne of biomass in the accelerated process under controlled conditions. As degradation in a landfill occurs over a much longer time, more complete decomposition is likely to occur than in the biogasification process, and there are also likely to be some process losses in biogasification. **A figure of 250 Nm^3 landfill gas per tonne of biodegradable waste (organic, paper and textile fractions) is therefore considered realistic and used in the LCI model.**

Waste fraction	Landfill gas production Nm³/tonne (wet material)	Data type	Source
MSW	372	Theoretical calculation	Gendebien et al. (1991)
MSW	229	Theoretical calculation	Ehrig (1991)
MSW	270	Calculated from Italian data	Ruggeri et al. (19910
MSW	120–160	Laboratory scale experiments	Ehrig (1991)
MSW	190–240	Measured at landfills	Ham et al. (1979)
MSW	60–180	Measured at landfills	Tabasaran (1976)
MSW	222	Mean UK landfill yield	Richards and Aitchison (1991)
MSW	135	Estimated average	IFEU |(1992)
MSW	200	Estimated average	de Baere et al. (1987)
MSW	100–200	Estimated average	Carra and Cossu (1990)
Food waste	191–344	Laboratory scale experiment	Ehrig (1991)
Grass	176	Laboratory scale experiment	Ehrig (1991)
Newspaper	120	Laboratory scale experiment	Ehrig (1991)
Magazines	100–225	Laboratory scale experiment	Ehrig (1991)
Cardboard	317	Laboratory scale experiment	Ehrig (1991)
Composted MSW	133	Laboratory scale experiment	Ehrig (1991)
Composted organic fraction	176	Laboratory scale experiment	Ehrig (1991)

Table 21.2 Production of landfill gas from MSW and selected waste fractions

Landfill gas from biologically treated material

Where biological treatment has been used to reduce the volume of waste for disposal, or where markets cannot be found for composted or anaerobically digested residues, these materials are also landfilled, and biochemical degradation will continue within the landfill. There are few estimates of the amount of landfill gas produced by such material. In laboratory studies, partially composted MSW ('a few days in a technical compost reactor') produced 133 Nm^3 of landfill gas (compared to 160 Nm^3 for untreated MSW in the same test), while a partially composted organic fraction produced 176 Nm^3 (Ehrig, 1991). For residues from biogasification and composting, IFEU (1992) used an estimate for landfill gas production of 20Nm^3 per tonne of restwaste entering the biogas/composting process. If, as in some biogasification processes (Chapter 11) the residue represents 20% of the input, this would give a gas production of 100 Nm^3 per tonne of residue landfilled. **In the LCI model, this figure of 100 Nm^3 per tonne of residue landfilled will be used for gas production from landfilled residues from both composting and biogasification processes.**

Landfill gas from ash

Given complete combustion, any ash entering a landfill should contain no organic carbon. Therefore no landfill gas should be generated. Not all combustion processes will completely remove the carbon, **but it will be assumed for the purposes of this model that no landfill gas is produced from ash.**

Landfill gas composition

Landfill gas is produced by the anaerobic decomposition of biodegradable organic material. As different anaerobic reactions proceed at varying rates, the composition of the gas released will vary through the different phases of the active life of a landfill site, as shown in Figure 21.3. It will also vary with the type of waste contained, but a typical landfill gas composition is given in Table 21.3. The major component is methane, which usually comprises 50–55%, followed by carbon dioxide which makes up most of the remaining volume. In addition more than 100 different

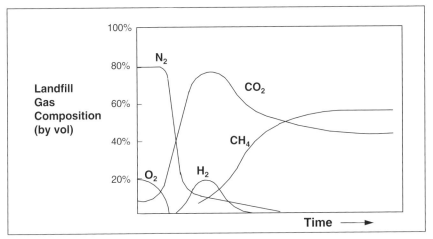

Figure 21.3 Production pattern for components of landfill gas. Adapted from DOE (1989).

Component	% by volume	
Methane	52.8	
Carbon dioxide	44.10	
Oxygen	0.5	
Nitrogen	2.00	
	Mean value ppm*	Maximum value ppm*
Actylonitrile	0.41	26.10
Benzcne	2.81	144.03
Carbon tetrachloride	ND	0.89
Chlorobenzene	0.93	40.0
1,2 -Dichloroethane	0.12	3.6
1,1,1-Trichloroethane	0.41	7.70
1,1-Dichloroethane	4.72	30.00
1,1,2,2-Tetrachloroethane	ND	3.8
Chloroethane	1.81	22.0
Chloroform	ND	3.11
1,1-Dichloroethene	0.14	2.00
Trans-1,2-dichloroethene	3.81	27.0
Ethylbenzene	8.16	69.94
Methylene chloride	21.15	366.0
Chloromethane	1.33	17.4
Flurotrichloromethane	1.21	15.8
Dichlorodifluromethane	15.10	160.0
Tetrachloroethene	7.18	33.4
Toluene	51.66	659.39
Trichloroethene	3.65	31.0
Vinyl chloride	6.75	42.0
Total xylene isomers	17.05	144.0
Methyl ethyl ketone	9.09	86.0
Methyl isobutyl ketone	1.43	13.0
Acetone	6.51	56.0
Chlorodifluromethane	3.03	32.0
Dichlorofluromethane	4.23	50.8
Hydrogen sulphide	21.0	100.0
Ethane	850.68	1780.0
Propane	24.16	328.0
Butane	4.93	36.0
Pentane	5.74	133.0
Hexane	7.21	35.55

Table 21.3 Landfill gas analysis.
*42 sites, 142 samples. Source: WMI (1994)

IWM2 Model Guide

volatile organic compounds have been identified as trace components, many of which are known to be toxic or carcinogenic. The actual trace components vary according to the landfill and the landfilled waste, but the major ones found are hydrogen sulphide, vinyl chloride, benzene, toluene, trichlorethane and mercaptans (Willumsen, 1991).

If no collection system exists for gas control or energy recovery, all of the landfill gas will eventually leak out of the site and enter the atmosphere. Where gas collection occurs, around 40% will typically be recovered (see above), whilst the remaining 60% still enters the atmosphere.

Gas control and energy recovery

In the simplest form of gas control, the collected gas is flared on the site to destroy both the combustible fractions of the gas and most of the organic trace components. Any methane, carbon monoxide and hydrogen in the landfill gas should be converted to carbon dioxide and water if combustion is complete. Comparing the performance of three different flaring systems, Baldwin and Scott (1991) found a large difference in combustion efficiency, however, with one system releasing 16% of the flare exhaust as unburned methane. In an efficient flare, however, all of the methane was shown to be burned. An efficient flare was also shown to significantly reduce, but not remove completely, levels of trace components from 4427 mg/m^3 to 32.8 mg/m^3 (a 99.3% reduction) (Table 21.4). Analysis of the flared gas from several systems also showed low levels (< 10 mg/m^3) of components not present in the unburned gas, such as methyl cyanide, nitromethane, acrolein, ethylene oxide and some alkynes, which must be formed *de novo* in the flaring process.

Landfill gas combustion control is enforced by regulatory requirements in Germany, The Netherlands, Switzerland, France and some states in the USA. Enclosed flare technology was developed in response to these requirements. Although enclosed flaring costs approximately four times as much as open flaring, combustion is more controlled and can be monitored more effectively (ENDS, 1999). The UK Environment Agency guidelines stipulate combustion at a minimum of 1000°C for at least 0.3 seconds. Table 21.5 summarises the landfill gas flare regulatory requirements in Germany and Switzerland.

Where the collected landfill gas is used for energy recovery, this can involve either heating applications (steam raising via boilers, kiln firing or space heating), or power generation systems. Power generation systems can involve spark-ignition or dual-fuel diesel engines, or gas turbines (in increasing order of generation capacity). Before the landfill gas is burned in these engines it is normally compressed and dewatered, which also removes most of the trace contaminants to protect the engines from acid gases and particulate matter. Comparison of the exhaust emissions from the different types of gas engines suggests that levels of contaminants can differ widely, as in Table 21.6.

The composition of landfill gas (with a typical methane content of around 55%) is similar to that of biogas (Chapter 11), so has a similar heat content. Given a calorific value of 37.75 MJ/Nm^3 for methane (Perry and Green, 1984), landfill gas has an energy content of 20.8 MJ/Nm^3. The UK Department of Environment (1989) gives a figure of 15–21 MJ/Nm^3 of landfill gas, depending on its methane content. **A heat content of 18 MJ/Nm^3 will be assumed for this analysis**. This amount of thermal energy is released on combustion; the amount of useful energy resulting will depend on whether the gas is used for heating or power generation purposes. **For the purposes of this study, it will be assumed that where energy recovery**

Total concentration (mg/m^3)

Components	Flare A LFG before flaring	Flare A Flared gas (adjusted for dilution)	Flare B LFG before flaring	Flare B Flared gas (adjusted for dilution)	Flare C LFG before flaring	Flare C Flared gas (adjusted for dilution)
Alkanes	920	0	370	39	510	1.2
Alkenes	400	0	170	1.1	230	18
Alcohols	180	2.1	1.3	55	7.3	6.1
Amines	0	0	0	0	0	0
Aromatic hydrocarbons	1600	6.1	380	58	350	2.6
Alkynes	0	0	0	0	0	3.9
Cycloalkanes	43	0	0	0	5.1	0
Carboxylic acids	0.8	0	0.96	0	0	0
Cycloalkenes	530	0	120	5.6	79	0.96
Dienes	1.7	0	0	0.14	10	0
Esters	290	0	0.2	0	0.5	0
Ethers	1.8	0	0.4	0	0.2	0.32
Halogenated organics	320	1.9	32	18	39	3
Ketones	120	1.6	2.9	2.2	1	0.43
Organosulphur compounds	19	21	2.6	6.7	3.9	4.6
Others	1.4	0	0.5	2.5	0.73	7.29
Totals	4427	32.8	1100	190	1200	48

Table 21.4 Landfill gas control by the use of flares: trace components in landfill gas before and after flare combustion at three different sites. Source: Baldwin and Scott (1991)

	UK EA guidelines	German TA Luft	Swiss regulations
Carbon monoxide (mg/Nm^3)	50	50	60
Nitrogen oxides (mg/Nm^3)	150	200	80
Unburnt hydrocarbons (mg/Nm^3)	10	10	20
Dust (mg/Nm^3)	–	10	20
Sulphur dioxide (mg/Nm^3)	–	50	50
Hydrochloric acid (mg/Nm^3)	–	30	20
Cadmium (mg/Nm^3)	–	0.05	0.1
Mercury (mg/Nm^3)	–	0.05	0.1
Dioxins and furans (TEQ) (ng/m^3)	–	0.18	–

Table 21.5 Emissions controls on landfill gas flares. Source: ENDS (1999)

Site Power generation Plant type	A Dual-fuel diesel engine	B Spark-ignition gas engine	C Gas turbine
Gaseous component	4.3	125	9
Particulates (mg/Nm^3)	800	c. 10,000	14
CO (mg/Nm^3)			
Total unburnt HC (mg/Nm^3)	22	>200	15
NOx (mg/Nm^3)	795	c. 1170	61
HCl (mg/Nm^3)	12	15	38
SO$_2$ (mg/Nm^3)	51	22	6
Dioxins (ng/Nm^3)	0.4	0.6	0.6
Furans (ng/Nm^3)	0.4	2.7	1.2

Table 21.6 Emissions from power generation plants using landfill gas. Notes: Differences in emissions reflect both differences between gas engine types and differences in the quality of the incoming landfill gas. At sites A and B the gas was dried, filtered, compressed and cooled before use. At site C the gas was passed through a wet scrubber to remove acid gases, then compressed, cooled, filtered and heated to 70°C before combustion. All three sites were in the UK. Source: Young and Blakey (1991)

IWM2 Model Guide

occurs it involves the burning of landfill gas in a gas engine to generate electricity, which is then exported into the grid. A conversion efficiency of 30% will be assumed (ETSU, 1995); this is in line with the value used for the biogas engine in Chapter 11 (Schneider, 1992; Schön, 1992), giving an electrical energy recovery of 1.5 kWh per Nm³ of landfill gas collected.

Leachate

As with landfill gas, it is difficult to provide 'typical' figures for the generation of leachate from landfilled wastes, since both the amount and composition of leachate will depend on many factors, including the nature of the waste landfilled, the landfilling method and level of compaction, the engineering design of the landfill, and the annual rainfall of the region.

Leachate production

The amount of leachate produced within a landfill will depend mainly on the rainfall of the area, how well the landfill is sealed (especially the cap), and the original water content of the waste deposited. Data on the amount of leachate produced by actual landfill sites are not commonly reported, but IFEU (1992) estimates that around 13% of the rainfall on a landfill site emerges as leachate. For sites in Germany with an average annual rainfall of 750 mm, this would produce around 100 litres of leachate per square metre of landfill site, per year. Using an estimated 20-metre depth of landfilled waste, with an approximate density of 1 tonne/m³, this gives a leachate production of 5 litres per tonne of landfilled waste per year. If the active period for leachate production is around 30 years, the total amount of leachate produced would be 150 litres per tonne of waste. *The model therefore assumes that 150 litres of leachate is produced per tonne of waste landfilled.*

Leachate composition

More data are available on the composition of landfill leachate than on the volume produced, but since the leachate composition depends mainly on the nature of the waste landfilled, reported leachate data differ widely. As with landfill gas, leachate composition also varies with the stage of decomposition of the waste: the initial acidification stage is characterised by low pH, along with high levels of organic matter (high Biological Oxygen Demand (BOD) and Chemical Oxygen Demand (COD) values), calcium, magnesium, iron and sulphate, which all decline as the methanogenic stage is reached (Spinosa et al., 1991).

A range of composition data from Municipal Solid Waste leachate are given in Table 21.7, from which the complexity of leachate mixtures can be seen. A range of volatile organic materials plus up to 46 non-volatile organic substances have been analysed from a single landfill site (Öman and Hynning, 1991). Since in many cases the biochemical pathways involved in the creation of leachate substances are not known, it is not possible to allocate individual pollutants to the different fractions of Municipal Solid Waste. One exception is the organic content of the leachate (BOD/COD), which is derived from the biodegradable fractions, i.e. from the putrescible organic, paper and textile fractions. *For the purposes of modelling, therefore, it is assumed that all of the BOD produced originates from these three fractions. All other leachate components are assumed to arise equally from all of the MSW material fractions that are landfilled, since it is not possible to identify their source with any degree of certainty.*

Component	MSW restwaste	Compost/biogas residues	MSW bottom ash
Aluminium	2.4	2.4	0.024
Ammonium	210	10	0.06
Antimony	0.066	0.051	0.051
Arsenic	0.014	0.007	0.001
Beryllium	0.0048	0.0048	0.0005
Cadmium	0.014	0.001	0.0002
Chlorine	590	95	75
Chromium	0.06	0.05	0.011
Copper	0.054	0.044	0.06
Fluorine	0.39	0.14	0.44
Iron	95	1.0	0.1
Lead	0.063	0.012	0.001
Mercury	0.0006	0.00002	0.001
Nickel	0.17	0.12	0.0075
Zinc	0.68	0.3	0.03
AOX	2.0	0.86	0.011
BOD	1900	1900	24
1,1,1-trichloroethane	0.086	0.0086	0.00086
1,2-dichloroethane	0.01	0.001	0.0001
2,4-dichloroethane	0.13	0.065	0.0013
Benzo (a) pyrene	0.00025	0.00013	0.0000025
Benzene	0.037	0.0037	0.00037
Chlorobenzene	0.007	0.0035	0.00007
Chloroform	0.029	0.0029	0.00029
Chlorophenol	0.00051	0.00025	0.0000051
Dichloromethane	0.44	0.044	0.0044
Dioxins/furans (TEQ)	0.32 ng	0.16 ng	0.0032 ng
Endrin	0.00025	0.00013	0.0000025
Ethylbenzene	0.058	0.029	0.00058
Hexachlorobenzene	0.0018	0.00088	0.000018
Isophorone	0.076	0.038	0.00076
PCB	0.00073	0.00036	0.0000073
Pentachlorophenol	0.045	0.023	0.00045
Phenol	0.38	0.1	0.005
Tetrachloromethane	0.2	0.02	0.002
Toluene	0.41	0.041	0.0041
Toxaphene	0.001	0.0005	0.00001
Trichloroethene	0.043	0.0043	0.00043
Vinyl chloride	0.04	0.004	0.0004

Table 21.7 Composition of landfill leachates from MSW, ash and biologically treated material (mg/litre, except for dioxins/furans). Source: IFEU (1992)

IWM2 Model Guide

Leachate composition data for the other two major types of material from the Municipal Solid Waste stream entering landfills, biologically treated material (composting and biogasification residues) and thermally treated material (ash), are also given in Table 21.7. Note that the residues from thermal treatment will include both bottom ash (clinker), which is relatively inert, and fly ash, which often contains high levels of inorganic pollutants such as heavy metals and salts. In some countries the fly ash is classified as a hazardous waste and subjected to stricter controls during its disposal. No data were available for the leachate composition from fly ash-containing landfills, however. *For the calculation of leachate resulting from ash in landfills in the LCI model, the fly ash and bottom ash amounts are combined, and the leachate composition from bottom ash is used.*

Leachate collection and treatment

Landfills operated on a 'dilute and disperse' basis will release all of the leachate generated into the surrounding soil and rock strata, where the constituent materials may be further broken down by soil micro-organisms, adsorbed onto soil particles or may enter the groundwater system. Most large modern landfills are lined by a geomembrane or layer of compacted clay, however, and operate on a 'containment' basis. The leachate produced within the sealed landfill can either be recirculated to accelerate the process of decomposition, or drained/ pumped out for leachate treatment. *Both recirculation and leachate treatment will consume energy, though these are not included in the present model due to lack of suitable data (if the user has data, this energy should be added to the total energy input for the landfill section).*

Leachate treatment can involve a range of physical (neutralisation, evaporation, drying, etc.) and biological (anaerobic digestion, bio-oxidation) processes to produce an effluent that can be discharged to municipal sewage systems or surface waters. Depending on the process used, the treatment of leachate from the methanogenic phase of landfill activity can produce from 9 to 22 kg of solid residue for every cubic metre treated (Weber and Holz, 1991). These residues can themselves be treated by incineration or landfilling, in which case they will produce further residues and emissions. *For simplicity in the LCI model, any leachate treatment residues will be added to the total amount of final solid waste.*

As was discussed in Chapter 13, it is generally accepted that most landfill liners will eventually leak, so that part of the leachate will be discharged directly into the underlying strata, from where it can contaminate the groundwater. For a lined site, therefore, It is difficult to estimate the proportion of the leachate generated that will be collected and treated, as opposed to leaking into the substrata. The level of leakage will depend on many factors, including the type of liner used (single versus multi-layered, mineral versus synthetic membrane, etc.), the geology of the site (permeability of underlying strata), and the efficiency of any leachate collection and treatment system. Although there is evidence that many lined landfills have leaked, there is little empirical data on leakage amounts. *For the purposes of the LCI model the default value for leachate collection is set to zero; this represents a landfill with no leachate collection system. Users modelling landfills with leachate collection systems should input their own data or can enter an estimated figure.* This is clearly an area where reliable data are urgently needed.

Final inert solid waste

Although it does not physically leave the site, one of the primary outputs of the solid waste management systems described in this book is the final solid waste left in a landfill at the end of all decomposition processes. This will not be the same weight of waste as was originally land-filled, since some of the waste has been degraded and will be released from the landfill as land-fill gas or leachate. The weight of each type of solid waste (mixed MSW, waste sorting residues, biological treatment residues, ash) that enter landfill is known, but the amount remaining after decomposition will depend on how extensive the degradation process has been. **Rather than attempt to predict the weight loss of the waste while in the landfill, the input ton-nages will be used as amounts of final solid waste**.

In any case, the important attribute of the final solid waste is its volume, rather than its weight, since landfill sites fill up rather than get too heavy. Using the specific densities of the different waste materials (Table 21.8) and the known input tonnages, it is possible to calculate the volume of material that is consigned to landfill. Whilst some further compaction and settling of the landfilled material may occur as decomposition occurs, **this volume will approximate**

Material/ fraction	Density (tonnes/m^3)	Specific volume in landfill (m^3/tonne)	Source
MSW	0.9	1.11	Bothmann (1992)
Paper/board	0.95	1.05	Habersatter (1991)
Glass	1.96	0.51	Habersatter (1991)
Metal – aluminium	1.08	0.93	Habersatter (1991)
Metal – ferrous	3.13	0.32	Habersatter (1991)
Plastic	0.96	1.04	Habersatter (1991) (average of resin types)
Textiles	0.7	1.43	(estimated)
Organic	0.9	1.11	(estimated)
Other (MSW)	0.9	1.11	(average for MSW used)
Bottom ash (MSW)	1.5	0.67	Bothmann (1992)
Filter ash/dust	0.6	1.67	Habersatter (1991)
Compost residue	1.3	0.77	Bothmann (1992)
Industrial waste	1.5	0.67	(assumed to be mainly ash)

Table 21.8 Specific densities of MSW fractions and waste materials in landfills

IWM2 Model Guide

to the final volume of solid waste resulting from landfilling, and will be the figure used for the output of the LCI model.

The environmental consequence of the remaining final solid waste is land consumption (or land generation if landfilling is used as a means of land reclamation). If we assume an average depth of waste in a landfill (IFEU, 1992, assuming a depth of 20 m), it is possible to calculate from final solid waste volume to space consumption by landfilling. Since landfills vary widely in geometry, however, depending on whether they are used to reclaim former quarries, clay pits, or as above-ground structures, this conversion will not be attempted, and the environmental burden of producing solid waste will be quoted as a volume requirement in this LCI model.

Economic costs

As with the other waste treatment options discussed earlier, the economic costs of landfilling vary widely across Europe and North America. The variability reflects geographical differences mainly in land costs, landfill design and engineering requirements and labour costs. A detailed model for the evaluation of the economics of landfill gas recovery is available (Milke, 1998), but a far less detailed approach is applied in this model to keep it simple.

The economic costs of landfilling should include the cost of the land, capital costs of site con-struction, operating costs, closure costs and long-term post-closure monitoring and aftercare costs. It is unlikely that most costs currently quoted fully account for all of these costs, in partic-ular those for post-closure monitoring and aftercare. The duration of post-closure monitoring and aftercare may be extensive. In the USA, a 30-year post-closure monitoring period is man-dated, although there can be significant leachate emissions for considerably longer periods. As discussed above, leakages are also quite likely to occur: a survey found that 18% of 1000 UK landfill sites studied had suffered 'significant' pollution incidents or failures (ENDS, 1992). Remedial costs following leakages can also be very expensive. Escapes of landfill gas accounted for most (48%) of the pollution incidents in the above survey, typically costing 65,000–130,000 euro to remediate. Surface water contamination accounted for 27% of incidents, with a typical cost of 6500–20,000 euro, whilst ground water pollution, accounting for 15% of incidents, had costs typically ranging from 65,000 to 1.3 million euro. Several instances of groundwater pollu-tion cost over 1.3 million euro to rectify.

As a result of the above potential costs for remediation work, it is likely that the real cost of landfilling is higher than currently quoted figures. A report by Pearce and Turner (1993) came to the same conclusion and suggested that additional external costs of between 1 and 5 euro per tonne should be added to the current figures. The same is likely to hold, to varying degrees, in other European countries.

Landfilling

Tab 1 Process Input (Screen 24)

As in the other **Process Input** windows this screen is included for transparency.

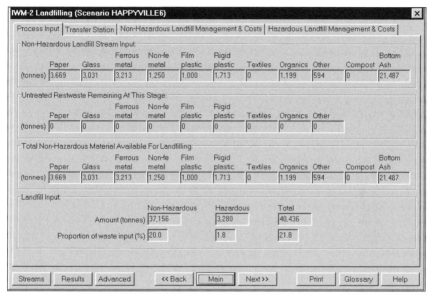

Screen 24 Landfilling – Process Input.

Non-Hazardous Landfill Stream Input:
(tonnes)

The model calculates and displays both the total amount and composition of residues from the collection, central sorting, biological treatment and thermal treatment processes.

Untreated Restwaste Remaining At This Stage:
(tonnes)

The model displays the composition of the remaining restwaste that has not been diverted into any of the previous treatments; this material is added to the **Landfill** stream.

Total Non-Hazardous Material Available for Landfilling:
(tonnes)

The model displays the total amount of non-hazardous material to be landfilled.

Landfill Input:

| Amount (tonnes) | Non-Hazardous | Hazardous | Total |
| Proportion of waste input (%) | Non-Hazardous | Hazardous | Total |

The model calculates the total amounts of non-hazardous and hazardous material entering the landfill. These are the totals used to calculate the air and water emissions and the final solid waste volume. Hazardous waste (fly ash and other gas-cleaning residues from thermal treatment) is accounted for separately, and it is assumed that all of this residual waste will be landfilled directly.

Tab 2 Transfer Station (Screen 25)

Any processing operations associated with transfer stations can be described on this screen.

IWM-2 Landfilling (Scenario HAPPYVILLE6)				☒

Process Input | Transfer Station | Non-Hazardous Landfill Management & Costs | Hazardous Landfill Management & Costs

Landfill Transfer Station:

	Non-Hazardous	Hazardous
Landfill material sent to transfer station (% of landfill input)	0.0	0.0
Landfill material sent to transfer station (tonnes)	0	0
Electrical energy consumption of transfer station (kWh/tonne input)	0.0	0.0
Diesel fuel consumption of transfer station (litres/tonne input)	0.0	0.0
Distance to landfill site from transfer station (km each way)	0.0	0.0

Streams	Results	Advanced		<< Back	Main	Next >>		Print	Glossary	Help

Screen 25 Landfilling – Transfer Station.

Landfill Transfer Station:
Landfill material sent to transfer station (% of landfill input)
Landfill material sent to transfer station (tonnes)
Electrical energy consumption of transfer station (kWh/tonne input)
Diesel fuel consumption of transfer station (litres/tonne input)
Distance to landfill from transfer station (km one way)

The collection and pre-sorting module covered transport of the residual waste to either a local landfill site or a transfer station if a distant landfill site is used. If a transfer station is used, the user must input the amount of energy/fuel used at the station per tonne of waste managed. The model calculates the total amounts of energy/fuel consumed to handle all of the waste, and adds this to the fuel/energy consumption totals within the model.

For onward transport to the landfill, the user enters the one-way distance to the landfill site. The model calculates the fuel consumed, assuming a 40-tonne truck capacity and that no return load is carried.

IWM2 Model Guide

Tab 3 Non-Hazardous Landfill Management and Costs (Screen 26)

Non-hazardous landfilling burdens and costs are accounted for in this window; landfill gas generation and subsequent energy generation along with leachate collection and treatment are also modelled.

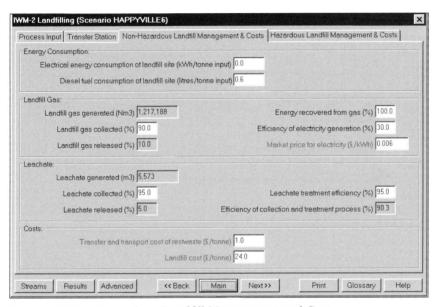

Screen 26 Non-Hazardous Landfill Management and Costs.

Energy Consumption:
   Electrical energy consumption of landfill site (kWh per tonne input)
   Diesel fuel consumption of landfill site (litres per tonne input)
Using this information, the model calculates the total energy/fuel consumption of the landfill operation and adds these amounts to the energy and fuel consumption totals for the overall system.

Landfill Gas:
   Landfill gas generated (Nm^3)   **Calculated**
   Landfill gas collected (%)
   Landfill gas released (%)   **Calculated**
   Energy recovery from gas (%)
   Efficiency of electricity generation (%)
   Market price for electricity (£/kWh)
If landfill gas is not collected, the appropriate amount of gas is added to the overall total air emissions. If gas is collected, the portion that diffuses out of the site (100% minus collection efficiency percentage) is added to the air emissions total. The collected portion is assumed to be burned, whether in a flare, furnace or gas engine, with the resultant air emissions. If energy

is recovered, the appropriate amount of electrical energy is added to the system energy production total.

Leachate:
    Leachate generated (m³)   **Calculated**
    Leachate collected (%)
    Leachate released (%)   **Calculated**
    Leachate treatment efficiency (%)

If the site is unlined, or lined but with no leachate collection and treatment system, it is assumed that all of the leachate produced leaks from the site and enters the substrata. The amounts of the leachate materials are therefore added to the totals for water emissions. If the site is lined and leachate collected and treated, the amount collected is calculated using the collection efficiency estimate, and the resulting effluent and residues added to the respective totals for water emissions and solid waste. The amount not collected is again assumed to leak from the site, and is added to the water emissions totals.

Costs:
    Transfer/transport cost of restwaste (£/tonne)
    Landfill cost (£/tonne)

The user inserts appropriate local unit costs for these operations, per tonne of waste material handled. The cost of landfilling should be inclusive of site purchase, construction, operation, gas sales (if appropriate), gas and leachate treatment, site closure and subsequent monitoring and aftercare. The model multiplies the unit costs by the amounts landfilled, and adds the total to the overall system cost.

Tab 4 Hazardous Landfill Management and Costs

This tab is the same as the previous screen except that the model assumes that no landfill gas is generated by a hazardous landfill.

CHAPTER 22

Materials Recycling

Summary

Energy consumption and emissions for recycled material manufacture are quantified where possible. These are then compared with the energy consumption and emissions associated with the production of an equivalent amount of the virgin material, so that overall savings or additional costs can be calculated.

Defining the system boundaries

The inputs and outputs of the recycling process are presented in Figure 22.1. The recycling processes are included within the system boundaries of this waste management LCI, but the processing of recycled materials into recycled products is outside the system boundaries of this model.

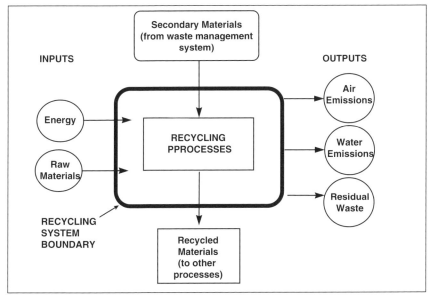

Figure 22.1 System boundaries of the recycling process.

Inputs

Several reports have been produced providing data on the energy consumption and emissions resulting from materials recycling (e.g. Habersatter, 1991; Henstock, 1992; Ogilvie, 1992; Porteous, 1992; Boustead, 1993a, b; BUWAL, 1998). Before using such data in the context of this chapter, however, it is necessary to determine their relevance.

In many cases the aim of the studies (e.g. Henstock, 1992) has been to compare the environmental burdens of producing recycled versus virgin material. This comparison is made on a 'cradle-to-produced material' basis: for virgin material this is from extraction of raw materials, whilst for recycled materials the cradle is often defined as the start of the collection process for the waste materials. This is not the appropriate comparison to determine the environmental savings or costs that can be attributed to the recovered materials leaving an IWM system. The environmental burdens of collection and sorting are included in such reports, but in the IWM-2 model developed in this book, they have already been accounted for within the waste management system boundaries. Therefore, what the present study requires are data for just the transport to the reprocessors and for the reprocessing itself. These can be added to the burdens of the defined waste management system. If any relative savings/costs are to be calculated, the relevant comparison would be between the transport and reprocessing burdens for the recycled material, and the total burdens for the virgin material, from raw material extraction to produced material (Figure 22.2).

A second note of caution is needed when interpreting the burdens of materials recycling. The burdens associated with material production (whether virgin or recycled) are usually presented on the basis of per kilogram (or per tonne) material produced, i.e. per unit of output. This is not the form of data relevant for an LCI of solid waste. The ultimate function of an IWM system is to manage a given amount of waste in an environmentally and economically sustainable way, not to produce recycled material (Chapter 5). The functional unit of the whole system is therefore the amount of waste entering the system, not the amount of recovered or recycled material leaving the system. If the reprocessing stage is to be taken into account in the overall system, therefore, data on environmental burdens need to be in the form of per tonne of recovered material sent for recycling, i.e. per unit of input, rather than output.

In the following sections covering the environmental burden of reprocessing each of the recovered materials, data will be presented for the following:

1. Energy consumption and emissions associated with production of the virgin materials that this material could replace (starting from raw material extraction).
2. Energy consumption and emissions associated with the reprocessing of recovered materials into recycled material.
3. The potential saving (or addition) of energy consumption and emissions for every tonne of recovered material sent for reprocessing.
4. Energy consumption and emissions associated with transporting the recovered material from the collection/sorting facility to the reprocessing facility. This needs to be subtracted from any potential savings to determine the actual savings (or additional costs) likely.

When considering the potential savings associated with materials recycling, the following notes of caution should be borne in mind.

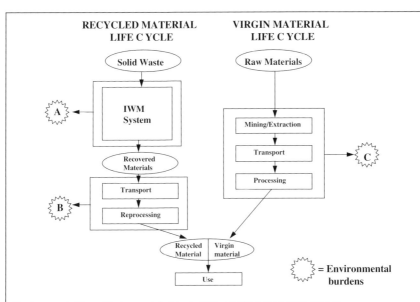

RECYCLED MATERIAL LIFE CYCLE — **VIRGIN MATERIAL LIFE CYCLE**

Solid Waste → IWM System → A
Recovered Materials → Transport → Reprocessing → B
Raw Materials → Mining/Extraction → Transport → Processing → C
Recycled Material | Virgin material → Use

⛤ = Environmental burdens

Environmental benefits (or costs) of including recycling in an IWM system

Comparing the overall burdens of an IWM system with recovery plus reprocessing of materials (Ar + B), with a system without materials recovery, where all products use virgin materials (Anr + C), the difference in the overall (system) environmental burden due to recycling is (Ar + B) – (Anr + C), where Ar is the burden of an IWM system that includes collection and sorting of recyclables, and Anr is the burden of a waste system that does not include collection and sorting of recyclables.

Difference in overall burden due to recycling $= Ar + B - Anr - C = \Delta A + B - C$

Where $\Delta A =$ difference between burden of IWM system with collection and sorting of recyclable, and system with only disposal of these materials, i.e. $(Ar - Anr)$

There will be an overall environmental benefit (reduced burden) so long as $\Delta A + B < C$

Choosing between recycled and virgin material

In product or package manufacture, there will be an environmental reason for choosing recycled over virgin material if the burdens of producing the recycled material are less than those of producing an equivalent amount of the virgin material,

i.e. $\Delta A + B < C$ for that material,

where $\Delta A =$ the difference in burden between collecting and sorting the material and collecting and disposing of the material by incineration and/or landfilling,
B = the burden of transporting and reprocessing the material
C = the burden of production from virgin raw materials.

Figure 22.2 Calculating the environmental benefits of recycling versus the use of virgin materials.

The data presented below tend to be generic (average data) or taken from a specific study of individual processes. Whilst the purpose is to give a broad indication of the available savings, the data will not be universally applicable.

IWM2 Model Guide

It is assumed that recycled material performs equally well and can replace an equal quantity of virgin material. This is not always the case since some high grade materials, e.g. writing papers, etc. cannot be replaced with recycled materials of equal quality.

Transport burdens

The energy consumption and emissions (per tonne) associated with transporting the recovered materials from the collection or sorting site (at which point they leave the basic I CI model developed in this book) to the reprocessing site will obviously vary with the distance involved. The fuel consumption data in Chapter 9 and the fuel production and use data in Chapter 5 are used to calculate the energy consumption and emissions associated with this transport. The calculation assumes that 20-tonne (payload) trucks are used and that the trucks carry loads in both directions.

Feed-stock energy

There has been considerable discussion as to whether feed stock energy should be included in comparisons between virgin and recycled materials. Habersatter (1991) includes the feed stock or inherent energy of both virgin and recycled materials, whereas Henstock (1992), in a comparison of recycled and virgin low-density polyethylene, includes the inherent energy of the raw materials used to make the virgin resin, but not the inherent energy of the recovered plastic used to produce the recycled resin. This convention significantly increases the apparent energy savings due to recycling. The decision must reflect the aim of the study, which is defined in the goal definition (Chapters 4 and 5). The aim of this study is to predict the energy consumption and emissions associated with managing the solid waste of an area in an environmentally and economically sustainable way.

The inherent energy contained in the waste entering the system is not considered as contributing to the energy consumption. **Therefore the inherent energy of the recovered material will not be considered as part of the energy consumption of the recycling process. For consistency, therefore, the inherent energy of the virgin raw materials will also not be included as part of the energy consumption of virgin materials.** It can be argued that the inherent energy of the material has not actually been consumed; it has merely been locked up and can be released at a later time, e.g. by burning as a fuel. The net effect of this convention is to give conservative estimates of the energy savings due to recycling. The processing energy savings and inherent energy savings of recycling are both given in Table 22.1, for all materials, so the higher value can be calculated if required.

Paper

A wide range of data has been reported for the energy consumption and emissions associated with the production of paper, whether from recovered or virgin materials. The figures differ according to the type of pulp or paper that is being produced (e.g. newsprint or bleached

Material	Process energy saved by recycling (GJ/tonne)	Inherent energy saved by recycling (GJ/tonne)	Air emissions for recycling	Water emissions for recycling	Solid waste reduced (increased) by recycling (kg/tonne)	Comments
Paper	5.6	30.3	Generally lower	Generally lower	(198)	Pulp and paper making included
Glass	3.5	–	Generally lower	Generally lower	29	Data for 100% virgin extrapolated as all glass-making uses some used cullet
Metal – Fe (tinplate)	18.6	–	Generally lower	Generally lower	57	Data for tinplate recycling up to production of new tinplate
Metal – Al	174.6	–	Generally lower (except HCl)	Generally lower	986	
Plastic (LDPE)	15.4	47.7	Generally lower (except CO_2)	Little data	(92)	Incomplete data for reprocessing of LDPE
Plastic (HDPE)	25.6	47.7	Generally lower	Poor data, may be higher for recycled	(184)	Incomplete data for reprocessing of HDPE
Textiles	52–59	little data	Little data	Little data	Little data	Energy range for woven and knitted wool only

Table 22.1 Summary of environmental benefits and costs of recycling. Table gives the calculated effect on energy consumption and emissions of reprocessing recovered materials, compared to the production of virgin material. Results are per tonne of recycled material produced.

Note: figures are indicative only and vary with processes and equipment used. Figures give the difference between reprocessing burdens for producing 1 tonne of recycled material and the burdens of producing 1 tonne of virgin material. The burdens of collecting and sorting the recovered material, and transporting it to the reprocessors are not included. Similarly, diversion of recovered material from landfill is not included in the solid waste savings

sulphite paper), and the boundaries chosen for the calculations (e.g. whether the data are for pulp production only, or for complete paper production; whether de-inking is included for recycled paper or not, etc.). What is important when making comparisons, is that similar products are compared and that comparable boundaries are drawn.

For newsprint production using 100% recovered paper, for example, the primary energy consumption for pulping, de-inking, paper making and effluent treatment has been calculated at 14.5 GJ per tonne produced. (This does not include collection, sorting and transport.) Equivalent production of newsprint from virgin wood consumes 21.0 GJ per tonne, giving a primary energy saving due to recycling of 6.5 GJ per tonne produced (Pulp and Paper, 1976). Similar savings have been calculated when energy consumption has been averaged across the different grades of paper and board produced (Figure 22.3). The average primary energy requirements for virgin and recycled paper processing were reported as 25.1 GJ (range 20–28) per tonne and 18.0 GJ per tonne, respectively in 1985, giving an average saving of 7.1 GJ per tonne (Porter and Roberts, 1985). It is likely that energy consumption has fallen since that time, due to more efficient processing techniques. The requirement for virgin paper making in the UK in 1989, for example, was around 21 GJ per tonne produced (Ogilvie, 1992).

Considerably higher energy consumption figures for both virgin and recycled paper have been quoted by some sources. Habersatter (1991), using data from Swedish and Swiss sources, gives energy consumptions of 53.0 GJ per tonne for production of virgin unbleached sulphite paper and 29.7 GJ per tonne for 100% recycled paper. The reason for this apparently higher energy consumption is that the inherent energy (i.e. calorific value) of the wood or recovered paper feed stock is also included in these totals. The inherent or feed stock energy of wood or waste paper is around 15 GJ per tonne and 2.02 tonnes of air-dried wood or 1.02 tonnes of recovered paper are needed to produce 1 tonne of unbleached sulphite pulp or 1 tonne of recycled fibre pulp, respectively (Habersatter, 1991). Subtracting the appropriate feed-stock energies (30.3 GJ/tonne and 15.3 GJ/tonne, respectively) leaves a processing energy requirement for unbleached sulphite paper of 22.7 GJ/tonne and 14.4 GJ/tonne for recycled paper. These accord with the data quoted above, giving an energy saving due to the processing associated with recycling of 8.3 GJ per tonne of paper produced.

Before leaving the subject of energy consumption in paper production, it is necessary to add a cautionary note on using primary energy totals alone to compare the environmental benefits of paper recycling. As well as the total energy consumption, it is necessary to know how the energy was produced, and in particular, whether it came from renewable or non-renewable

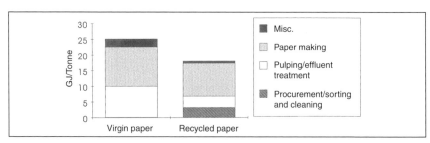

Figure 22.3 Process energy requirements for virgin and recycled paper production. Sources: Porter and Roberts (1985); Porteous, (1992).

sources. Many pulp and paper mills in Sweden, for example, generate their own electricity and steam on site using by-products from the pulping process (bark, process liquors, etc.). Some plants produce excess steam or electricity, which is exported from the site. Such use of biomass (a renewable energy resource) will result in no net production of some emissions, such as carbon dioxide, since these were absorbed during the growing of the trees in the first place. By contrast, recycling processes will tend to use power generated from fossil fuel with a net production of carbon dioxide and depletion of finite fossil fuel reserves (though equally power could be generated from on-site boilers fuelled by waste paper). The point here is that energy consumption is not an environmental burden in itself; what is important are the environmental burdens resulting from the generation of the energy used, i.e. the emissions and the depletion of finite resources.

Carbon balance

In the IWM-2 model, to allow the allocation of the environmental burdens of paper recycling, paper production using virgin material (wood) is allocated a CO_2 credit (a negative burden) as it requires grown trees to be harvested. It is assumed the wood comes from forests managed in a sustainable manner and young trees are re-planted as part of the usual forest management process. The growth period of trees is when they take up and fix CO_2 most rapidly, so paper production from virgin material results in a net uptake of CO_2.

Paper recycling on the other hand, does not require trees to be cut down and therefore the rapid CO_2 uptake associated with new growth does not occur. As paper recycling uses energy, the overall process is a net CO_2 producer (but many other air and water emissions are avoided by paper recycling) (Figure 22.4).

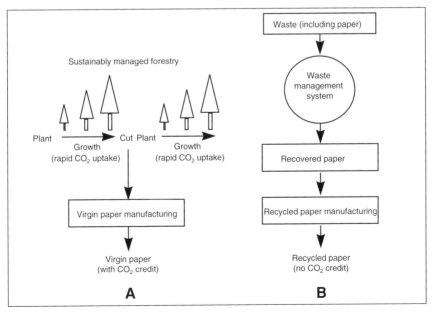

Figure 22.4 Carbon balance of virgin paper compared to recycled paper.

	Virgin paper/tonne (sulphite, bleached)	Recycled paper (100%, de-inked)	Saving/tonne recycled paper
Source	BUWAL (1998)	BUWAL (1998)	Calculated
Energy consumption (GJ)*:	27.84	22.25	5.59
Air Emissions (g):			
Particulates	141	743	−602
CO	297	386	−89
CO_2	155,000	354,000	−199,000
CH_4	155	630	−475
NOx	1590	1920	−330
N_2O	1.73	12.8	−11.07
SOx	4,950	2630	2320
HCl	9.33	24.9	−15.57
HF	0.67	2.45	−1.78
H_2S		3.71	−3.71
HC			
Chlorinated HC			
Dioxins/furans			
Ammonia	0.122	94	−93.878
Arsenic			
Cadmium	0.006	0.00682	−0.00082
Chromium			
Copper			
Lead	0.0203	0.0354	−0.0151
Mercury	0.0011	0.00457	−0.00347
Nickel	0.166	0.3	−0.134
Zinc	0.0396	0.103	−0.0634
Water emissions (g):			
BOD	1300	1630	−330
COD	56,000	8340	47660
Suspended solids	1100	4130	−3030
Total organic compounds	10.1	2950	−2939.9
AOX	500	17.2	482.8
Chlorinated HCs	0.0008	0.00329	−0.00249
Dioxins/furans (TEQ)			
Phenol	0.132	0.493	−0.361
Aluminium	17.3	67.8	−50.5
Ammonium	2.36	17.4	−15.04
			(continued)

Table 22.2 Energy consumption and emissions from recycled and virgin paper production.
*Not including inherent energy of feed stock materials

IWM2 Model Guide

	Virgin paper/tonne (sulphite, bleached)	Recycled paper (100%, de-inked)	Saving/tonne recycled paper
Arsenic	0.0341	0.13	–0.0959
Barium	3.73	13.5	–9.77
Cadmium	0.0022	0.00796	–0.00576
Chloride	1340	17200	–15860
Chromium	0.173	0.659	–0.486
Copper	0.0833	0.317	–0.2337
Cyanide	1340	0.0143	1339.98
Fluoride			
Iron	18.2	59.1	–40.9
Lead	0.152	0.545	–0.393
Mercury	0.0000366	0.000182	–0.0001454
Nickel	0.0859	0.326	–0.2401
Nitrate	1.07	837	–835.93
Phosphate	0.978	24.1	–23.12
Sulphate	496	8660	–8164
Sulphide	0.0306	0.106	–0.0754
Zinc	0.177	0.671	–0.494
Solid waste (kg) :	22.34	220.1	–197.76

Table 22.2 (continued) Energy consumption and emissions from recycled and virgin paper production.
*Not including inherent energy of feedstock materials

In the IWM-2 model paper production from virgin material is allocated a CO_2 credit of 1.833 tonnes CO_2 per tonne virgin wood consumed (BUWAL, 1998) (the US EPA model uses a figure of 2.2 tonnes CO_2 per tonne virgin wood consumed, from Kramer and Kozlowski, 1979). This credit is added to the total CO_2 burden from virgin paper production presented in Table 22.2. It is expressed in the model by subtracting the emissions associated with recycled paper production from the total emissions associated with virgin paper production, represented as A–B in Figure 22.4.

The emissions (to air, water and as solid waste) associated with production of recycled and virgin paper are presented in Table 22.2. Note that these data relate to conditions in Switzerland where steam production for both virgin and recycled paper production uses fossil fuels, and electricity production is assumed to be according to the general UCPTE model introduced in Chapter 5. If virgin pulps were made in integrated pulp mills such as in Sweden, fuelled by biowaste by-products, the emission levels associated with virgin pulp production would be generally lower.

To calculate the energy consumption and emissions per tonne of recovered paper sent for reprocessing, rather than per tonne of recycled paper produced requires data on material losses during the process. The actual process losses will depend on the quality of the input fibres, the level of filler in the recovered material, the type of recycling process employed and the quality of recycled product required. For recycled newsprint manufacture, a loss of 15–18% is typical (Claydon, 1991; US EPA, 1997), and across different paper grades a general loss of 15–20% applies (Cathie and Guest, 1991). Taking the former figure and an energy saving of 5.6 GJ per tonne produced (above), the energy saving amounts to 4.8 GJ per tonne of recovered paper sent for reprocessing. There are few savings in emissions and in the overall amount of solid waste generated as the pulping process has improved significantly in recent years, whereas, the environmental burdens associated with paper recycling (mainly from the de-inking process) have remained unchanged.

Glass

The use of recovered glass cullet in glass making has the advantage of lowering the furnace temperature needed to melt the other raw material ingredients. These energy savings can be estimated by the simple equation:

Energy savings (%) = 0.25 × % of scrap glass used (BUWAL, 1998).

Considering the process of glass recycling up to the production of the hot 'gob' of glass, energy consumption and emissions data per tonne of 100% recycled glass produced are presented in Table 22.3.

Corresponding data for the production of 100% virgin glass are not available, since all glass production processes incorporate some recovered glass, either collected post-consumer material or production scrap, because of the furnace energy savings available. Extrapolating linearly from the data for recycling rates of 100%, 65% and 60% given by BUWAL (1998), however, it is possible to estimate the energy consumption and associated emissions for glass production from virgin materials (Table 22.3). Using this estimate, it can be seen that the recycling of glass saves around 3.46 GJ per tonne of glass produced. There are savings in the amounts of those emissions that are associated with this energy production, but there are slight increases in the amounts of some air emissions, notably NOx, hydrogen fluoride (HF) and lead, associated with recycling.

Contaminants (metals, ceramics, paper) are reported to constitute 2.7% of the feed stock (Habersatter, 1991). The default substitution ratio (the amount of virgin product replaced by 1 tonne of recycled product) in the model assumes that contamination is around this 3% level in feed-stock cullet.

Metal

Metal – ferrous

Typical primary energy consumption and emissions associated with the full production process for tinplate from 100% recovered material are presented in Table 22.4, along with data for the production of tinplate from virgin materials. These figures, from BUWAL (1998), suggest an energy saving of around 18.6 GJ per tonne of tinplate produced, when recovered tinplate is used. There are also savings in most, but not all emissions.

	Virgin glass	Recycled glass (100%) /tonne produced	Savings/tonne recycled glass
Source	Extrapolated*	BUWAL (1998)	Calculated
Energy consumption (GJ):	14.5	11.04	3.46
Air emissions (g):			
Particulates	1488	704	784
CO	1197	222	975
CO_2	145600	57000	88600
CH_4	827	767	60
NOx	1500	2880	−1380
N_2O	2.12	1.66	0.46
SOx	2970	728	2242
HCl	117	58.5	58.5
HF	8.2	23.4	−15.2
H_2S			
HC			
Chlorinated HC			
Dioxins/furans			
Ammonia	25.45	16.5	8.95
Arsenic	64.4	2.61	61.79
Cadmium	0.0118	0.009	0.0028
Chromium			
Copper			
Lead	−11.7	35.5	−47.2
Mercury	0.00172	0.002	0.00028
Nickel	0.4565	0.362	0.0945
Zinc	0.258	0.155	0.103
Water emissions (g):			
BOD	0.57	0.374	0.196
COD	11.64	7.41	4.23
Suspended solids	7760	796	6964
Total organic compounds	68.475	80.7	−12.225
AOX	0.0358	0.0287	0.0071
Chlorinated HCs	0.00876	0.0075	0.00126
Dioxins/furans (TEQ)			
Phenol	1.46	1.18	0.28
Aluminium	24.1	16.5	7.6
Ammonium	42.2	10.3	31.9
Arsenic	0.0584	0.038	0.0204
Barium	28.2	22	6.2
			(continued)

Table 22.3 Energy consumption and emissions from recycled and virgin glass production.
*Calculated by extrapolation from data for 100%, 65% and 60% recycled glass

	Virgin glass	Recycled glass (100%) /tonne produced	Savings/tonne recycled glass
Cadmium	0.087	0.0099	0.0771
Chloride	99900	8410	91490
Chromium	0.338	0.227	0.111
Copper	0.143	0.0918	0.0512
Cyanide	0.041	0.032	0.009
Fluoride			
Iron	28.65	19.2	9.45
Lead	0.368	0.0151	0.3529
Mercury	-0.000001	0.000198	-0.000199
Nickel	0.153	0.102	0.051
Nitrate	7.1	5.64	1.46
Phosphate	1.6	1	0.6
Sulphate	773	480	293
Sulphide	0.316	0.253	0.063
Zinc	0.346	0.232	0.114
Solid waste (kg):	74	44.97	29.03

Table 22.3 (continued) Energy consumption and emissions from recycled and virgin glass production.
*Calculated by extrapolation from data for 100%, 65% and 60% recycled glass

The amount of recovered material input that is lost in the reprocessing of tinplate is reported as 8.2% by Habersatter (1991); Porteous (1992) suggests a loss of 5% in the process up to the production of de-tinned washed steel. Taking the former value would mean that each tonne of recovered tinplate delivered to the reprocessors would produce 918 kg of recycled tinplate. The typical primary energy consumption associated with this production would therefore be 18.36 GJ per tonne of recovered tinplate used, or an energy saving (when compared to the use of virgin materials) of around 12.4 GJ per tonne (using the BUWAL data). Comparisons of the amounts of emissions (including solid waste) associated with the processing of 1 tonne of recovered tinplate scrap versus production of an equivalent amount of tinplate from virgin are also presented in Table 22.4.

For the large part of ferrous scrap not in the form of tinplate, reprocessing is simpler and consists only of removal of contaminants and then remelting. For iron, the energy needed to remelt is 1.8 GJ per tonne whilst iron production from ore requires 7.92 GJ per tonne. This would give an energy saving of 6.12 GJ per tonne of recycled iron produced. For steel, the energy saving of using electric arc melting of scrap versus virgin steel production using a blast furnace is around 15.8 GJ per tonne (Ogilvie, 1992). Assuming the same material loss as used above (8.2%), the possible savings from recycling of iron and steel, per tonne of recovered material used, would be 5.0 GJ and 12.9 GJ, respectively.

IWM2 Model Guide

	Virgin tinplate/tonne	Recycled tinplate/tonne	Savings/tonne recycled tinplate
Source	BUWAL (1998)	BUWAL (1998)	Calculated
Energy consumption (GJ):	35.77	17.18	18.59
Air emissions (g):			
Particulates	1,410	1130	280
CO	18500	4330	14170
CO_2	297,000	1090000	1880000
CH_4	10800	1920	8880
NOx	4560	1730	2830
N_2O	9.6	4.55	5.05
SOx	6230	2730	3500
HCl	86.4	130	–43.6
HF	11	14.8	–3.8
H_2S	9.9	0	9.9
Total HC			
Chlorinated HC			
Dioxins/furans			
Ammonia	1.97	1.81	0.16
Arsenic	0	0	0
Cadmium	0.11	0.009	0.101
Chromium	0.14	0.19	–0.05
Copper	0.26	0.53	–0.27
Lead	4.59	9.47	–4.88
Mercury	0.02	0.024	–0.004
Nickel	1.79	0.24	1.55
Zinc	0.27	0.17	0.1
Water emissions (g):			
BOD	170	170	0
COD	465	460	5
Suspended solids	395	175	220
Total organic compounds	150	127	23
AOX	0.52	0.0013	0.5187
Chlorinated HCs	0.01	0.499	–0.489
Dioxins/furans(TEQ)			
Phenols	0.65	0.08	0.57
Aluminium	1920	285	1635
Ammonium	8.01	2.65	5.36
Arsenic	3.85	0.57	3.28
Barium	165	23.4	141.6
Cadmium	0.1	0.0168	0.0832
Chromium	19.6	3.14	16.46
			(continued)

Table 22.4 Energy consumption and emissions from recycled and virgin tinplate production

	Virgin tinplate/tonne	Recycled tinplate/tonne	Savings/tonne recycled tinplate
Copper	9.61	1.71	7.9
Cyanide	0.03	0.0064	0.0236
Iron	790	567	223
Lead	9.74	1.65	8.09
Mercury	0.02	0.02	0
Nickel	9.69	1.73	7.96
Zinc	19.4	2.85	16.55
Chloride	14200	4500	9700
Fluoride	32	32	0
Nitrate	6.08	7.03	-0.95
Phosphate	146	47.9	98.1
Sulphate	8950	2970	5980
Sulphide	0.14	0.01	0.13
Solid waste (kg):	67	10.2	56.8

Table 22.4 (continued) Energy consumption and emissions from recycled and virgin tinplate production

Metal – aluminium

There are clear energy advantages in the use of recovered material to produce aluminium, due to the large energy requirement of production from virgin materials (bauxite). Since the production of virgin aluminium relies on electrolysis and consumes large amounts of electrical energy, the total primary energy consumption and emissions are heavily dependent on the method used for electricity generation. Data from Switzerland for the energy consumption and emissions associated with the production, per tonne, of both virgin and recycled aluminium are presented in Table 22.5. Average European values have been used for primary aluminium production, rolling and recycling. As only 60% of the aluminium used in Switzerland comes from Europe, the actual places of origin of the metal (Canada and Iceland) have been considered for the remaining 40%. Production in these countries is solely by means of hydroelectric power. The data here use a 'Western world' model (UCPTE, 1994) for electricity generation.

The data presented here show that energy savings associated with aluminium recycling can be in the order of 175 GJ per tonne of recycled aluminium produced. There are also large savings in most of the associated emissions to both air and water, and in the overall amount of solid waste produced.

In the model a material loss of 5% during the recycling process is assumed (a default value of 95% substitution ratio), due to the removal of any contaminants in the recovered material feed stock, and that this material becomes an additional residue for landfilling. The savings in primary energy and emissions per tonne of recovered aluminium sent for reprocessing are calculated by the model.

IWM2 Model Guide

	Virgin aluminium/tonne*	Recycled aluminium/tonne*	Savings/tonne aluminium
Source	BUWAL, 1998	BUWAL, 1998	Calculated
Energy consumption (GJ):	182.8	8.24	174.56
Air emissions (g):			
Particulates	21,300	235	21,065
CO	61,500	123	61,377
CO_2	7,640,000	403,000	7,237,000
CH_4	16,700	847	15853
NOx	16,000	893	15107
N_2O	41.6	1.91	39.69
SOx	54,600	1520	53,080
HCl	699	20.7	678.3
HF	72.7	12.2	60.5
H_2S			
Total HC			
Chlorinated HC			
Dioxins/furans			
Ammonia	13.6	0.365	13.235
Arsenic			
Cadmium	0.263	0.00371	0.25929
Chromium			
Copper			
Lead	1.03	0.0283	1.0017
Mercury	0.107	0.00842	0.09858
Nickel	8.37	0.205	8.165
Zinc	2.18	0.0596	2.1204
Water emissions (g):			
BOD	3.37	0.0907	3.2793
COD	83	1.67	81.53
Suspended Solids	4780	200	4580
Total organic compounds	743	113	630
AOX	0.217	0.0056	0.2114
Chlorinated HCs	0.0587	0.00294	0.05576
Dioxins/furans (TEQ)			
Phenol	9.04	0.248	8.792
Aluminium	2440	54.1	2385.9
Ammonium	76.8	2.19	74.61
Arsenic	4.94	0.109	4.831
Barium	350	8.29	341.71
			(continued)

Table 22.5 Energy consumption and emissions from recycled and virgin aluminium production.
*Using a 'Western world' scenario for electricity generation

IWM2 Model Guide

	Virgin aluminium/tonne*	Recycled aluminium/tonne*	Savings/tonne aluminium
Cadmium	0.197	0.00474	0.19226
Chloride	51300	1210	50090
Chromium	24.8	0.557	24.243
Copper	12.2	0.269	11.931
Cyanide	0.288	0.00859	0.279
Fluoride	2.71	0	2.71
Iron	915	71	844
Lead	13.5	0.326	13.174
Mercury	0.00444	0.000414	0.004026
Nickel	12.4	0.274	12.126
Nitrate	91.9	2.29	89.61
Phosphate	145	3.2	141.8
Sulphate	17500	611	16889
Sulphide	1.96	0.0517	1.9083
Zinc	25	0.553	24.447
Solid waste (kg):	995	9.3	985.7

Table 22.5 (continued) Energy consumption and emissions from recycled and virgin aluminium production.
*Using a 'Western world' scenario for electricity generation

Plastics

There have been numerous reports calculating the energy consumption and emissions associated with the production of specific virgin plastic resins (e.g. Kindler and Mosthaf, 1989; Lundholm and Sundström, 1986; Habersatter, 1991; PWMI, 1993 [see review in Ogilvie, 1992]). Comparable detailed data are not available, however, for the process of plastics recycling, probably due to its relatively recent introduction and rapid rate of development. Clark and New (1991) suggest that the energy savings from plastics recycling vary from 27 to 215 GJ/tonne, depending on the resin type, but there is no detail on how these figures are obtained. A more detailed study by Henstock (1992) reports the energy consumption and some emissions associated with the reprocessing of low-density polyethylene film collected from supermarkets into recycled LDPE granules and then into recycled polyethylene bags. The primary energy consumption of the reprocessing from recovered LDPE film into recycled LDPE granules (excluding sorting at the stores and transport) is presented as between 25.4 and 33.2 GJ per tonne of recycled LDPE produced. The air emissions that result from the electrical power and propane consumption during the process are presented in Table 22.6, but no details of emissions to water are provided.

For reprocessing of rigid plastic bottles (HDPE), Deurloo (1990) gives a figure of 2.88 GJ of electricity per tonne of recycled HDPE produced (equivalent to 7.6 GJ thermal energy/tonne, using the UCPTE generation efficiency of 37.8% – Chapter 5). Again, the only air emissions included are those for the electricity generation, but data for water emissions were given (Table 22.6).

Source	Virgin LDPE/ tonne produced BUWAL (1998)	Recycled LDPE/ tonne produced Henstock (1992)	LDPE savings/ tonne recycled ‡Calculated	Virgin HDPE/ tonne produced BUWAL (1998)	Recycled HDPE/ tonne produced Deurloo (1990)	HDPE savings/ tonne recycled Calculated	Virgin PP/ tonne BUWAL (1998)	Virgin GP PS/tonne BUWAL (1998)
Energy consumption (GJ):	40.828	25.4	15.42	33.258	7.62	25.63	32.308	40.048
Air emissions (g):								
Particulates	3000			2000	158	1842	2000	1700
CO	900			600	280	320	700	1100
CO_2	2,320,000	1,299,900	1,020,100	2,060,000	353,325	1,706,675	1,800,000	2,600,000
CH_4	4400			3700			3400	11,000
NOx	12,000	6390	5610	10,000	989	9011	10,000	12,000
N_2O	6.7			4.9	56	-51.1	5.7	5.2
SOx	9000	13,870	-4870	6000	2002	3998	11,000	11,000
HCl	70			50			40	26
HF	5			1	0.01	0.99	1	2.8
H_2S								
Total HC					1690			
Chlorinated HC								
Dioxins/furans								
Ammonia	1.1			0.66			0.83	0.44
Arsenic								
Cadmium	0.018			0.013			0.018	0.019
								(continued)

Table 22.6 Energy consumption and emissions from recycled and virgin plastic production.

*Assuming a material loss of 5%; †assuming a material loss of 15%; ‡using best case data; §feed-stock energy not included; ¶calculated from power consumption and material losses

	Virgin LDPE/ tonne produced	Recycled LDPE/ tonne produced	LDPE savings/ tonne recycled	Virgin HDPE/ tonne produced	Recycled HDPE/ tonne produced	HDPE savings/ tonne recycled	Virgin PP/ tonne	Virgin GP PS/tonne
Chromium								
Copper								
Lead	0.1			0.068			0.088	0.068
Mercury	0.036			0.03			0.023	0.022
Nickel	1			0.72			0.96	0.99
Zinc	0.4			0.31			0.42	0.5
Water emissions (gJ):								
BOD	200			100	2365	-2265	60	51
COD	1500			200	4620	-4420	400	370
Suspended Solids	500			200			200	290
Total organic compounds	100			150			300	94
AOX	0.067			0.055	24.2	-24.145	0.075	0.1
Chlorinated HCs	0.023			0.019			0.021	0.028
Dioxins/furans (TEQ)								
Phenol	2.4			1.9	0.55	1.35	2.6	5
Aluminium	160			98			120	66
Ammonium	5			10			10	9
Arsenic	0.34			0.21	0.1	0.11	0.26	0.15
Barium	56			43			58	70
Cadmium	0.028			0.021	0.055	-0.034	0.028	0.032
Chloride	130			800	97.9	702.1	800	5600
Chromium	1.8			1.1	0.33	0.77	1.4	0.9

(continued)

Table 22.6 (continued) Energy consumption and emissions from recycled and virgin plastic production.

*Assuming a material loss of 5%; †assuming a material loss of 15%; ‡using best case data: §feed-stock energy not included; ¶calculated from power consumption and material losses

	Virgin LDPE/ tonne produced	Recycled LDPE/ tonne produced	LDPE savings/ tonne recycled	Virgin HDPE/ tonne produced	Recycled HDPE/ tonne produced	HDPE savings/ tonne recycled	Virgin PP/ tonne	Virgin GP PS/tonne
Copper	0.83			0.51	2.31	-1.8	0.65	0.37
Fluoride					1.07			
Iron	220			140			170	94
Lead	1			0.61	0.11	0.5	0.77	0.43
Mercury	0.002			0.0017	0.006	-0.0043	0.0013	0.0014
Nickel	0.85			0.53	0.22	0.31	0.67	0.4
Nitrate	5			10	1.06	8.94	20	2
Phosphate	5			1			13.5	4.1
Sulphate	2100			1300			1700	160
Sulphide	0.54			0.44	0.55	-0.11	0.61	0.9
Zinc	1.8			1.1			1.4	0.89
Solid Waste (kg):	39.1	132.0¶	-92.9	31.9	216	-184.1	31.2	30.19

Table 22.6 (continued) Energy consumption and emissions from recycled and virgin plastic production.
*Assuming a material loss of 5%; †assuming a material loss of 15%; ‡using best case data: §feed-stock energy not included;
¶calculated from power consumption and material losses

For the production of virgin LDPE and HDPE, data (averaged across Europe) give total primary energy consumptions of 88.55 GJ and 80.98 GJ per tonne, respectively (PWMI, 1993). These include the inherent energy of the feed-stock material used (47.73 GJ/t), and it can be argued whether this should be included when comparisons are made. As discussed above, using the recovered material does result in the saving of the raw material that contains this amount of energy, but using the convention stated above, the inherent energy of the plastic feed stock will not be included in the energy consumption of using virgin plastic. Taking the data given above for recycling, this results in a potential (processing) energy saving due to LDPE reprocessing of 7.6–15.4 GJ per tonne produced, and a saving of 25.7 GJ per tonne of recycled HDPE produced.

Table 22.6 also gives the energy consumption and emissions reported for the production of other virgin plastic resin types. Although no recycling data are available for these, if the processing consists of flaking, washing, drying and granulating, then they may well be similar to those reported for LDPE and HDPE. More data are clearly needed in this area, however.

The material loss during the recycling process is given as 5% for LDPE film (Henstock, 1993), and 15% for HDPE (Deurloo, 1990), though this will depend on how well the material has been sorted and is likely to vary between different resin types. Using these values, however, the potential energy saving associated with recycling will be in the order of 7.2–14.6 GJ per tonne of recovered LDPE, and 21.8 GJ per tonne of recovered HDPE, reprocessed.

One of the assumptions made when calculating possible savings due to materials recycling is that the recycled material performs in exactly the same way as the virgin material. This is not always the case, especially with regard to some plastics. For example, bags made from the recycled LDPE described above had to be 30 μm thick, compared to 20 μm for a bag made of virgin HDPE (Henstock, 1992). Similarly use of recycled material is reported to result in higher wastage rates than virgin material (an increase of 3.5%) during bag production. Thus while the figures calculated and used here will give a broad estimate of potential savings, their accuracy must be viewed with some caution.

On the other hand, it has been possible to use recycled HDPE material successfully in laundry detergent bottles. By co-extruding a layer of recycled plastic between two outer layers of virgin material, over 25% of recycled material can be used, without the need to change the weight, performance or aesthetics of the bottle. In these cases, therefore, a straight comparison of the energy consumption and emissions of recycled versus virgin material production is valid.

Textiles

Information on the energy consumption and emissions of textile recycling processes is very limited. One study on the woollen industry reported that the energy consumption of producing woven cloth of virgin wool was 115.61 GJ per tonne, compared to 56.61 GJ per tonne for cloth with 100% recycled content, giving a saving of 59 GJ per tonne produced (Lowe, 1981). For knitted products, virgin wool use consumed 108.28 GJ per tonne, compared with 56.61 GJ per tonne for recycled material (a saving of around 52 GJ per tonne). Note that these figures are per tonne produced, not per tonne reprocessed; no data are available on material losses during processing. The figures for virgin wool do not include the initial scouring process, so the actual savings associated with the use of recycled content are likely to be larger (Ogilvie, 1992).

Economic costs

Just as there can be savings or additional costs in energy consumption and emissions, there will be economic savings or costs associated with the production of recovered materials. The additional costs will be the transport cost to the reprocessors, and the cost of the reprocessing operation. Since the price paid for the recovered material by the reprocessors has already been included in the income of the basic model, it must be included within the processing costs.

Additional income to the system when reprocessing is included comes from the sale of recycled material, values for which can be found in current commodity market prices.

Thus if the transport and processing costs (including recovered material prices) exceed the value of the recycled material, there will be an extra cost incurred by reprocessing the recovered material. Alternatively, overall cost savings will occur if the value of the recycled material produced exceeds the transport and reprocessing costs.

In the final analysis, recycled material will only sell if it is priced competitively compared to virgin. How close the recycled material price comes to the virgin price will depend on whether it has equal performance for the intended use (e.g. in the case of glass) or whether there is a fall in performance that requires a compensating discount. In either event the price for recycled material is pegged to virgin material prices and is thus relatively fixed. For the recycling industry to expand, therefore, it is necessary for the transport and reprocessing costs to be competitively less than the virgin material price. For materials where there are large energy savings resulting from reprocessing compared to virgin production (e.g. aluminium, steel), this will hold. If there are high reprocessing costs and/or small associated energy savings, it may not be possible to produce the recycled material for a competitive price, as is often stated for plastics. In such cases, other options are still possible. Rattray (1993) conducted an exercise with different sectors in the plastics industry in the USA to determine where cost savings could be found in the recycling of HDPE, so that the cost of recycled resin could be reduced below that of virgin resin. By optimising the system rather than individual operations within it, the exercise generated ideas to reduce costs by 20 cents per pound. Likely savings would be more in the region of 6–8 cents per pound, but this would be enough to allow recycled material to compete well with virgin material for many end markets.

The alternative way to encourage materials recycling would be to lower the price the reprocessors pay for the recovered materials, either by improving the efficiency of the waste management system or by increasing the charges levied against users of the system.

Model data

The model uses the data presented in Table 22.7, compiled from Tables 22.2–22.6, energy consumption and emissions savings due to recycling.

IWM2 Model Guide

	Paper	Glass	Metal – ferrous	Metal – aluminium	Plastic – film (LDPE)	Plastic – rigid (HDPE)	Textiles
Energy consumption (GJ)*:	5.59	3.46	18.59	174.56	15.42	25.63	52.0
Air emissions (g):							
Particulates	−602	784	280	21,065		1842	
CO	−89	975	14170	61,377		320	
CO_2	−199,000	88.600	1,880.000	7,237.000	1,020,100	1,706,675	
CH_4	−475	60	8880	15,853			
NOx	−330	−1380	2830	15,107	5610	9011	
N_2O	−11.07	0.46	5.05	39.69		−51.1	
SOx	2320	2242	3500	53,080	−4870	3998	
HCl	−15.57	58.5	−43.6	678.3			
HF	−1.78	−15.2	−3.8	60.5		0.99	
H_2S	−3.71		9.9				
HC							
Chlorinated HC							
Dioxins/furans							
Ammonia	−93.878	8.95	0.16	13.235			
Arsenic		61.79	0				
Cadmium	−0.00082	0.0028	0.101	0.25929			
Chromium			−0.05				

Table 22.7 Energy consumption and emissions savings due to recycling used in the LCI model. Data are per tonne of recovered material sent for reprocessing

(continued)

	Paper	Glass	Metal – ferrous	Metal – aluminium	Plastic – film (LDPE)	Plastic – rigid (HDPE)	Textiles
Copper			-0.27				
Lead	-0.0151	-47.2	-4.88	1.0017		-2265	
Mercury	-0.00347	-0.00028	-0.004	0.09858		-4420	
Nickel	-0.134	0.0945	1.55	8.165			
Zinc	-0.0634	0.103	0.1	2.1204			
Water emissions (g):							
BOD	-330	0.196	0	3.2793			
COD	47660	4.23	5	81.53			
Suspended solids	-3030	6964	220	4580			
Total organic compounds	-2939.9	-12.225	23	630			
AOX	482.8	0.0071	0.5187	0.2114		-24.145	
Chlorinated HCs	-0.00249	0.00126	-0.489	0.05576			
Dioxins/furans (TEQ)							
Phenol	-0.361	0.28	0.57	8.792		1.35	
Aluminium	-50.5	7.6	1635	2385.9			
Ammonium	-15.04	31.9	5.36	74.61			
Arsenic	-0.0959	0.0204	3.28	4.831		0.11	
Barium	-9.77	6.2	141.6	341.71			
Cadmium	-0.00576	0.0771	0.0832	0.19226		-0.034	
Chloride	-15860	91490	16.46	50090		702.1	

(continued)

Table 22.7 (continued) Energy consumption and emissions savings due to recycling used in the LCI model. Data are per tonne of recovered material sent for reprocessing

	Paper	Glass	Metal – ferrous	Metal – aluminium	Plastic – film (LDPE)	Plastic – rigid (HDPE)	Textiles
Chromium	-0.486	0.111	7.9	24.243			0.77
Copper	-0.2337	0.0512	0.0236	11.931			-1.8
Cyanide	1339.98	0.009	223	0.279			
Fluoride			8.09	2.71			
Iron	-40.9	9.45	0	844			
Lead	-0.393	0.3529	7.96	13.174			0.5
Mercury	-0.0001454	-0.000199	16.55	0.004026			-0.0043
Nickel	-0.2401	0.051	9700	12.126			0.31
Nitrate	-835.93	1.46	0	89.61			8.94
Phosphate	-23.12	0.6	-0.95	141.8			
Sulphate	-8164	293	98.1	16889			
Sulphide	-0.0754	0.063	5980	1.9083			-0.11
Zinc	-0.494	0.114	0.13	24.447			
Solid waste (kg):	-197.76	29.03	56.8	985.7	-92.2	-184.1	

Table 22.7 (continued) Energy consumption and emissions savings due to recycling used in the LCI model. Data are per tonne of recovered material sent for reprocessing

Materials recycling

The materials recycling screen of the LCI model is shown in Screen 27.

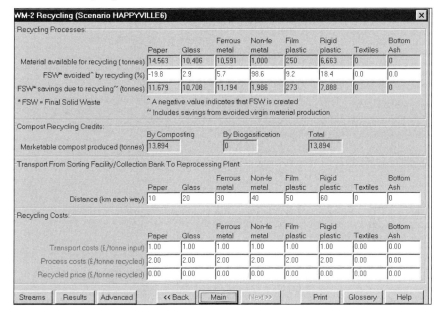

Screen 27 Recycling.

Recycling Processes:
   Material available for recycling (tonnes)
   FSW * avoided ^ by recycling (%)
   FSW * savings due to recycling ~(tonnes)
        *FSW= Final Solid Waste
        ^ A negative value indicates that FSW is created
        ~ Includes savings from avoided material production

The model calculates and displays the total amount of material collected by the waste management system for recycling. The user can edit the **FSW * avoided ^ by recycling (%)** default values given in the model if more accurate local data are available. This term **FSW * avoided ^ by recycling,** simply accounts for the avoidance or creation of solid waste during the recycling process. The model calculates the final tonnage of recycled materials and subtracts the environmental burdens associated with these materials from the overall environmental burdens of the waste management system being modelled.

Compost Recycling Credits:
   Marketable compost produced (tonnes)   By Composting   By Biogasification   Total

The model calculates and displays the total amount of compost recycled as a soil improver and adds the appropriate recycling credits to the recycling section of the model.

IWM2 Model Guide

Transport from Sorting Facility/Collection Bank to Reprocessing Plant:
Distance one way (km)

The user inputs the average distance (one way) in km to the reprocessing plant used for each material. The model calculates the energy consumption and emissions. The model then calculates the savings in energy consumption associated with the recycling of the amount of recovered material predicted in the rest of the model. The energy consumption and emissions from the transport to the reprocessing plant are subtracted to give the actual savings likely from the recycling process.

Recycling Costs:
    Transport costs (£/tonne input)
    Process costs (£/tonne recycled)
    Recycled price (£/tonne recycled)

The user inserts these economic data and the model calculates the additional cost or saving attributable to the recycling of the recovered materials. Default values in the **Recycling Tab** of the **Advanced Variables** screen specify the material loss associated with the recycling of each material, if region-specific data are available they should be entered here.

CHAPTER 23

Advanced Variables

Summary

The **Advanced Variables** screen, accessed by the Advanced button on the Main Screen, allows expert users to change some of the default variables contained in the basic set up of the model.

Fuels & Electricity

Tab 1 Fuels & Electricity (Screen 28)

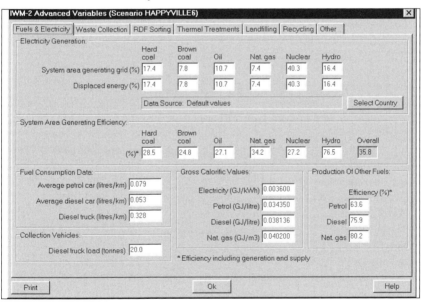

Screen 28 Advanced Variables – Fuels & Electricity.

Electricity Generation:

System area generating grid (%) Hard coal, Brown coal, Oil, Nat. Gas, Nuclear, Hydro

Displaced energy (%): Hard coal, Brown coal, Oil, Nat. Gas, Nuclear, Hydro

Select Country

This group box allows the user to select a suitable electricity supply grid for the model. The default values are based on the UCPTE model (1994). Users can select generic country data from the list in the **Select Country** button or can enter their own values.

457

User note: the energy emissions data include the emissions from the generation and supply of electricity.

The user can select the breakdown of the energy displaced by any energy recovered within the modelled waste management system (from thermal treatment, biogasification and landfill gas). The model defaults to the current breakdown of the generating grid, but this can be edited by the user to displace a single energy type or a mixture of energy types.

System area generating efficiency (%)* Hard coal, Brown coal, Oil,   Nat.   Gas, Nuclear, Hydro

*Efficiency including generation, production and supply

The model provides a default set of System area generating efficiency values; this accounts for the losses in energy during the generation, production and supply of electricity. Again users can enter their own values if they have more accurate data.

Fuel Consumption Data:
 Average petrol car (litres/km)
 Average diesel car (litres/km)
 Diesel truck (litres/km)

This fuel consumption data applies to the whole waste management collection system, the transport of materials between collection, processing and final disposal. Default values are supplied but these can be replaced by the user if more accurate data are available.

Collection Vehicles:
 Diesel truck load (tonnes)

The model defaults to a 20-tonne truck being used for collection and transport of all materials within the waste management system. This value can be altered here if necessary, but any change applies to all truck transport throughout the whole system.

Gross calorific value:
 Electricity (GJ/kWh)
 Petrol (GJ/litre)
 Diesel (GJ/litre)
 Nat. Gas (GJ/m^3)

These values are used to calculate the total energy consumption of each section of the waste management system (Energy consumption = Gross calorific value x 100/efficiency of generation or production and supply of fuel).

Production Of Other Fuels:
 Efficiency (%)*
 Petrol
 Diesel
 Nat. gas

*Efficiency including generation, production and supply

Default values for the efficiency of production and supply of Petrol, Diesel and Natural Gas are given here; these figures can be edited by the user if more accurate country-specific data is available.

Waste Collection

The **Waste Collection** tab contains variables relating to the **Kerbside Collection** systems, the **Material Bank Collection** systems, Collection Bags (both plastic and paper) and Commercial waste collection.

Tab 2 Waste Collection – Kerbside Collection System (KCS) #1 (Screen 29)

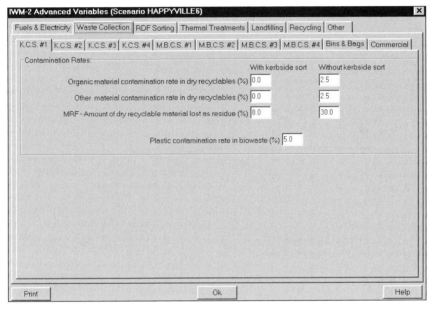

Screen 29 Advanced Variables – Waste Collection – KCS#1.

Contamination Rates:	With kerbside sort	Without kerbside sort
Organic material contamination rate in dry recyclables (%)		
Other material contamination rate in dry recyclables (%)		
MRF – Amount of dry recyclable material lost as residue (%)		

Default values are provided for each of the contamination rates; the values are taken from available literature and communications with actual operators. The availability of local data to use in place of the default values will improve the accuracy of the model. The **Kerbside Sort** option is available in each of the KCS tabs in the **Collected Household Waste** tab in the **Waste Collection** screen. The contamination rate in kerbside collection systems is very low (default value of 0) where the additional kerbside sort is carried out by the operator. This relatively uncontaminated material can be well sorted in an MRF and less material is lost as residue (8%) compared to a similar kerbside collection system operated without a kerbside sort (30%).

IWM2 Model Guide

Plastic contamination rate in biowaste (%)

The plastic contamination rate in biowaste defaults to 5% (2.5% rigid plastic and 2.5% film plastic); this material is subtracted from the **Plastic** stream and added as plastic contamination to the **Organic** stream, the same amount of Organic material is subtracted from the **Organic** stream and added to the **Restwaste** stream to balance the material flow.

The three other KCS tabs in the **Advanced Variables** screen are exactly the same as the above. The default data set being used by each will depend on whether **Kerbside Sort** or **No Kerbside Sort** has been selected for each of the kerbside collection systems described in the model.

Tab 2 Waste Collection – Material Bank Collection System (MBCS) #1 (Screen 30)

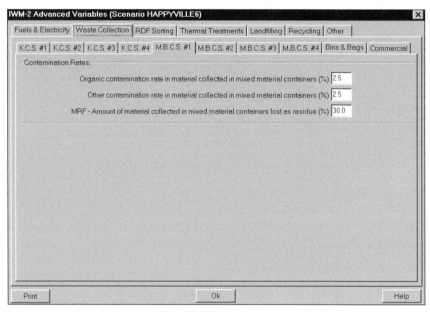

Screen 30 Advanced Variables – Waste Collection – MBCS.

Contamination Rates:

Organic contamination rate in material collected in mixed material containers (%)
Other contamination rate in material collected in mixed material containers (%)
MRF–Amount of material collected in mixed material containers lost as residue (%)

Again the default values for contamination rates that are supplied in the model are based upon available literature and operator experience. Actual local data will, as always, improve the accuracy of the model.

The three other MBCS tabs in the **Advanced Variables** screen are exactly the same as the above.

Tab 2 Waste collection – Bins & Bags (Screen 31)

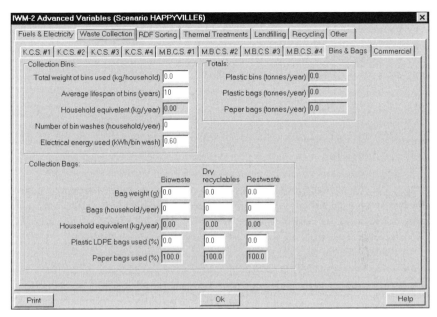

Screen 31 Advanced Variables – Waste Collection – Bins & Bags.

Collection Bins:
Total weight of bins used (kg/household)
Average lifespan of bins (years)
Household equivalent (kg/year)   Calculated
Number of washes (household/year)
Electrical energy used (kWh/bin wash)

For comparison with bag systems, the effects of bin use are calculated. The user inserts the (average) total weight of bins used per household, and the average life span of the bins used. This is converted into an equivalent total usage of material per year for the system area, and then into energy consumption and emissions. It is assumed that the bins are made from injection-moulded polypropylene.

Bin washing accounts for additional energy consumption and emissions due to the heating of the water used. The user inserts the average number of bin washes carried out per house-hold per year. The model calculates the total number of bin washes per year in the area, and converts this to electricity consumption using the estimated figure of 0.6 kWh per bin washed. This figure can be edited by the user if more accurate data is available. Electricity usage is totalled throughout the solid waste Life Cycle and converted to primary energy consumption and emissions using generic data.

IWM2 Model Guide

Collection Bags:

Bag weight (g)	**Biowaste**	**Dry Recyclables**	**Restwaste**
Bags (household/year)			
Equivalent to (kg/household/year)	Calculated	Calculated	Calculated
Plastic LDPE bags used (%)			
Paper bags used (%)	Calculated	Calculated	Calculated

This section accounts for the upstream impacts of the use of bags (i.e. during raw material acquisition, processing, bag manufacture and transport). The user is required to insert the average weight of the bags, the bag material (paper or plastic), and the number of bags used per household per year, for each fraction collected in this way. The model calculates the total amount of paper and plastic required, and converts this to energy consumption and emissions by multiplying by the generic data for production of paper and plastic given in the materials impacts section of the model. *It is assumed that the paper bags are made from 100% recycled paper, and that the plastic bags are made from low-density polyethylene (LDPE) although many refuse sacks are made at least partly from recycled LDPE; reliable data for such processes are scarce.*

The model assumes that the amounts of waste defined previously already include refuse sacks and/or bins. This modelling approach allows the user to compare the effects (with respect to air and water emissions only) of using bags versus bins.

Tab 2 Waste Collection – Commercial (Screen 32)

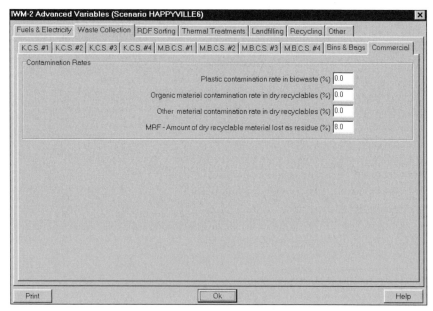

Screen 32 Advanced Variables – Waste Collection – Commercial.

IWM2 Model Guide

Contamination Rates:

Plastic contamination rate in biowaste (%)

Organic material contamination rate in dry recyclables (%)

Other material contamination rate in dry recyclables (%)

MRF – Amount of dry recyclable material lost as residue (%)

As the composition and quality of commercial waste is completely region-specific no contamination rate default values are provided with the model. Low contamination rates are to be expected in commercial wastes as they are normally of a more homogeneous nature than household waste. This low contamination rate again results in optimum sorting efficiency in an MRF and therefore low (8%) amounts of material lost as residue.

RDF Sorting

The **RDF Sorting** tab contains variables relating to both the sorting and processing of cRDF and dRDF fuels.

Tab 3 RDF Sorting – cRDF (Screen 33)

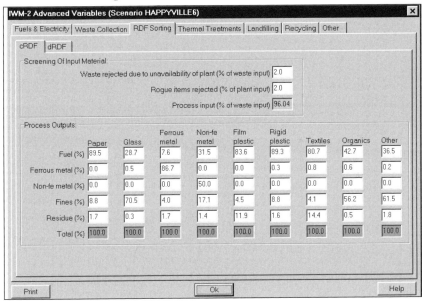

Screen 33 Advanced Variables – RDF Sorting – cRDF.

Screening Of Input Material:

Waste rejected due to unavailability of plant (% of waste input)

Rogue items rejected (% of plant input)

Process input (% of waste input)

Two screening processes are available to the user; the default values contained in the model are based on best available data.

Process Outputs:
  Fuel (%)
  Ferrous metal (%)
  Non-fe metal (%)
  Fines (%)
  Residue (%)
  Total (%)   Calculated

These default values describe the effect of the RDF Sorting process on the incoming waste fractions, and are based on the best available data (to see reference in model double click any piece of data on this screen).

Tab 3 RDF Sorting – dRDF

The **dRDF** tab is the same as the **cRDF** tab except that the default values are different and there is a default value given for Loss of moisture due to drying and pelletising (% of fuel fraction).

Thermal Treatments

The **Thermal Treatments** tab contains variables relating to process operating data, emissions data (based on old and new facilities), Refuse-Derived Fuel burning processes and Paper and Plastic-Derived Fuel burning processes.

Tab 4 Thermal Treatments – Incineration Process #1 (Screen 34)

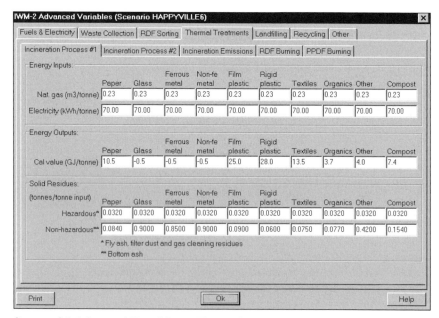

Screen 34 Advanced Variables – Thermal Treatments – Incineration Process #1.

Energy Inputs:
Nat. Gas (m³/tonne)
Electricity (kWh/tonne)

The energy input per tonne of the process is required to balance against the energy generated and recovered from the process.

Energy Outputs:
Cal. value (GJ/tonne)

This allows the model to calculate the calorific value of the waste stream entering the incinerator.

Solid Residues (tonnes/tonne input):
Hazardous*
Non-hazardous**
*Fly ash, filter dust and gas cleaning residues
**Bottom ash

The **Solid Residue** stream is split into hazardous and non-hazardous material to allow the user to determine the impact of any particular system on available landfill space for both types of landfill. A waste management system designed around large volumes of material being incinerated is unsustainable if adequate hazardous landfill space is not available.

Tab 4 Thermal Treatments – Incineration Process #2

The **Incineration Process #2** tab is exactly the same as the **Incineration Process #1** tab, but can be modified to allow two different (new and old) incinerators to be modelled within the same waste management system.

Tab 4 Thermal Treatments – Incineration Emissions (Screen 35)

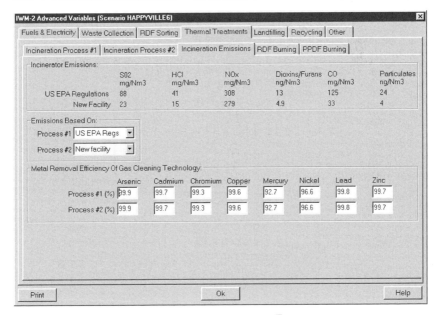

Screen 35. Thermal Treatments – Incineration Emissions.

Incinerator Emissions:

	SO$_2$ mg/Nm^3	HCl mg/Nm^3	NOx mg/Nm^3	Dioxins/Furans ng/Nm^3	CO mg/Nm^3	Particulates ng/Nm^3
US EPA Regulations	88	41	308	13	125	24
New Facility	23	15	279	4.9	33	4

The emissions from a facility complying with the US Federal Standard (1995) for Municipal Waste Combustors and the emissions from a new facility are presented on this screen. From these figures the user can decide which data set most closely matches the incineration process they are attempting to model.

Emissions Based On:
  Process #1   **US EPA Regs**
  Process #2   **New facility**

Here the user can select to model each incineration process based upon the performance data given above.

Metal Removal Efficiency of Gas Cleaning Technology:
        Arsenic   Cadmium   Chromium   Copper   Mercury   Nickel   Lead   Zinc
        Process #1 (%)
        Process #2 (%)

The default values given for the efficiency of metal removal by the gas-cleaning technology are based on average operating figures from a new facility. The user can edit these values if site-specific data are available.

Tab 4 Thermal Treatments – RDF Burning (Screen 36)

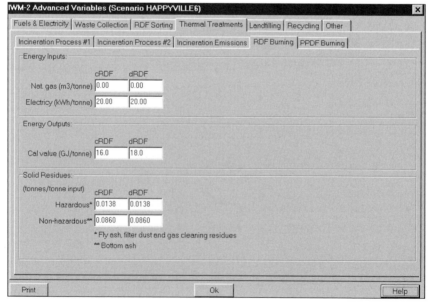

Screen 36 Advanced Variables – Thermal Treatments – RDF burning.

Energy Inputs:
  Nat. Gas (m3/tonne)   **cRDF**   **dRDF**
  Electricity (kWh/tonne) **cRDF**   **dRDF**

  Energy Outputs:
  Cal. value (GJ/tonne)   **cRDF**   **dRDF**

  Solid Residues (tonnes/tonne input):
  Hazardous*   **cRDF**   **dRDF**
  Non-hazardous**   **cRDF**   **dRDF**

*Fly ash, filter dust and gas cleaning residues
**Bottom ash

This tab is the same as the **Incineration Process #1 and #2** tabs, except that the default values for cRDF and dRDF are presented on this screen.

Tab 4 Thermal Treatments – PPDF Burning

The **PPDF** tab is again the same as the **RDF** tab, except that the default values for paper and plastic are presented, rather than cRDF and dRDF.

Landfilling

The **Landfill** tab contains variables relating to landfill gas, landfill leachate and the volume of material entering the landfill.

Tab 5 Landfilling (Screen 37)

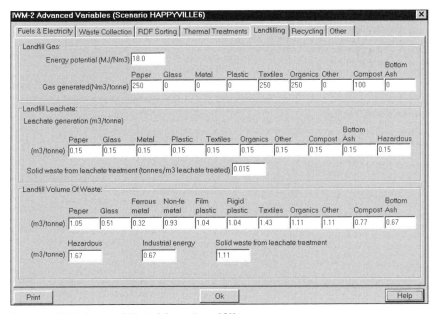

Screen 37 Advanced Variables – Landfilling.

Landfill Gas:
  Energy potential (MJ/Nm^3)
  Gas generated (Nm^3/tonne)

As with all the previous default values, the figures above are from available literature. Actual operating data for the energy potential of the landfill gas will again improve the accuracy of the overall model.

Landfill Leachate:
  Leachate generation (m^3/tonne)
  Solid waste from leachate treatment (tonne/m^3 leachate treated)

Using the default values shown here the model calculates the total volume of leachate generated by the material entering the landfill. The model also calculates the amount of solid waste residue generated during the treatment of this leachate and adds this amount to the total amount of material to be landfilled. Both figures can be edited by the user if more accurate data are available.

Landfill Volume of Waste:
  (m^3/tonne)

As landfills run out of space rather than become too heavy, the model calculates the final volume of the material landfilled in each scenario. Default values are provided (based on available literature). These values can be edited by the user if necessary.

Recycling

The **Recycling** tab contains variables that relate to the amount of Material Lost during the recycling process. These figures are used by the model to calculate the revenue per tonne of output from the recycling section.

Tab 6 Recycling (Screen 38)

Material losses (%):
  These figures are used to calculate the air and waste emissions and the recycling process costs (per tonne output) and the price for recycled materials (per tonne output).

Default values are provided for paper, glass, metals and plastics based upon figures calculated from BUWAL (1998).

Tab 6 Recycling   469

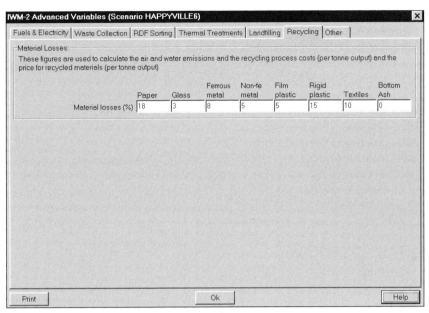

Screen 38 Advanced Variables – Recycling.

Other Variables

Tab 7 Other Variables (Screen 39)

Screen 39 Advanced Variables – Other.

SI Conversion Factors:
   Giga Joules (GJ) to Kilowatt Hours (kWh)
   Mega Joules (MJ) to Kilowatt Hours (kWh)

The values given here cannot be changed, as they are Système International d'Unités (SI) Conversion Factors.

Global Warming Potentials:
(Over a 100-year time horizon)

CO_2 (relative to CO_2)	**1.0**
CH_4 (relative to CO_2)	**21.0**
N_2O (relative to CO_2)	**310.0**

The Global Warming Potential system based on CO_2 equivalents over a 100-year time horizon as proposed by the IPCC (1996) is used in the model. The user can change the time horizon if required by entering the appropriate figures (GWP 20, CO_2 = 1.0, CH_4 = 62.0, N_2O = 290.0. GWP 500, CO_2 = 1.0, CH_4 = 7.5, N_2O = 180.0.)

IWM2 Model Guide

CHAPTER 24

Waste Flow

Waste System Flow

The Waste System Flow (Screen 40) is a schematic representation of the flow of materials in the model. This has been included to allow users to view (and print) a simple diagram that summarises each scenario, enabling quick and easy comparisons.

The Waste System Flow represents the entire waste management system: the flow of material is from the top left (inputs) to the bottom right (outputs). Figures are given for Collected Household Waste, Delivered Household Waste and Commercial Waste; the sum of these is the Total Waste input to the system.

Single-material collection banks result in the first output from the system with the collected material being sent directly to reprocessors (Materials column).

Of the remaining waste, fractions can be sent to Sorting, PPDF burning or RDF burning, Biological treatment, Incineration and finally Landfill.

Sorting waste results in the production of residue that can be incinerated or landfilled or waste suitable for Paper and Plastic-Derived Fuel burning. This burning process results in a fraction of the waste being combusted and the resulting ash being landfilled.

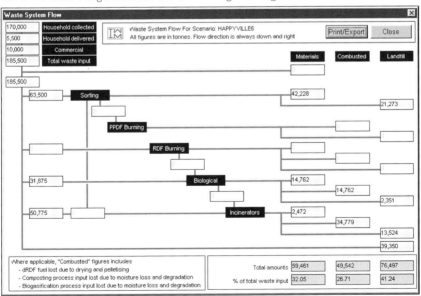

Screen 40 Waste Flow Diagram.

The waste fraction sent to RDF processing can result in the production of an organic fraction that can be sent for biological treatment and the recovery of materials (such as metals and glass) from a pre-sort. The RDF fuel that is produced and subsequently combusted results in ash that must be disposed of to landfill.

The waste fraction sent to Biological treatment can be pre-sorted to improve the quality of the final compost. The pre-sort can result in a residue that can be incinerated and materials (such as metals and glass) that can be sent for recycling. The material lost due to evaporation and biodegradation is accounted for in the Combusted column (due to space restrictions another column could not be added to this diagram). The product of Biological treatment, and Compost, is accounted for in the Materials column in the flow diagram.

The waste fraction sent for Incineration (including the resides from Sorting and Biological treatment) can also be subjected to a pre-sort, resulting in recyclable materials. The remaining waste that is combusted results in ash that must be disposed of to Landfill.

Any waste that has not been treated by one of the previous methods is sent directly to landfill.

The Waste System Flow presents the figures for the Total amount of materials recovered by the system, the Total amount of waste combusted by the system and the Total amount of waste landfilled by the system. These figures are then presented as the percentage of the Total waste input to the system.

As the figures presented in this schematic diagram are the waste flow of the scenario, they do not match the figures in the **Final Solid Waste** tab in the results section of the model. The figures in the **Final Solid Waste** tab take into account the Life Cycle of the scenario and therefore include solid waste arisings from energy production and leachate treatment and savings in solid waste from energy production and recycling.

CHAPTER 25

Streams Button

Streams

The **Streams** button on the main screen opens the **Streams** window that shows a detailed mass flow of all components in the Life Cycle Inventory (LCI) model. This screen has 11 tabs that detail the mass flow of the entire **Restwaste** stream, the **Sorting** process, the Paper and Plastic-Derived Fuel (**PPDF**) process, the **cRDF** (coarse Refuse-Derived Fuel) and **dRDF** (densified Refuse-Derived Fuel) processes, the **Biological** processes, the **Thermal** processes, the **Landfill** processes, the recovered **Materials** flow, the **Fuel** flow (electricity, natural gas, petrol and diesel) and the **Costs** breakdown.

Clicking the **Hide unused entries** box in the top left of the screen instructs the model not to display any rows that have no data in them.

The **Restwaste** tab is shown in Screen 41.

Description	Dest.	Paper	Glass	Ferrous Metal	Non-fe Metal	Film Plastic	Rigid Plastic	Textiles	Organics	Other	Total
Collected Household Waste	-	62,900	15,300	10,710	1,190	8,500	8,500	3,400	32,300	27,200	170,000
Delivered Household Waste	-	n/a	1,250	2,500	1,250	250	250	n/a	Zero	Zero	5,500
Collected Commercial Waste	-	2,000	1,000	1,800	200	500	500	500	2,500	1,000	10,000
Delivered Garden Waste	B	n/a	n/a	n/a	n/a	n/a	n/a	n/a	Zero	n/a	Zero
KCS1 Biowaste	B	-14,939	n/a	n/a	n/a	-797	-797	n/a	-15,342	n/a	-31,875
KCS1 Dry Recyclables	S	-35,625	-11,875	-7,125	-950	Zero	-4,750	Zero	-1,588	-1,588	-63,500
KCS2 Biowaste	B	Zero	n/a	n/a	n/a	Zero	Zero	n/a	Zero	n/a	Zero
KCS2 Dry Recyclables	S	Zero	Zero	Zero	Zero	Zero	Zero	Zero	Zero	Zero	Zero
KCS3 Biowaste	B	Zero	n/a	n/a	n/a	Zero	Zero	n/a	Zero	n/a	Zero
KCS3 Dry Recyclables	S	Zero	Zero	Zero	Zero	Zero	Zero	Zero	Zero	Zero	Zero
KCS4 Biowaste	B	Zero	n/a	n/a	n/a	Zero	Zero	n/a	Zero	n/a	Zero
KCS4 Dry Recyclables	S	Zero	Zero	Zero	Zero	Zero	Zero	Zero	Zero	Zero	Zero
MBCS1 Single Material Containers	M/B	Zero	Zero	Zero	Zero	Zero	Zero	Zero	Zero	n/a	Zero
MBCS1 Mixed Material Containers	S	Zero	Zero	Zero	Zero	Zero	Zero	Zero	Zero	Zero	Zero
MBCS2 Single Material Containers	M/B	Zero	Zero	Zero	Zero	Zero	Zero	Zero	Zero	n/a	Zero
MBCS2 Mixed Material Containers	S	Zero	Zero	Zero	Zero	Zero	Zero	Zero	Zero	Zero	Zero
MBCS3 Single Material Containers	M/B	Zero	Zero	Zero	Zero	Zero	Zero	Zero	Zero	n/a	Zero
MBCS3 Mixed Material Containers	S	Zero	Zero	Zero	Zero	Zero	Zero	Zero	Zero	Zero	Zero
MBCS4 Single Material Containers	M/B	Zero	Zero	Zero	Zero	Zero	Zero	Zero	Zero	n/a	Zero
MBCS4 Mixed Material Containers	S	Zero	Zero	Zero	Zero	Zero	Zero	Zero	Zero	Zero	Zero
Bulky Household Waste Recovered	M	n/a	Zero	Zero	Zero	Zero	Zero	n/a	n/a	n/a	Zero
Bulky Household Waste Residue	T/L	n/a	-1,250	-2,500	-1,250	-250	-250	n/a	n/a	Zero	-5,500
Commercial Biowaste	B	Zero	n/a	n/a	n/a	Zero	Zero	n/a	Zero	n/a	Zero
Commercial Dry Recyclables	S	Zero	Zero	Zero	Zero	Zero	Zero	Zero	Zero	Zero	Zero
Input To Coarse RDF Production	C	Zero	Zero	Zero	Zero	Zero	Zero	Zero	Zero	Zero	Zero
Input To Densified RDF Production	D	Zero	Zero	Zero	Zero	Zero	Zero	Zero	Zero	Zero	Zero
Material To Biological Stream	B	Zero	Zero	Zero	Zero	Zero	Zero	Zero	Zero	Zero	Zero
Material To Thermal Stream	T	-8,602	-2,655	-3,231	-264	-4,922	-2,072	-2,340	-10,722	-15,968	-50,775
Material To Landfill Stream	L	-5,734	-1,770	-2,154	-176	-3,281	-1,381	-1,560	-7,148	-10,645	-33,850

All figures in tonnes. Dest = Destination (S=Sorting,B=Biological,C=cRDF,D=dRDF,T=Thermal,L=Landfill,M=Materials)

Screen 41 Streams – Restwaste.

IWM2 Model Guide

In the table above, the variables used in the model are presented in the first column titled Description. The first three rows of the grid (Collected Household Waste, Delivered Household Waste and Collected Commercial Waste) describe the composition and present the total amount of waste material entering the system. In the table, the figures in black represent inputs to the system, while the negative figures in red represent material being transferred to another process or out of the system. The destination of each of the transferred materials is shown by the abbreviation in the second column of the table (the key is at the top of the tab).

Groups of rows can be subtotalled by clicking the **Sub-Total** button (select the rows to be sub-totalled and click the **Apply** button); this is helpful when investigations into the flows of individual waste fractions are being made. The sub-total of all the columns in the **Restwaste** tab should come to zero if the waste management system is complete.

The calculation used to obtain the figures in the table can be viewed by double-clicking the number itself. A **Variable Information** window opens, which displays the full calculation for every variable used in the model and gives reference details where appropriate.

For example, the calculation used to obtain the value for Collected Household Waste – Paper is (double-click the box at row 3 column 3):

S1 Population * (S1 HouseholdWastePersonYear/1000) * (S1 HouseholdPaperPercent/100)

The letter and number combinations before each variable are part of the internal identification system of the model and can be ignored. Some of the variables shown in the **Calculation** tab are calculated from other calculated variables. These secondary calculations can also be viewed by the user by double-clicking the full variable name within the **Calculation** tab.

For example, KCS1 Dry recyclables – paper, the figure is calculated using:

– S2KCS1 NumberHouseholds * (S2KCS1 DryRecyclablesPaper/1000) * ((100 –
S2KCS1 OrganicContamination – S2KCS1 OtherContamination)/100)

and the term Organic Contamination is calculated from:

if S2KCS1 KerbsideSortChoice = "Yes" then
SO2KCS1 OrganicContaminationRateWithSort else
SO2KCS1 OrganicContaminationRateWithoutSort

This ability to drill down into the model and view every calculation for every piece of data in the model is the only way true transparency can be achieved.

CHAPTER 26

Results Button

Results

The **Results** window includes five tabs: **Costs**, **Fuels**, **Final Solid Waste**, **Air Emissions** and **Water Emissions**. Each tab displays a table of results broken down into **Collection**, **Sorting**, **Biological**, **Thermal**, **Landfill**, **Recycling** and **Total**.

Throughout the results section, resources used (fuel and costs) and emissions generated (solid waste, air emissions, water emissions) are in black type, while offsets (revenue, energy produced, reductions/savings in solid waste, air emissions, water emissions) are in red type. Rows showing **Totals** (for **Costs**, **Fuels/Energy** and **Final Solid Waste**) are highlighted with yellow text on a blue background. On the **Air Emissions** tab, GWP (Global Warming Potential) is calculated and presented as an example of an internationally agreed Impact category (see Life Cycle Impact Assessment (LCIA) in Chapter 4); to highlight that this is different from the other results in the table the row has black text on a green background.

The drop-down menu labelled **Number display** allows the user to select the format of the results tables. The data can be presented as the **Nearest whole number**, between 1 to **4 decimal places** or as **Scientific notation**.

By clicking the **Print** button, the option to print, copy or save each results table becomes available.

Tab 1 Results – Costs (Screen 42)

The **Costs** table presents the **Costs** (i.e. expenditures), **Revenue** and **Total costs** of each phase of the waste management system as well as providing a **Cost/household** and a **Cost/person**. Double-clicking any data box in the results table opens the **Variable Information** screen containing the calculation details and reference information. The absolute costs presented in the model should not be considered as 100% accurate (e.g. for determining fees for waste management services), but the relative differences in costs between different scenarios can be used as simple guidelines.

Tab 2 Results – Fuels (Screen 43)

The **Fuels** tab again displays numbers as black, representing energy consumption and red, representing energy production. The gross calorific values used to convert kWh of electricity, litres of petrol and diesel and m^3 of natural gas into Giga Joules are in the **Advanced Variables** screen on the **Fuels** tab.

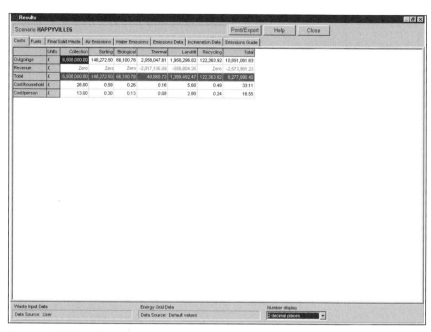

Screen 42 Results – Costs.

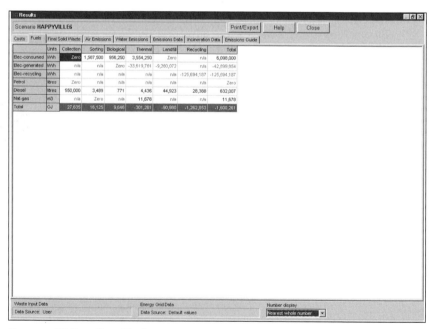

Screen 43 Results – Fuels.

Tab 3 Results – Final Solid Waste (Screen 44)

Screen 44 Results – Final Solid Waste.

The **Final Solid Waste** tab presents data on the amounts of waste being sent for final disposal to landfill. This is broken down into **Non-hazardous** and **Hazardous material** (fly ash from incineration), solid waste from the production of energy used in the waste management system (**Industrial-energy** in the table), solid waste from the manufacturing process for bin bags and collection bins, solid waste from leachate treatment, a total tonnage and a total volume. As before, black numbers represent generation, negative red numbers represent off-sets to resources used and emissions generated.

User Note: the **Total** column of the results section presents the net results of each Life Cycle Inventory, i.e. the global burdens of each waste management system modelled. When recycling is carried out, the associated burdens or savings in burdens are added to the burdens of the local waste management system (collection, treatment and disposal) giving a global perspective of the burdens associated with the waste management system rather than a local one. To calculate the local burdens of a waste management system, export the data from the **Results** tables into a spreadsheet and calculate a new total that excludes the Recycling figures from the calculation.

Tab 4 Results – Air Emissions (Screen 45)

The **Air Emissions** tab presents the emissions to air (in grams) for each phase of the waste management system. As before, black numbers represent generation, negative red numbers represent savings.

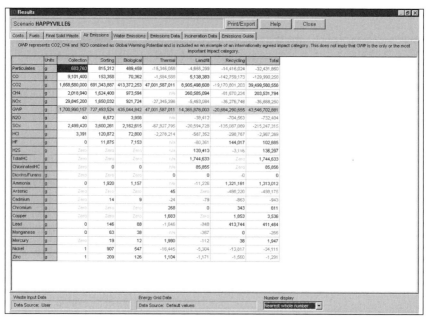

Screen 45 Results – Air Emissions.

Global Warming Potential (GWP) is included in this results table and is calculated from the Carbon Dioxide equivalents (over a 100-year time horizon) presented in the **Advanced Variables** screen on the **Other** tab. These figures are based upon the Intergovernmental Panel on Climatic Change (IPCC, 1996) report on climate change. GWP is calculated and presented as an example of an internationally agreed impact category, but this does not imply that GWP is the only (or the most important) impact category. To calculate other impact categories, the Print/Export function allows the user to copy the results table into a spreadsheet where further data manipulation can be carried out.

Tab 5 Results – Water Emissions

The **Water Emissions** tab presents the emissions to water (in grams) for each phase of the waste management system. As before, black numbers represent generation, negative red numbers represent savings. This screen is not shown in this guide document as it is very similar to the **Air Emissions** screen.

Tab 6 Results – Emissions Guide (Screen 46)

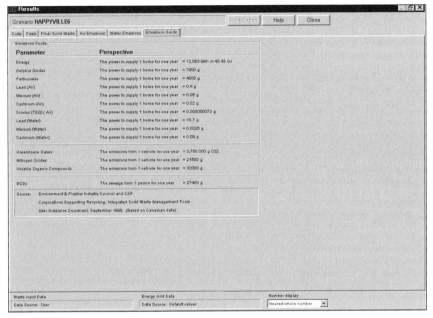

Screen 46 Results – Emissions Guide.

The purpose of the **Emissions Guide** is to provide some guidance for users that may not be familiar with either the parameters or the units (with respect to scale) used in the **Air** and **Water Emissions** tabs of the **Results screen**. Some selected parameters are put into perspective by defining them with respect to emissions from the activities of daily life. For example, the power to supply one home for 1 year produces 0.02 g of cadmium, while the emissions from one vehicle for 1 year produces 3,750,000 g of carbon dioxide. This perspective will allow users to better interpret the results of the model and to better understand the relevance of the emissions from the whole waste management system.

CHAPTER 27

Scenario Comparisons

Compare Scenarios

This feature of the IWM2 model allows the user to compare the results of two and up to eight saved scenarios. This is achieved by clicking the **Utilities** menu on the main screen and selecting the **Compare Scenarios** function. The model then requires the user to select which scenarios to compare (double-click to select, click **Next** button to begin the comparison) and calculates the results of all the scenarios. This takes a little time so please be patient. The results of this are presented in the **Compare Scenarios** window.

The model contains seven theoretical scenarios; the results of these are compared in the text below. The Happyville scenarios are as follows:

- Happyville I – Baseline scenario, all waste arisings to landfill without landfill gas collection or leachate collection.
- Happyville2 – All waste arisings to landfill with landfill gas collection and energy recovery and leachate collection and treatment.
- Happyville3 – Separate collection of 25% paper and 50% organic material for composting, restwaste as 2 above.
- Happyville4 – Material bank collection system for dry recyclables, restwaste as I above.
- Happyville5 – Kerbside collection of dry recyclables plus Material bank collection system for dry recyclables, restwaste as 2 above.
- Happyville6 – Separate collection of biowaste (paper and organics), Kerbside collection of dry recyclables plus Material bank collection system for dry recyclables, 25% of restwaste incinerated with energy recovery, remaining restwaste as 2 above.
- Happyville7 – Separate collection of biowaste (paper and organics), Kerbside collection of dry recyclables plus Material bank collection system for dry recyclables, all restwaste incinerated with energy recovery. All residues to landfill with landfill gas collection and energy recovery and leachate collection and treatment.

In the figures the cost comparisons of the Happyville scenarios are not shown as the generic numbers used are not representative of real waste management costs at any single location and could therefore be viewed as misleading. Cost data should always be entered by the user and should reflect realistic costs of waste management under the prevailing (or proposed) local conditions.

The box in the top left corner of the screen contains the results (**Costs**, **Fuels**, **Final Solid Waste**, **Air Emissions** and **Water Emissions**) that have been compared by the model. Select a set of results such as **Final Solid Waste** by double-clicking, then select a single

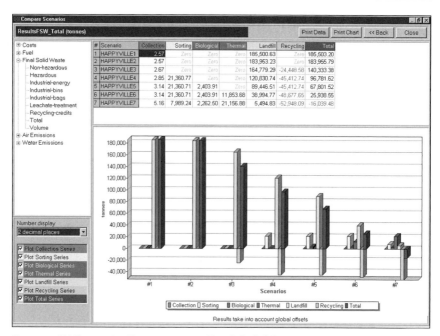

Screen 47 Scenario Comparison – Final Solid Waste.

variable within this set (e.g. **Total**) by clicking. The results of the user's selected scenarios are presented in the table in the top right of the screen and this data is presented graphically below. In the bottom left of the screen the user can select (click on and off) which phases of the waste management system are graphed. The model allows the user to select the scale on the graphs by using the **Units** drop-down menu at the top of the screen; this ensures that the graphs can be switched between grams, kilograms and tonnes depending on the variable being studied.

User Note: the scenario comparison section (like the results section) presents the net results of each Life Cycle Inventory, i.e. the global burdens of each waste management system modelled. When recycling is carried out, the associated burdens or savings in burdens are added to the burdens of the local waste management system (collection, treatment and disposal) giving a global perspective of the burdens associated with the waste management system, rather than a local one. This is why in Scenario #7 above the total amount of material landfilled is negative. Of course at a local level this is not the case: residues from collection, sorting, composting and thermal treatment still need to be landfilled locally. But, when the savings in solid waste production due to the use of recycled materials rather than virgin materials are added to the amount of material requiring landfill, the overall figure can sometimes be negative. This situation occurs in waste management systems with high recycling and diversion rates.

The user can select from the list of **Air Emissions** and compare these emissions between each of the scenarios as in Screen 48, which presents data on Global Warming Potential.

User Note: Global Warming Potential is included in the **Air Emissions** as an example of an Impact Category, as the methodology for calculating this as been agreed by the Intergovernmental Panel on Climate Change (IPCC, 1996).

IWM2 Model Guide

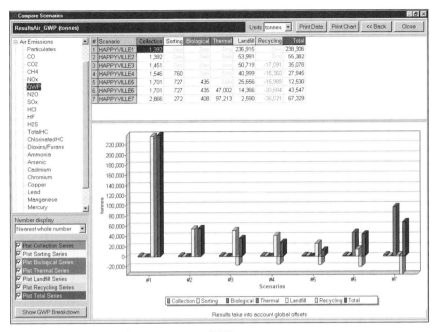

Screen 48 Scenario Comparison – GWP.

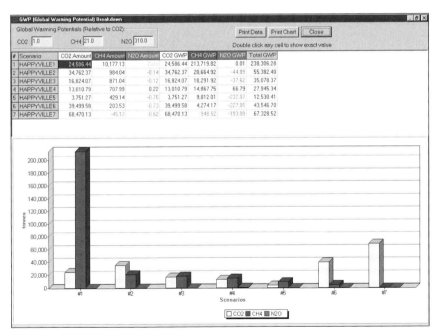

Screen 49 Scenario comparison – GWP breakdown.

If, as in the above example, Global Warming Potential (GWP) is graphed, an extra option becomes available: the **Show GWP Breakdown** button. This option simply presents the breakdown of the GWP figures into carbon dioxide equivalents for CO_2, CH_4 and N_2O, as presented in Screen 49.

Making comparisons

This section looks at the results that can be obtained from a Life Cycle Inventory of solid waste, and considers ways in which such results can be interpreted and used. Life Cycle Inventories for solid waste, or for any other product or service, are mainly used in a comparative way (Figure 27.1). The absolute environmental performance (in terms of energy consumption and emissions) or economic cost is less informative than a comparison between different options to see which is preferred. This comparison may be between completely different systems or

How to use the LCI tool?

Approach 1: Assume waste stream is constant, and compare the performance of different Integrated Waste Management systems in dealing with this waste.

- valid comparison

Approach 2: Assume the waste management system is constant, and look at the effect of altering the amount and composition of waste that the system handles.

- valid comparison if comparing geographical differences in waste
- caution needed if looking at changing the materials entering the waste stream in a given area, since these materials performed a function before they became waste. Alternative materials for this function may have environmental burdens elsewhere than in the waste management system. A product-specific LCI is needed to find the best material for any function, throughout the Life Cycle, not just in the waste management system.

How to choose between options?

Single criterion   – where there is a single over-riding concern (e.g. lack of landfill space)

Multiple criteria   – where more than one issue is important (e.g. energy consumption and landfill space)
                       – 'less is better' – where one option is lower in all categories
                       – Impact Assessment – can combine some parameters that contribute to the same environmental effect such as global warming (must be a transparent process).

Figure 27.1 Making Comparisons using LCI Results.

products, or between an existing one and a potential improvement. This section looks at how LCI results for solid waste can be used effectively in a comparative way. But first it is necessary to determine which comparisons are valid (Figure 27.1).

The objective of the LCI of solid waste was defined earlier as attempting to predict the environmental performance (in terms of emissions and energy consumption) and economic costs of an integrated waste system, which can manage the waste of a given area. A valid comparison, therefore, is between alternative IWM systems that deal (in different ways) with the specified waste stream of the area in question (Approach 1, Figure 27.1).

Rather than keeping the waste stream entering the system constant and looking at the effect of altering the waste management system, an alternative approach is to keep the waste management system constant, and to investigate the effect of altering the amount and composition of the waste stream entering the system (Approach 2, Figure 27.1). This could take the form of asking "What would be the effect of removing all plastic (or paper, glass, metal, etc.) from the waste stream?" These comparisons, however, need to be treated with considerable caution. The LCI model can predict how present waste management systems would perform with changed waste inputs, and if particular materials were eliminated from the waste stream, but that is not the end of the story. If materials are eliminated from the waste stream by not using

Environmental factors:

Energy –	Electricity consumed by each part of the waste management system
	Electricity generated by each part of the waste management system
	Electricity balance of recycling
	Petrol used by each part of the waste management system
	Diesel used by each part of the waste management system
	Natural gas used by each part of the waste management system
	Total energy used by each part of the waste management system
Air Emissions –	Amount of each individual material listed (24 parameters)
Water Emissions –	Amount of each individual material listed (27 parameters)
Final Solid Waste –	Weight of non-hazardous material landfilled
	Weight of hazardous material landfilled
	Weight of material from industrial energy production
	Weight of material from industrial bin production
	Weight of material from industrial bag production
	Weight of material from leachate treatment
	Weight of material from recycling credits (offset burdens)
	Total weight of material landfilled
	Total volume of material landfilled

Economic costs:

	Outgoings of each part of the waste management system
	Revenue of each part of the waste management system
	Total cost of each part of the waste management system
	Cost per household of each part of the waste management system
	Cost per person of each part of the waste management system

Figure 27.2 Outputs of the Life Cycle Inventory.

IWM2 Model Guide

them in products or packaging, then there will also be environmental and economic effects earlier in the Life Cycle of the product or package itself, i.e. before it becomes waste. Thus while the burdens of waste management may be reduced, there may be greater Life Cycle burdens elsewhere, such as in the sourcing, manufacture or transport of any replacement material. Choosing which material to use for a given product or package requires a product-specific LCI, rather than the LCI for waste that has been developed in this book. Again this shows the important linkage between product-specific LCIs and the LCI for solid waste; both are important as they answer different questions.

Given that a valid comparison is being made, how can the user choose between different waste management system options on the basis of LCI results? The full inventory for each option will consist of all the different parameters presented in Figure 27.2; how can useful comparisons be made between such extensive lists?

1. Single criteria. If there is a single over-riding environmental or cost consideration, then the basis of choice between alternative waste management systems is straightforward. If the aim is to minimise landfill requirement (as in countries such as Japan), then the system that meets this objective can be chosen. Similarly, if cost is the only important factor, the cheapest option can be selected. There are few instances, however, where a single factor alone is important. Indeed, if the objective of Integrated Waste Management is environmental and economic sustainability, then both environmental burdens and economic costs must be considered. Similarly, within the realm of environmental burdens, there are many individual factors to consider, rather than just one.

2. Multiple criteria. Where many categories, both environmental and economic, are considered important, choosing between alternative waste management options will be straightforward if one option has lower burdens, in all respects, than the other. Given the number of parameters calculated in this LCI for solid waste, it is generally unlikely that the energy consumption, individual emissions to air and water, final inert waste and cost of one system will all be lower than those of an alternative system. If all burdens are lower, selecting the preferred system is simple; if, as is more likely, some burdens are higher whilst others are lower, a method to trade-off these difference is needed.

Choosing between different waste management alternatives on the basis of environmental and economic performance is facilitated if the number of parameters to be considered can be reduced by aggregation. As discussed in Chapter 4, various methods have been used to aggregate inventory data into a smaller number of impact categories. The method that has been most widely accepted is the aggregation of parameters that contribute to the same environmental effect. Methane and carbon dioxide emissions, for example, both contribute to the greenhouse effect and so will lead to global warming. Using equivalence factors according to their relative effect on global warming, it is possible to aggregate emissions of these gases together into a global warming impact category. The release of 1 kg of methane will cause an equivalent contribution to global warming as 21 kg of carbon dioxide, taken over a 100-year time-span (IPCC, 1996), giving a Global Warming Potential (GWP) for methane of 21 (CO_2 is taken as the reference global warming gas). Similar Impact Assessment methodologies have been suggested for some, but not all, of the environmental impact categories listed in Table 4.2 (Potting et al., 1998; Finnveden and Potting, 1999; De Haes et al., 1999). Further development of this important stage of a Life Cycle Assessment is ongoing.

The objective of the IWM-2 computer model is limited to producing a Life Cycle Inventory, so a full Impact Assessment is not attempted. In the model however, the impact category of Global Warming Potential is calculated as an example of how Impact Assessment can be applied.

Identifying improvement opportunities

In discussions of Integrated Waste Management in this book, there has been considerable emphasis on the use of Total Quality Management techniques. One of the key concepts of Total Quality is continuous improvement: the process of continually monitoring performance and looking for ways to improve a system. As it provides a way to monitor performance, the LCI of waste can not only be useful in choosing between different options, as demonstrated above; it can also be used within any waste management system to identify areas for potential improvement.

The importance of operations in the home

To identify where the greatest potential for improvement exists, it is first necessary to determine where the largest burdens occur, and then find ways of reducing them. Essentially, this consists of running a sensitivity analysis on the system, to show which possible alterations will result in significant improvements. When this is done, some unexpected results can occur. Clearly, if the environmental performance of the waste treatment processes is improved, there will be overall system improvements. If, for example, more of the landfill gas could be collected and burned with energy recovery, or more landfill sites were lined with leachate collection and treatment, there would be fewer potentially harmful emissions to air and water. What is perhaps surprising, however, is the significant effect that the behaviour of householders can have on the overall system.

Taking a basic waste collection system, where all household waste is collected (on a weekly basis, where 30 litres fuel is used to collect waste from 1000 households) co-mingled, the effect of householders' behaviour can be predicted. If each household was to wash out its bin with warm water every month, the energy consumption of the collection system would increase by approximately 16%. Similarly, if each household in the system was to use an extra 5 g LDPE plastic collection bag per week, there would also be a 16% increase in the energy consumption of the collection system. If every household was to make one special trip of 2 km each way by car each week to a material bank site to drop off recyclable materials, this would have a dramatic effect on the energy consumption of the collection system, which would increase by 190%.

The importance of the actions of the householders should perhaps not come as a complete surprise. After all, the household is the source of the waste that needs to be managed, and the place that the initial sorting of waste can occur. Householders can have a very dramatic effect on the burdens of waste management by reducing the amount of waste that they put into the waste management system in the first place, by choosing to use, for example, concentrated or compact products that contribute less waste themselves, and use less packaging per use. Using refill packs or light-weighted containers will also tend to reduce the overall generation of household waste. If households were to generate less waste the overall environmental burden of managing the total amount of waste would fall accordingly. Similarly, for the waste that they still generate, households can ensure that as much as possible is valorised by effectively separating the waste into the different categories requested.

One of the most important opportunities for overall environmental improvement, therefore, is likely to occur within each household. The household's behaviour in generating, and separating the waste (if required) and also in its use of car transport, collection bags and bin washing will have major effects on the overall system performance. Any burdens that occur at the household stage will repeated in every household in the system, so can end up having a highly significant overall effect.

This importance of consumer behaviour in determining the overall environmental burden of both products and services has been highlighted before. The well-known LCA practitioner Dr Ian Boustead has often told the story that the overall energy consumption for producing a bottle of whisky can be doubled simply by the consumer driving a short distance to the shop to purchase this item alone. Hindle et al. (1993) have similarly shown that it is by focusing on the usage of a product by consumers that the largest potential environmental improvements can be identified. In waste management too, householders' behaviour has a critical effect on the overall environmental burden. Communicating with these vital players in the system is essential. Investing time and money in educating households about the effect their own actions in generating and handling waste can have on the environment would seem to be time and money well spent.

System improvements

Whilst the case studies described above may not themselves be strictly applicable to a particular area, they do provide some general pointers as to how an Integrated Waste Management system can be optimised.

Optimising collection – Collection of the waste usually represents a significant part of the economic cost of a waste management system, and, as the above examples show, it can also be a significant source of environmental burdens. Optimising the collection system will therefore improve the overall performance of most systems, in both economic and environmental terms. This means collecting all of the waste from households in the form necessary for the chosen waste management options in as few visits as possible. At the same time the comfort level of the householders must be maintained for full participation and co-operation to be achieved.

Including paper in the biobin – There are several advantages in adopting a wider definition of biowaste, when source-separated organic material is collected for composting, non-recyclable paper can be included in the biowaste. The major advantage is the diversion of more material from landfill (i.e. less final solid waste produced, along with the emissions that landfilling produces) along with the production of more useful product (compost). There are other benefits as well, including the need to rinse out the collection bin less often, the production of less leachate during collection and composting, and the production of a compost with a higher organic matter content (see Chapter 11 for full details).

Benchmarking waste treatment processes and identifying outliers – When the model is used, the most useful results will come from inserting locally sourced data for the region and system in question. In such cases the generic data supplied in the model can be considered as a benchmark against which the local process performance can be assessed. If the level of energy consumption or emission generation differs widely from the benchmark figure, this will act as a prompt to re-check the data, and perhaps to seek ways of carrying out the relevant process more efficiently.

CHAPTER 28

What Parameters Have Changed?

What's Changed?

The final function of the model allows a detailed examination of the changes within one scenario resulting from the adjustment of one or more variables. The **What's Changed?** function in the **Utilities** menu enables users to quickly view the effects of changing variables as a scenario is being developed or to view them in a previously saved scenario (Screen 50).

For example, a fundamental change such as increasing the number of people per household in the Waste input screen – **System Area** tab results in major changes in the mass flow throughout the model (more than 200 variables change, the exact number depending upon the existing scenario), but a small change such as increasing the landfill gas recovery rate only results in two variables changing in the model.

As with previous tables, double-clicking any data box opens the **Variable Information** screen containing the appropriate calculation details and ISO reference information. As before, black numbers represent generation, negative red numbers represent savings. This function

Variable Name	Input Section	Current Value	Saved Value	Difference	Type
S02PlasticBins	#0 Advanced Variables	50.0	55.6	-5.6	Calculat
S02PlasticBags	#0 Advanced Variables	400.1	444.6	-44.5	Calculat
S02PaperBags	#0 Advanced Variables	414.4	460.4	-46.0	Calculat
S1AveragePersonsPerHousehold	#1 Waste Inputs	2.0	1.8	+0.2	User En
S1NumberOfHouseholdsServed	#1 Waste Inputs	100,000	111,111	-11,111	Calculat
S1TotalGlass	#1 Waste Inputs	20,186	20,242	-56	Calculat
S1TotalFerrousMetal	#1 Waste Inputs	22,216	22,305	-89	Calculat
S1TotalNonFerrousMetal	#1 Waste Inputs	3,278	3,290	-12	Calculat
S1TotalFilmPlastic	#1 Waste Inputs	10,354	10,365	-11	Calculat
S1TotalRigidPlastic	#1 Waste Inputs	14,854	14,865	-11	Calculat
S1TotalOrganics	#1 Waste Inputs	80,008	80,141	-133	Calculat
S1TotalOther	#1 Waste Inputs	64,868	64,968	-100	Calculat
S1TotalTotal	#1 Waste Inputs	237,300	237,711	-411	Calculat
S2KCS1NumberHouseholds	#2 Waste Collection	10,000	11,111	-1,111	Calculat
S2KCS2NumberHouseholds	#2 Waste Collection	20,000	22,222	-2,222	Calculat
S2KCS3NumberHouseholds	#2 Waste Collection	30,000	33,333	-3,333	Calculat
S2KCS4NumberHouseholds	#2 Waste Collection	40,000	44,444	-4,444	Calculat
S2KCSTotalNumberHouseholds	#2 Waste Collection	100,000	111,111	-11,111	Calculat
S2MBCS1NumberHouseholds	#2 Waste Collection	5,000	5,556	-556	Calculat
S2MBCS2NumberHouseholds	#2 Waste Collection	10,000	11,111	-1,111	Calculat
S2MBCS3NumberHouseholds	#2 Waste Collection	1,000	1,111	-111	Calculat
S2MBCS4NumberHouseholds	#2 Waste Collection	10,000	11,111	-1,111	Calculat
S2MBCSTotalNumberHouseholds	#2 Waste Collection	26,000	28,889	-2,889	Calculat

Whats Changed

Variables changed since the last save: 246. Double click any variable for more info.   Print   Help   Close

Screen 50 What's Changed?

IWM2 Model Guide

has been designed as a method to help users quickly identify in detail the consequences of their decisions within the model. This is likely to be helpful during the process of optimising waste management systems, as it can be used as a very simple sensitivity analysis.

References

Adan, B.L., Cruz, V.P. and Palalay, M. (1982) *Solid Waste Management Study: Scavenging Study.*

ADAS (1997) *Composting Report.* Available from http://www.adas.co.uk/

Adur District Council (1996) West Sussex County, UK. Mr P. Olver, personal communication.

AEA (1999) *Methane Emissions from UK Landfills.* Final report for the Department of the Environment, Transport and the Regions, UK. AEA Technology Environment, Culham, Abingdon, Oxfordshire, UK.

AFPA (1997) *1997 Statistics on Paper, Paperboard & Woodpulp.* American Forest & Paper Association, Washington DC, USA.

AFR (1997) AFR Report 154, *Composting.*

AMG (1992) *Metal Waste Reclamation.* AMG Resources Brochure, AMG Ltd, Harborne, Birmingham, UK.

Anon. (1994). *Modern Plastics,* Vol. 71, No. 1, p. 73.

Anon. (1995) *Fiber Organon* Vol. 66 Fibre Economics Bureau, Washington, DC.

Anon. (1998) *Getting Green Dotted: The German Recycling Law in Plain English.* Raymond Communications, Riverdale, MD 20737, USA.

Anon. (1999) Recycling plastic bottles. *Resource Recycling,* Vol. 18, No. 2, pp. 11–19.

ANRED (1990) *Sorting and Composting of Domestic Waste.* ANRED Publication 27, ANRED, France.

ANS (1999) Arbeitskreis für die Nuutzbarmachung von Siedlungsabfällen eV. *Müll und Abfall,* February, pp. 103–104.

APC (1998) New Generation Municipal Waste Combustion Facility meets NSPS. *Air Pollution Consultant,* Vol. 8, No. 1, pp. 1.2–1.4.

Archer, D.B. and Kirsop, B.H. (1990) The microbiology and control of anaerobic digestion. In: *Anaerobic Digestion: A Waste Treatment Technology,* Ed. A. Wheatley, Elsevier Applied Science, Oxford.

Area Metropolitana de Barcelona (1997) *Programa Metropolita de Gestio de Reisus Municipals (1997–2006).*

Atkinson, W. and New, R. (1993a) *An Overview of the Impact of Source Separation Schemes on the Domestic Waste Stream in the UK and Their Relevance to the Government's Recycling Targets.* Warren Spring Laboratory Report LR 943, February, 1993. Warren Spring Laboratory, Stevenage, UK.

Atkinson, W. and New, R. (1993b) *Kerbside Collection of Recyclables from Household Waste in the UK – A Position Study.* Warren Spring Laboratory Report LR 946, May, 1993. Warren Spring Laboratory, Stevenage, UK.

Augentein, D. and Pacey, J. (1991) Modelling landfill methane generation. In: *Biogas Disposal and Utilisation, Choice of Material and Quality Control, Landfill Completion and Aftercare, Environmental Monitoring.* Third International Landfill Symposium, Sardinia, 14–18 October 1991, pp. 115–148.

Axt, P.G. (1998) *Comparison of the Cost Efficiency of Packaging Recovery Systems in Germany and The Netherlands (Verwertung ausgedienter Verpackungen: Kosten/Leistungsvergleich am Beispiel von Deutschland und den Niederlanden).* Diploma from University of Bayreuth, Institute for Applied Economics, Supervisor: Professor Dr Andreas Troge, Honorary Professor for Environmental Economy.

Baccini, P., Belevi, H. and Lichtensteiger, T. (1992) *Die Deponie in einer kologisch orientierten Volkswirtschaft.* GAIA No. 1, pp. 34–49.

Baldwin, G. and Scott, P.E. (1991) Investigations into the performance of landfill gas flaring systems in the UK. In: *Biogas Disposal and Utilisation, Choice of Material and Quality Control, Landfill Completion and Aftercare, Environmental Monitoring.* Third International Landfill Symposium, Sardinia, 14–18 October pp. 301–312.

Barres, M., Grenet, Y., Millot, N. and Meisel, A. (1990) Sanitary landfilling in France. In: *International Perspectives on Municipal Solid Wastes and Sanitary Landfilling*, Eds J.S. Carra and R. Cossu, Academic Press, London, pp. 78–93.

Barton, J. (1986) *The Application of Mechanical Sorting Technology in Waste Reclamation: Options and Constraints*. Warren Spring Laboratory. Paper presented at the Institute of Waste Management and INCPEN Symposium on Packaging and Waste Disposal Options, London.

Barton, J.R., Poll, A.J., Webb, M. and Whalley, L. (1985) *Waste Sorting and RDF Production in Europe*. Elsevier Applied Science, Oxford.

Bartone, C.R. (1999) *Financing Solid Waste Management Projects in Developing Countries: Lessons from a Decade of World Bank Lending*. Proc. Organic Recovery and Biological Treatment, ORBIT 99, Part 3. Rhombos, Berlin, pp. 757–765.

Bechtel, P. and Lentz, R. *Landfilling and Incineration: Some Environmental Considerations and Data Concerning the Disposal of Municipal Solid Waste*. P&G Internal report, unpublished.

Beckerman, W. (1995) *Small is Stupid: Blowing the Whistle on the Greens*. Duckworth, London, Chapter 4, Section 2, p. 51.

Beker, D. (1990) Sanitary landfilling in The Netherlands. In: *International Perspectives on Municipal Solid Wastes and Sanitary Landfilling*, Eds. J.S. Carra and R. Cossu. Academic Press, London, pp. 139–155.

Benitez, J. (1995) *Process Engineering and Design for Air Pollution Control*. Prentice Hall, Englewood Cliffs, NJ, pp. 333–371.

Bergmann, M. and Lentz, R. (1992) *Vorstudie zu einer Ökobilanz über biologische Abfallbehandlungsverfahren*, Internal report. Procter & Gamble GmbH, Schwalbach, Germany.

Bernache-Perez, G., Sanchez-Colon, S. and Garmendia, A.M. (1999) *Solid Waste Characterization Study in the Guadalajara Metropolitan Zone (GMZ), Mexico*. See *Warmer Bulletin*, No. 68, pp. 4–5 for summary.

Berndt, D. and Thiele, M. (1993) *Status des Dualen Systems und seine Kosten*. Verpackungs-Rundschau, October 1993, pp. 84–88.

Beyea, J., DeChant, L., Jones, B. and Conditt, M. (1992) Composting plus recycling equals 70 percent diversion. *Biocycle*, May 1992.

BfE (1990) Bundesamtes für Energiewirtschaft. *Vergärung biogener Abfälle aus Haushalt, Industrie und Landschaftspflege*. Schriftreihe des Bundesamtes für Energiewirtschaft Studie Nr 47.

BGK Bundesgütegeinschaft Kompost (1998) *Kompostverwertung in den verschiedenen Bundesländern (Stand 1997)*. Briefliche Mitteilung und Marketpotential von Kompost im 1996. Bundesgütegeinschaft Kompost eV, Köln.

Biala, J. (1998) Economic implications of separate collection and treatment of organic waste materials. *Waste Disposal and Water Management in Australia*, March/April 1998, pp. 10–25.

Biffa (1992) Biffa Waste Services, personal communication, May 1992.

Biffa (1994) Biffa Wastes Services, personal communication, January 1994.

Biocycle (1992) Co-collection at curbside. *Biocycle*, September 1992, pp. 56–57.

Birley, D. (1993) Does the Blue Box have a future? *Warmer Bulletin*, Vol. 35, pp. 10–12.

Birtley, D. (1996) Kerbside boxes revisited: why kerbside boxes after all? *Waste Management*, August 1996, pp. 16–17.

Boelens, J., De Wilde, B. and De Baere, L. (1996) Comparative study on biowaste definition: effects on biowaste collection, composting process and compost quality. *Compost Sci. and Utiliz.*, Vol. 4, No. 1, pp. 60–72.

Bongartz, T. and von Dörte Naumann, M. (1991) *Aktuelle Entsorgungskosten in der Bundesrepublik Deutschland*. EntsorgungsPraxis, March 1991, pp. 84–86.

Boswell, J.E.S. and Charters, G.J. (1997) *Scavenging, Salvaging and Recycling on Landfills*. Proc. Sardinia 97, 6th International Landfill Symposium, Vol. 5, pp. 403–414.

Bothmann, (1992) Cited in *IFEU*, 1992.

Boustead, I. (1992) *The Relevance of Re-use and Recycling Activities for the LCA Profile of Products*. Procs. Third CESIO International Surfactants Congress and Exhibition, London, 1–5 June, 1992. pp. 218–226.

Boustead, I. (1993a) *Resource Use and Liquid Food Packaging. E.C. Directive 85/339: UK Data 1986–1990*. A report for INCPEN, May 1993.

Boustead, I. (1993b) *Aerosols and Other Containers*. A Report for the British Aerosol Manufacturers' Association, January, 1993.

Brisson, I.E. (1997) *Assessing the Waste Hierarchy – a Social Cost–Benefit Analysis of Municipal Solid Waste Management in the European Union*. AKF, Institute of Local Government Studies, Denmark. Available at www.akf.dk/eng

Brown, P. (1991) Gas risk to homes from 1,000 rubbish tips. *The Guardian*, 4 April 1991.

Brunner, P.H. (1998) *Material Flow Analysis as a Tool for Sustainable Regional Materials Management*. International NOW Conference Beyond Sustainability, Amsterdam 1998.

Bund (1992) *Bioabfallkompostierung vorrangige Abfallverwertung*. Report by E. von Lossau, M. Krauß, and Neidhardt, R. Bund Hessen.

Bundesamtes für Energiewirtschaft (1991) *Vergärung biogener Abfälle aus Haushalt, Industrie und Land-schaftspflege*. Schriftreihe des Bundesamtes für Energiewirtschaft Studie, Nr 47.

Bundesgütegemeinschaft Kompost (1997) *Kompostanlagen in Deutschland 1995*. Kompost-Information Nr 104. Abfall Now Verlag, Stutgart.

BUWAL 250/II (1998) *Life Cycle Inventories for Packagings*, Vol. II. Swiss Agency for the Environment, Forests and Landscape (SAEFL). Environmental Series No.250/II. CH-3003 Berne.

Camacho, L.N. (1995) *Recycling in Manila*. UNEP Industry and Environment, January–March, p. 81.

Campbell, D.J.V. (1991) An universal approach to landfill management acknowledging local criteria for site design. In: *Biogas Disposal and Utilisation, Choice of Material and Quality Control, Landfill Completion and Aftercare, Environmental Monitoring*. Third International Landfill Symposium, Sardinia, 14–18 October, pp. 15–32.

Canterbury, J. (1998) How to succeed with pay as you throw. *BioCycle*, December.

Cardinale, P. (1998) *Análisis de ciclo de vida: Una herramienta de gerenci ambiental*. Debates IESA. Instituto de Estudios Superiores de Administración. Venezuela, pp. 34–38.

Carra, J.S. and Cossu R. (1990) Introduction. In: *International Perspectives on Municipal Solid Wastes and Sanitary Landfilling*, Eds J.S. Carra and R. Cossu. Academic Press, London, pp 1–14.

Cathie, K. and Guest, D. (1991) *Wastepaper*. PIRA International, UK.

CEMPRE. (1999) Manual de Gerenciamento Integrado. Rua Anatalicia Ferreira Silva, 50, Santo Amaro – CEP: 04710–060 – São Paulo – SP, Brazil. http://www.cempre.org.br/

Centemero, M., Ragazzi, R. and Favoino, E. (1999) *Label Policies, Marketing Strategies and Technical Developments of Compost Markets in the European Countries*. Proc. Organic Recovery and Biological Treatment, ORBIT 99, Part 2. Rhombos, Berlin, pp. 355–362.

CEPI (1999) Confederation of European Paper Industries. 250, Avenue Louise, B-1050 Brussels, Belgium.

Chang, D.P.Y. (1996) Chlorine in waste combustion. *Hazardous waste & Hazardous Materials*, Vol. 13, No. 1, pp. 9–11.

Charters, G. (1996) *Waste Site Scavenging. A Viable Recycling Service or Dangerous Social Exploitation?* Proc. Wastecon 96, Durban Institute of Waste Management of Southern Africa, Parallel session paper 45.

Chem Systems *Life Cycle Inventory: Incinerator Build, Maintain, and Decommission*. A report prepared for the UK Environment Agency, unpublished.

Cheremisinoff, P.N. (1987) *Waste Incineration Handbook*. Reed International, UK, pp. 52–56.

Christensen, T.H. (1990) Sanitary landfilling in Denmark. In: *International Perspectives on Municipal Solid Wastes and Sanitary Landfilling*, Eds J.S. Carra and R. Cossu. Academic Press, London, pp.37–50.

Clark, R.M. (1978) *Analysis of Urban Solid Waste Services – A Systems Approach*. Ann Arbor Science Publishers, Ann Arbor, MI.

Clark, H. and New, R. (1991) *Current UK Initiatives in Plastics Recycling*. Paper presented to Plastics Recycling '91 European Conference, Copenhagen, 5–7 February.

Claydon, P. (1991) Recycled fibre – the major raw material in quality newsprint. *Paper Technology*, Vol. 32, No. 5, pp. 34–37.

Clayton E. (1991) *Review of Municipal Solid Waste Incineration in the UK*. Warren Spring Laboratory Report LR 776. (PA) Department of the Environment Research Programme.

Cointreau-Levine, S. (1997) *Occupational and Environmental Health Issues of Solid Waste Management*. International Occupational and Environmental Medicine, Mosby, St Louis, MO, Chapter 4, pp. 38-1–38-22.

Consoli, F. (1993) *Guidelines for Life-Cycle Assessment: A 'Code of Practice'*. SETAC, Brussels, Pensacola.

Coombs J. (1990) The present and future of anaerobic digestion In: *Anaerobic Digestion: A Waste Treat-ment Technology*. Elsevier Applied Science, Oxford, pp. 1–42.

Cooper, J. (1998) Integrating energy recovery and recycling. *Materials Recycling Week.*, 18 December, pp. 14–15.

Corbitt, R.A. (1998) *Standard Handbook of Environmental Engineering*, 2nd edition. McGraw-Hill, New York.

Corporation of the City of London (Ruksana Mirza) (1999) RFP 98–58, *Selected Programs Within London's Solid Waste Management System*. Technical Memorandum: *Environmental Evaluation of Recycling Packages*.

Cosslett, G. (1997) Bring and burn. *The Waste Manager*, May, pp. 21–23.

Council Report (1997) *Overview: The Recommended Waste Management Strategy for London/Middlesex*. City of London, Ontario, Canada.

DOE (1989) *Landfill Gas*. Waste management paper no. 27. Department of the Environment, London.

DOE (1990) *Our Common Inheritance*. Environmental Protection Act White Paper, Department of Environment, London.

DOE (1992) Ed. A. Brown, *The UK Environment*. Department of Environment and Government Statistical Office report, HMSO, London, Chapter 13.

DOE (1993) *Landfill Costs and Prices: Correcting Possible Market Distortions*. A Study by Coopers & Lybrand for the Department of the Environment HMSO, London.

DOE (1997) *Markets and Quality Requirements for Composts and Digestates from the Organic Fraction of Household Wastes*. Available from the Waste Management Information Bureau, F6 Culham Laboratory, Abingdon, Oxfordshire OX14 3DB, UK.

DOE/DTI (1991) Ecolabelling of washing machines. *Environmental Labelling*, Newsletter from the National Advisory Group on EcoLabelling. Issue 2, Winter.

DOE/DTI (1992) *Economic Instruments and Recovery of Resources from Waste*. Department of Trade and Industry and Department of the Environment, London.

De Baere, L. (1993) *DRANCO, A Novel Way of Composting Food and Wastepaper: Results of a First Full-Scale Plant*. Paper given at Second CIES Conference on the Environment, Barcelona, 15–18 June.

De Baere, L. (1999) *Anaerobic Digestion of Solid Waste: State of the Art*. Proc. Second International Symposium on Anaerobic Digestion of Solid Waste. Barcelona, 15–18 June.

De Baere, L., van Meenen, P., Deboosere, S. and Verstraete, W. (1987) Anaerobic fermentation of refuse. *Resources and Conservation*, Vol. 14, pp. 295–308.

De Wilde, B., Boelens, J. and De Baere, L. (1996) Results of laboratory and field studies on wastepaper inclusion in biowaste in view of composting. *The Science of Composting*, Part 2. Blackie Academic & Professional, Glasgow, pp. 803–812.

Deardorff, G.B. (1991) Construction inspection of Municipal landfill lining systems: a USA perspective. In: *Biogas Disposal and Utilisation, Choice of Material and Quality Control, Landfill Completion and Aftercare, Environmental Monitoring*. Third International Landfill Symposium, Sardinia, 14–18 October, pp. 741–752.

Denison, R.A. (1996) Environmental Life Cycle comparisons of recycling, landfilling and incineration: a review of recent studies. *Ann. Rev. Energy Environ.*, Vol. 21, pp. 191–237.

DETR (1997) *Marketing Guide for Producers of Waste-derived Compost*. Department of the Environment, Transport and the Regions Report, HMSO, London.

DETR (1998) *Monitoring and Assessment of Peat and Alternative Products for Growing Media and Soil Improvers in the UK – Results for 1996 and 1997*. Department of the Environment, Transport and the Regions Report, HMSO, London.

Deurloo, T. (1990) *Assessment of Environmental Impact of Plastic Recycling in P&G Packaging*. Procter & Gamble European Technical Center, internal report.

DHV Environment and Infrastructure. (1997) *Composting in the European Union*. Report for the European Commission.

Diaz, L.F. (1999) CalRecovery Inc, Hercules, California, USA, Personal communication.

Diaz, L.F. and Eggert, L.L. (1994), New MRF – material recycling facility in Mexico City. *BioCycle*, Vol. 35, No. 6, p. 53.

Doh, W. (1990) *Biologische Verfahren der Abfallbehandlung*, EF-Verlag, Berlin.

Doppenberg, A.A.T. (1998) Organisational trends in domestic waste collection in The Netherlands. *Wastes Management*, April. Institute of Waste Management, UK, pp. 19–21.

DPPEA (1997) *Analysis of the Full Costs of Solid Waste Management for North Carolina Local Governments.* North Carolina Division of Pollution Prevention and Environmental Assistance, USA.

DTI (1996) *Energy from Waste – Best Practice Guide,* revised edition. A guide for local authorities and Private Sector developers of municipal solid waste combustion and related projects. ETSU for the DTI. New and Renewable Energies Bureau, ETSU, Harwell, Oxfordshire, UK.

EC (1989a) 89/369/EEC. Council Directive on the Prevention of Air Pollution from New Municipal Waste Incineration Plants. *Official Journal of the European Communities,* 8 June.

EC (1989b) 89/429/EEC. Council Directive on the Prevention of Air Pollution from Existing Municipal Waste Incineration Plants. *Official Journal of the European Communities,* 21 June.

EC (1992a) *Green Paper on the Impact of Transport on the Environment – A Community Strategy for Sustainable Mobility.* European Commission, Brussels, 20 February.

EC (1992b) Ecolabelling regulation. *Official Journal of the European Community*: L99, April.

EC (1994) 94/62/EC Packaging and Packaging Waste Directive. *Official Journal of the European Communities.*

EC (1997a) *Economic Evaluation of the Draft Incineration Directive.* A report produced for the European Commission DGXI.

EC (1997b) *Mass Burn of MSW: Economic Evaluation of the Draft Incineration Directive.* A report produced for the European Commission DG XI, section 7, p. ix.

EC (1999a) 1999/31/EC. EC Landfill Directive. *Official Journal of the European Communities,* 16 July.

EC (1999b) *Discussion Paper on the Revision of the Packaging and Packaging Waste Directive 94/62/EC.*

EEWC (1993) *Waste to Energy – An Audit of Current Activity, January 1993.* Report prepared for the European Energy from Waste Coalition, private communication.

European Energy from Waste Coalition (1997) *What Is Energy From Waste?* Information literature, European Energy from Waste Coalition.

EEA (1998) Waste generation and management. *Environment in the European Union at the Turn of the Century,* European Environment Agency, Copenhagen, Office for Official Publications of the European Communities, Luxembourg, Chapter 3.7.

Egosi, N. and Weinberg, A. (1998) New technology overview: Single-stream processing. *Resource Recycling,* May, pp. 42–46.

Ehrig, H.J. (1991) Prediction of gas production from laboratory scale tests. In: *Biogas Disposal and Utilisation, Choice of Material and Quality Control, Landfill Completion and Aftercare, Environmental Monitoring.* Third International Landfill Symposium, Sardinia, 14–18 October, pp. 87–114.

Elkington, J. (1997) *Cannibals with Forks, The Triple Bottom Line of Sustainable Business.* Capstone, Centre for Innovation, Oxford, UK, pp. 57–58.

Elkington, J. and Hailes, J. (1988) *The Green Consumer Guide.* Victor Gollancz, London.

ENDS (1992a) *Landfill, Oil and Gas Industries Top Methane Emissions League.* Environmental Data Services Report 206, pp. 3–4.

ENDS (1992b) *Survey Puts Landfill Clean-Up Costs in Perspective.* Environmental Data Services Report 214, p. 11.

ENDS (1992c) *Subsidising the Dash to Burn Trash.* ENDS Report 211. Environmental Data Services, London, pp. 12–14.

ENDS (1993) *Catalyst Cuts Dioxin Emissions From Incinerators.* ENDS Report 216. Environmental Data Services, London, pp. 9–10.

ENDS (1997) *Making Compost From Unsorted Rubbish: The Saga Continues.* ENDS Report 274. Environmental Data Services, London, pp. 24–28.

ENDS (1998) *Germany Plans To End Domestic Waste Dumping.* Environmental Data Services Daily, 23 August.

ENDS (1999) *Progress on Recycling Incinerator Ash,* ENDS Report 290. Environmental Data Services, London, p. 16.

Enustun, M. (1999) Procter & Gamble, Turkey, personal communication.

Environment Canada (1988) *National Incinerator Testing and Evaluation Program: Environmental Characterisation of Mass Burn Incinerator Technology at Quebec City, Summary Report.* Environment Canada Report EPS 3/UP/5.

EPIC (1997) *A Review Of The Role Of Plastics In Energy Recovery.* Procter & Redfern Ltd, January. Available from EPIC, 5925 Airport Road, Suite 500, Mississauga, Ontario L4V IWI, Canada.

EPU (1998) *Lighter Pampers Package Introduced in Eastern Europe*. Procter & Gamble, Environmental Progress Update, p. 3.

ERL/UCD (1993) *Waste Management Statistics for Ireland*. Environmental Resources Ltd and Environmental Institute of University College Dublin.

Ernst, A-A, (1990) A review of solid waste management by composting in Europe. *Resources, Conservation and Recycling*, Vol. 4, pp. 135–149.

ERRA (1991) *Resource*. Report of European Recovery and Recycling Association, Autumn, 1991. ERRA, Brussels.

ERRA (1992a) *Nomenclature: Secondary materials*. Reference report of the ERRA Codification Programme, December 1992. Available from European Recovery and Recycling Association, 83 Ave E. Mounier, Box 14, Brussels 1200, Belgium.

ERRA (1992b) *Programme Ratios*. Reference report of the ERRA Codification Programme, Nov. 1992. Available from European Recovery and Recycling Association, 83 Ave E. Mounier, Box 14, Brussels 1200, Belgium.

ERRA (1993a) *Waste Analysis Procedure*. Reference report of the ERRA Codification Programme, March 1993. Available from European Recovery and Recycling Association, 83 Ave E. Mounier, Box 14, Brussels 1200, Belgium.

ERRA (1993b) *Terms and Definitions*. Reference report of the ERRA Codification Programme, March 1993. Available from European Recovery and Recycling Association, 83 Ave E. Mounier, Box 14, Brussels 1200, Belgium.

ERRA (1993c) Project summary sheets. European Recovery and Recycling Association, 83 Ave E. Mounier, Box 14, Brussels 1200, Belgium.

ERRA (1994a) *The West Sussex Recycling Initiative: Lessons from Adur and Worthing*, December 1994. Available from European Recovery and Recycling Association, 83 Ave E. Mounier, Box 14, Brussels 1200, Belgium.

ERRA (1994b) *Tres Años de Recogida Selectiva en Barcelona*. European Recovery and Recycling Association, 83 Ave E. Mounier, Box 14, Brussels 1200, Belgium.

ERRA (1996) *The Sheffield Kerbside Project 1989–1995*. European Recovery and Recycling Association, 83 Ave E. Mounier, Box 14, Brussels 1200, Belgium.

ERRA (1998) *Towards Integrated Management of Municipal Solid Waste*, Vol. 1: *Report Summary*. European Recovery and Recycling Association, 83 Ave E. Mounier, Box 14, Brussels 1200, Belgium.

ETSU (1992) *Production and Combustion of c-RDF for On-Site Power Generation*. Energy Technology Support Unit report no. B 1374, by Aspinwall and Company Ltd., Department of Trade and Industry, London.

ETSU (1993a) *An Assessment of Mass Burn Incineration Costs*. Energy Technology Support Unit report no. B R1/00341/REP, by W.S. Atkins, Consultants Ltd. Department of Trade and Industry, London.

ETSU (1993b) *Assessment of d-RDF Processing Costs*. Energy Technology Support Unit report no. B 1314, by Aspinwall and Company Ltd. Department of Trade and Industry, London.

ETSU (1995) *Energy From Landfill Gas: Appley Bridge Extended Renewable Energy Case Study*. ETSU, New and Renewable Energies Bureau, Harwell, Oxfordshire, UK.

ETSU (1996a) *Landfill Gas Development Guidelines*. ETSU for the DTI, London, p. 11.

ETSU (1996b) *Alternative Road Transport Fuels – A Preliminary Life-Cycle Study for the UK*, Vol. 2. ETSU, B156, New and Renewable Energies Bureau, Harwell, Oxfordshire, UK, p. 54.

Ettala, M.O. (1990) Sanitary landfilling in Finland. In: *International Perspectives on Municipal Solid Wastes and Sanitary Landfilling*, Eds J.S. Carra and R. Cossu, Academic Press, London, pp. 67–77.

Europe Environment (1998) *Recycling: Commission Anxious to Improve Competitiveness*, No. 528, September.

European Packaging and Waste Law (1999) *Germany – Spare Plastics Recycling Capacity*, March, pp. 26–27.

EUROPEN (1999) *Use of Life Cycle Assessment (LCA) as a Policy Tool in the Field of Sustainable Packaging Waste Management*. A EUROPEN discussion paper. European Organisation for Packaging and the Environment, Brussels. Available from http://www.europen.be/issues/lca/lca_revised.html

EUROSTAT (1994a) *Europe's Environment 1993: Statistical Compendium*. Statistical Office of the European Community, June.

EUROSTAT (1994b) Report of the Statistical Office of the European Community, News Release, 15 February.

Fichtner Ingenieurgesellschaft (1991) *Gegenüberstellung der Schadstofffrachten bei der thermischen Restmüll-behandlung und einer Restmülldeponie, im Auftrag Fichtner, Landesentwicklungsgesellschaft,* Ministerium für Umwelt, Baden-Württemberg.

Fickes, M. (1998) Getting the best bang for the waste buck. *World Wastes,* June, pp. 22–27.

Fielder, H. (1998) Thermal formation of PCDD/PCDF: a summary. *Env. Eng. Sci.,* Vol. 15, No. 1, pp. 49–58.

Finnveden, G. and Ekvall, T. (1997) On the usefulness of LCA in decision making – the case of recycling vs incineration of paper. In: *Presentation summaries of Fifth LCA Case Studies Symposium.* SETAC-Europe, Brussels, Belgium, pp. 9–17.

Finnveden, G. (1992) Landfilling – a forgotten part of Life Cycle Assessment. In: *Product Life Cycle Assessment – Principles and Methodology.* Nordic Council of Ministers, Norway, pp. 263–288.

Finnveden, G. (1995) *Treatment of Solid Waste in Life-cycle Assessment – Some Methodological Aspects. Life Cycle Assessment and Treatment of Solid Waste.* Proc. International workshop, September 1995, Stockholm, Sweden. AFR Report 98, pp. 107–114.

Finnveden, G. and Potting, J. (1999) Eutrophication as an impact category: state of the art and research needs. *Int. J. LCA,* Vol. 4, No. 6, pp. 311–314.

Forrest, P., Heaven, S. and Sandels, C. (1990) *Sorting at Source, Separation of Domestic Refuse.* Save Waste and Prosper, Leeds, UK.

Franke, M. (March, 1999) Visit to Bombay (Mumbai) City, guided and accompanied by Mr P.U. Panjwani, Consultant and Ex City Engineer Municipal Corporation of Greater Mumbai, personal communication.

Franke, M. (March, 1999) Visit to specific areas of Madras (Chennai) City, guided and accompanied by the Non-governmental Organization Exnora International, Chennai, India, personal communication.

Franke, M. (May, 1999) Visit to Buenos Aires "First International Seminar about Waste Minimization and Recycling" ARS (Association Para El Estudio De Los Residuos Solidos), personal communication.

Fricke, K. (1990) *Grundlagen der Kompostierung.* EF-Verlag, Berlin.

Fricke, K. and Vogtmann, H. (1992) *Biogenic Waste Compost, Experiences of Composting in Germany.* IGW, April, p. 33.

Frischknecht, R. *et al.* (1995) *Relevanz der Infrastructur in Ökobilanzen – Untersuchung anhand der Ökoinventare für Energiesysteme (The Relevance of Infrastructure in Life Cycle Assessments – An Investigation Using the Life Cycle Inventories for Energy Systems).* Federal Institute of Technology (ETH), Institute of Energy Technology (IET), Zurich.

Frischknecht, R. *et al.* (1996) *Ökoinventare für Energiesysteme (Life Cycle Inventories for Energy Systems).* Federal Institute of Technology (ETH), Institute of Energy Technology (IET), Zurich.

Furedy, C. (1991) Source separation in developing countries. *Warmer Bulletin,* No. 30, August, pp. 12–13.

Gabola S. (1999) *Towards a Sustainable Basis for the EU Packaging and Packaging Waste Directive.* A presentation given at the ERRA (European Recovery & Recycling Association) Symposium, Brussels. Available from the European Recovery and Recycling Association, 83 Ave E. Mounier, Box 14, Brussels 1200, Belgium.

Gaines, L.L. and Mintz, M.M. (1994) *Energy Implications of Glass Container Recycling.* Argonne National Laboratory, Argonne, IL, March.

Gandolla, M. (1990) Sanitary landfilling in Switzerland. In: *International Perspectives on Municipal Solid Wastes and Sanitary Landfilling,* Eds J.S. Carra and R. Cossu. Academic Press, London, pp. 190–198.

Garmendia, A.M. (January 1997) Visit to landfill sites in the Mexico City Metropolitan Zone. SUSTENTA, Mexico City, personal communication.

Gendebien, A., Pauwels, M., Constant, M. *et al.* (1991) Landfill gas: from enviroment to energy. State of the art in the European Community context. In: *Biogas Disposal and Utilisation, Choice of Material and Quality Control, Landfill Completion and Aftercare, Environmental Monitoring.* Third International Landfill Symposium, Sardinia, 14–18 October, pp. 69–76.

Gendzelevich, W., and Ramos, D. (1996) Remediation of the New Cabalan Landfill, Olongapo City, Philippines. *Geotechnical News,* September, p. 10.

German Government (Deutscher Bundestag) (1993) *Anaerobe Vergärung als Baustein der Abfallverwertung.* Drucksache 12, 4905, 12 May.

Glenn, J. (1997) MSW composting in the United States. *Biocycle,* November, pp. 64–70.

Goldstein, N. (1993) The curb and pile trial. *Biocycle,* August, pp. 37–39.

Goldstein, N. and Block, D. (1997) Biosolids composting holds its own. *Biocycle,* December, pp. 64–74.

Gomes, M.M., and Hogland, R.H.W. (1995) *Scavengers and Landfilling in Developing Countries*. Proc. Sardinia 95, Fifth International Landfill Symposium, Vol. 1, pp. 875–880.

Greenberg, M, Caruana, J. and Krugman, B. (1976) Solid-waste management: a test of alternative strategies using optimization techniques. *Environment and Planning*, Vol. 8, pp. 587–597.

Grüneklee, C.E. (1997) *Development of Composting in Germany*. Proc. Organic Recovery and Biological Treatment into the Next Millennium, ORBIT 97, pp. 313–316.

Gunaseelan, V.N. (1997) Anaerobic digestion of biomass for methane production: a review. *Biomass and Bioenergy*, Vol. 13, pp. 83–114.

Habersatter, K. (1991), *Oekobilanz von Packstoffen Stand 1990*, Bundesamt für Umwelt, Wald und Landschaft (BUWAL) Report No. 132, Bern, Switzerland.

Habig, G. (1992) *Arbeitskreis der Fachgemeinschaft Thermo Prozess – und Abfalltechnik im VDMA (Verband Deutscher Maschinen– und Anlagenbau)* Frankfurt, March.

Hackl, A.E. (1993) *Energy Aspects in Environment Protection*. Proc. of the CEFIC Conference: The Challenge of Waste, Vienna, 23–24 August, pp. 127–129.

Ham, R.K., Hekimian, K.K., Katten, S.L. et al. (1979) *Recovery, Processing and Utilisation of Gas from sanitary landfills*. U.S. Environmental Protection Agency Report EPA-600/2-79-001.

Härdtle, G. Marek, K., Bilitewski, B. and Kijewski, K. (1986) Recycling von Kunstoffabfällen. *Fachzeitschrift für Behandlung und Beseitigung von Abfällen*, Vol. 27.

Harris, R. (1992) *The groundwater protection policy of the NRA*. Proceedings of Landfilling waste – asset or liability, IBC Conference, London, 29 May, 1992.

Haskoning (1991) *Conversietechnieken voor GFT–afval*, NOH, 53430/0110.

Heijungs, R., (1996) On the identification of key issues for further investigation in Life Cycle Screening; the use of mathematical tools and statistics for sensitivity analysis. *Journal of Cleaner Production*, Vol. 4, pp. 159–166.

Helmut Kaiser Consultancy (1996) Incineration report. Available from Helmut Kaiser Consultancy, Philosophenweg 2, D-72076 Tübingen, Germany.

Henstock, M. (1992) *An Analysis of the Recycling of LDPE at Alida Recycling Limited*. Report by Nottingham University Consultants, Ltd, Nottingham, UK.

Hickey, T.J and Rampling, T.W.A. (1989) *The Hedon Boiler Trials*. Warren Spring Laboratory Report LR 747 (MR).

Hindle, P., White, P.R. and Minion, K. (1993) Achieving real environmental improvement using value: burden assessment. *Long Range Planning*, Vol. 26, No 3, pp. 36–48.

Hindle, P. and McDougall, F. (1997) *The Value of Biological Treatment*. Proc. Organic Recovery and Biological Treatment into the Next Millennium, ORBIT 97, pp. 11–17.

Homes, J. (1996) *The United Kingdom Waste Management Industry Report 1996*. Table 13.

Horton, T. (1998) Economics: Can the marriage of economics and the environment end happily ever after? *MSW Management*, December.

Hunt, R.G., Franklin, W.E., Welch, R.O., Cross, J.A., and Woodal, A.E. (1974) *Resource and Environmental Profile Analysis of Nine Beverage Container Alternatives*. Midwest Research Institute for U.S. Environmental Protection Agency, Washington DC, USA.

Hunt., R.G., Sellers, J.D., and Franklin, W.E. (1992) Resource and environmental profile analysis: a Life Cycle environmental assessment for products and procedures. *Environmental Impact Assessment Review*, Vol. 12, pp. 245–269.

Huppes. (1992) Allocating impacts of multiple economic processes in LCA. In: *Life-cycle Assessment*. Report of the SETAC Workshop, Leiden. Society of Environmental Toxicology and Chemistry, Brussels, December, pp. 57–70.

Huruska, J.P. (1998–2000) Vice President, Municipal Development, 26 Wellington Street East, Suite 601, Toronto, Ontario, M5E 1S2, Canada, personal communication.

Hutzinger, O. and Fielder, H. (1991) Formation of dioxins and related compounds from combustion and incineration processes. In: *Dioxin Perspectives: A Pilot Study on International Information Exchange on Dioxins and Related Compounds*, Elsevier Science BV, Amsterdam, Chapter 3, pp. 263–434.

Hutzinger, O., and Fielder, H. (1993) From source to exposure: some open questions. *Chemosphere*, Vol. 27, pp. 121–129.

Hutzinger, O., Blumich, M.J., Van Den Berg, M. and Olie, K. (1985) Sources and fate of PCDDs and PCDFs: an overview. *Chemosphere*, Vol. 14, pp. 581–600.

IAG (1997) *Municipal Solid Waste Incinerator Residues.* International Ash Working Group, Studies in Environmental Science 67, Elsevier Science, Oxford.

IEA Biotechnology (1998) *Biogas and More! Systems and Markets Overview of Anaerobic Digestion.* Resource Development Associates, Washington DC, USA.

IEM (1998) Special Report: Eco-efficiency, Part 2, Focus on Life Cycle Assessment. *Journal of the Institute of Environmental Management*, Vol. 5, Issue 3, June.

IFEU (1991) *Bewertung verschiedener Verfahren der Restmüllbehandlung in Wilhelmshaven.* Institut fur Energie–und Umweltforschung, Heidelberg, GmbH. Heidelberg, January.

IFEU (1992) *Vergleich der Auswirkungen verschiedener Verfahren der Restmüllbehandlung auf die Umwelt und die menschliche Gesundheit.* Institut für Energie-und Umweltforschung, Heidelberg, GmbH. Heidelberg, November.

IGD (1992) *Sustainable Waste Management: The Adur Project.* Report by the Institute for Grocery Distribution, Letchmore Heath, Watford, UK, p. 85.

IGD (1994) *Environmental Impact Management.* Report by A. Hindle for Policy Issues Programme of the Institute for Grocery Distribution, Watford, UK, p. 65.

INCPEN (1996) *Environmental Impact of Packaging in the UK Food Supply System.* The Industry Council for Packaging and the Environment, London.

INTEGRA (1998) *Recovery: Material Recovery Facilities, Portsmouth.* Information Booklet. Hampshire Waste Services, Otterbourne, UK.

IPCC (1994) *Radiative forcing of climate change.* The 1994 report of the Scientific Assessment Group of the Intergovernmental Panel on Climate Change, J. Houghton. Meteorological Office, Bracknell, UK.

IPCC (1996) *Intergovernmental Panel on Climate Change: Climate Change 1995; The Science of Climate Change*, Eds J.T. Houghton et al., Cambridge University Press, Cambridge.

ISO 14040 (1997) *Environmental Management – Life Cycle Assessment – Principles and Framework.* ISO/FDIS.

ISO 14041 (1998) *Environmental Management – Life Cycle Assessment – Goal and Scope Definition and Life Cycle Inventory Aanalysis.* ISO/FDIS.

ISO 14042 (1999) *Environmental Management – Life Cycle Assessment – Life Cycle Impact Assessment.* ISO/FDIS.

ISO 14043 (1999) *Environmental Management – Life Cycle Assessment – Life Cycle Interpretation.* ISO/FDIS.

ISO Technical Report 14049 (1999) *Environmental Management – Life Cycle Assessment – How to Apply ISO 14041 Goal and Scope Definition and Life Cycle Inventory Analysis.* ISO/FDIS.

Jespersen, L. (1991) *Source Separation and Treatment of Biowaste in Denmark.* Verwerkingsmogelijkheden en scheidingsregels van groente-, fruit- en tuinafval, Koninklijke Vlaamse Ingenieursvereniging (K.V.I.V.). Syllabus studiedag 7 Maart.

Juniper Consultancy Services (1995) *Municipal Solid Waste Incineration in Europe: A Market Study.* Juniper Consultancy Services, UK.

Kahn, F. (1996) Recycling as a matter of survival. *Conservation*, Vol. 2, No. 1, January–March. South African Department of Environmental Affairs and Forestry, pp. 16–17.

Keeling, J.B. (1991) *The Recyclers of Harare.* Source unknown.

Kerbside Dublin (1993) *Kerbside Dublin, One Year On.* Report of Kerbside Dublin, Dublin, Ireland.

Kern, M. (1993) Grundlagen verfahrenstechnischer Vergleiche von Kompostierungsanlagen. In: *Biologische Abfallbehandlung; Kompostierung –Anaerobtechnik – Kalte Vorbehandlung*, Eds. K. Wierner and M. Kern, Reihe (book series) Abfall-Wirtschaft – Neues aus Forschung und Praxis, University of Kassel, pp. 11–40.

Kern, M., Fulda, K. and Mayer, M. (1999) Stand der biologischen Abfall–behandlung in Deutschland. *Müll und Abfall*, Feburary, pp. 78–81.

Kilgroe, J.D. (1996) Control of dioxin, furan, and mercury emissions from municipal waste combustors. *Journal of Hazardous Materials*, Vol. 47, May, pp. 163–194.

Kindler, H. and Mosthaf, H. (1989) *Ökobilanz von Kunstoffverpackungen*, BASF- Ludwigshafen, 1989.

Kirkpatrick, N. (1992) *Choosing a Waste Disposal Option on the Basis of a Lifecycle Assessment.* Proceedings of PIRA Conference: Lifecycle Analysis – protecting your market share, 4 November, Gatwick, UK.

Koopmans, W.F. (1997) *The Quality of Compost from VGF Waste.* Proc. Organic Recovery and Biological Treatment into the Next Millennium, ORBIT 97, pp. 319–323.

Korz, J. and Frick, B. (1993) Das BTA–Verfahren. In: *Biologische Abfallbehandlung; Kompostierung – Anaerobtechnik – Kalte Vorbehandlung*, Eds. K. Wierner and M. Kern, Reihe (book series) Abfall–Wirtschaft – Neues aus Forschung und Praxis, University of Kassel, pp. 831–849.

Kramer, P.J. and Kozlowski, T.T. (1979) *Physiology of Woody Plants*. Academic Press, London.

Kreuzberg, G. and Reijenga, F. (1989) *Handboek Gescheiden Inzameling Groente–, Fruit–end Tuinafval*. Provinciale Waterstaat Noord-Holland.

Kuta, C.C., Koch, D.G., Hildebrant, C.C. and Janzen, D.C. (1995) Improvement of products and packaging through the use of Life Cycle analysis. *Resources, Conservation and Recycling*, Vol. 14, pp. 185–198.

La Manita, G.P. and Pilati, G. (1996) Introduction. *Polymer Recycling*, Vol. 2. No. 1, pp. 1–2.

Landbank. (1992) *WARM Report. A Proposal for a Model Waste Recovery and Recycling System for Britain*. A Gateway Foodmarkets report, prepared by the Landbank Consultancy.

Lashof, D.A. and Ahuja, D.R. (1990) Relative contributions of greenhouse gas emissions to global warming. *Nature*, 344, April.

Lentz, R. Carra, R. and Scherer, P. (1992) *Composite Hygiene Papers in Anaerobic Digestion of Biowaste – A Case Study at the Biogas Pilot Plant of BTA, Munich*. Proc. International Symposium of Anaerobic Digestion of Solid Waste, Italy.

Lewis, M.R. and Newell, T.A. (1992) *Development of an Automated Colour Sorting System for Recyclable Glass Containers*. Proc. National Waste Processing Conference. ASME, New York, pp. 197–207.

Loader, A. (1991) *Review of Pollution Control at Municipal Waste Incinerator Works*. Department of the Environment, Warren Spring Laboratories, Stevenage, UK.

Lopez-Real, J. (1990) *Agro-Industrial Waste Composting and its Agricultural Significance*. Paper presented at Fertiliser Society of London, 5 April 1990. Proc. of The Fertiliser Society, No. 293.

Lowe, J. (1981) *Energy Usage and Potential Savings in the Woollen Industry*. Wool Industry Research Association, Leeds, UK.

Lundholm, M.P. and Sundström, G. (1986) *Resource and Environmental Impact of Two Packaging Systems for Milk: Tetrbrik Cartons and Refillable Glass Bottles*. Mälmo, Sweden.

Maddox, B. (1994) Politics ahead of science. *Financial Times*, 2 March, p. 20.

Magnetic Separation Systems Inc. (1995) *Plastic Bottle Sorting*. Literature from Magnetic Separation Systems Inc., Nashville, USA.

Manser A.G.R., and Keeling A.A. (1996) *Practical Handbook of Processing and Recycling Municipal Waste*. Lewis , Englewood Cliffs, NJ.

McDougall, F.R. and White, P.R. (1998) *The Use of Lifecycle Inventory to Optimise Integrated Solid Waste Management Systems: A Review of Case Studies*. Paper presented at Systems Engineering Models for Waste Management, Gothenburg, Sweden.

Meadows, D.H., Meadows, D.L. and Randers, J. (1992) *Beyond The Limits: Global Collapse or Sustainable Future?* Earthscan Publications Ltd, London.

Meadows, D.H., Meadows, D.L., Randers, J. and Behrens, W.W III (1972) *The Limits to Growth: A Report to the Club of Rome's Project on the Predicament of Mankind*. Universe Books, New York.

Meyers, N. and Simon, J.L. (1994) *Scarcity Or Abundance? A Debate on the Environment*. W.W. Norton, New York.

Milke, M.W. (1998) Computer simulation to evaluate the economics of landfill gas recovery. *Wat. Sci. Tech.*, Vol. 38, No. 2, pp. 201–208.

Mirza, R. (1998–1999) Enviroshpere EMC Inc., personal communication.

MOPT (1992) *Residuos sólidos urbanos*. Report by Luis Ramon Otero Del Peral for Ministerio de Obras Publicas y Transportes, Madrid, Spain.

Nagpal, A., Gupta, S. and Mohan, M.K. (1999) Plastics Waste Recycling Trade in Delhi. *Warmer Bulletin*, Vol. 65, pp. 14–15.

Newell Engineering (1993) *Black Magic Non-ferrrous Separator*. Literature from Newell Engineering Ltd, Redditch, UK.

Nilsson-Djerf, J. (1999) *Social Factors in Integrated Waste Management: Measuring the Social Elements of Sustainable Waste Management*. Report available from European Recovery and Recycling Association, 83 Ave E. Mounier, Box 14, Brussels 1200, Belgium.

Noone, G.P. (1990) The treatment of domestic waters. In: *Anaerobic Digestion: A Waste Treatment Technology*, Ed. A. Wheatley. Elsevier Applied Science, Oxford, pp. 139–170.

NSWMA (1992) *Processing Costs for Residential Recyclables at Materials Recovery Facilities*. National Solid Wastes Management Association, Washington DC.

Nyns, E.-J. (1989) Methane fermentation. In: *Biomass Handbook*, Eds O. Kitani and C.W. Hall. Gordon and Breach, New York, pp. 287–301.

QECD (1997) *Environmental Data Compendium*. Organisation for Economic Cooperation and Development, Paris.

OWS Organic Waste Systems. (1998) Company Lit. OWS n.v., Dok Noord 7b, B-9000 Belgium.

O'Brien J. (1998) *Tools and Strategies for Competition*. Public Sector Solid Waste Systems. MSW Management, Elements.

Oakland, D. (1988) *Filtr. Sep.*, January/February, p. 39.

Oakland, J.S. (1989) *Total Quality Management*. Heinemann, Oxford.

Oakland, J.S. (1994) *Total Quality Management – The Route to Improving Performance*, 2nd edition. Butterworth-Heinemann, Oxford.

Oakland, J.S. (1995) *Total Quality Management – Text with Cases*. Butterworth-Heinemann, Oxford.

Obermeier, T. (1990) Reduction of dioxins and other toxic substances in emissions and combustion residues from waste incineration plants. *Warmer Bulletin*, Summer, p. 19.

QECD (1997) *Environmental Data Compendium*. Organisation for Economic Cooperation and Development, Paris.

Ogilvie, S.M. (1992) *A Review of the Environmental Impact of Recycling*. Report LR 911 (MR) Warren Spring Laboratories, Stevenage, UK.

Olie, K., Vermeulen, P.L. and Hutzinger, O. (1977) Chlorodibenzo-p-dioxins and chlorodibenzofurans are trace components of fly ash and flue gas of some municipal waste incinerators in the Netherlands. *Chemosphere*, Vol. 6, pp. 445–459.

Öman, C. and Hynning, P. (1991) Identified organic compounds in landfill leachate. In: *Biogas Disposal and Utilisation, Choice of Material and Quality Control, Landfill Completion and Aftercare, Environmental Monitoring*. Third International Landfill Symposium, Sardinia, 14–18 October, pp. 857–864.

ORCA (1991a) *The Role of Composting in the Integrated Waste Management System. Solid Waste Management, An Integrated System Approach*, Part 5. Organic Reclamation and Composting Association, Brussels.

ORCA (1991b) *Composting of Biowaste – The important Role of the Waste Paper Fraction. Solid Waste Management, An Integrated Approach*, Part 8. Organic Reclamation and Composting Association, Brussels.

ORCA (1992a) *Information on Composting and Anaerobic Digestion*. ORCA Technical Publication No. 1. Organic Reclamation and Composting Association, Brussels, p. 74.

ORCA (1992b) *A Review of Compost Standards in Europe*. ORCA Technical Publication No. 2. Organic Reclamation and Composting Association, Brussels.

Ovam (1991) *Ontwerp Afvalstoffenplan 1991–1995 van Vlaanderen*. (Waste Management Plan of the Flemish region for the period 1991 to 1995).

Owens, J.W. (1998) Life Cycle Impact Assessment: The use of subjective judgements in classification and characterisation. *Int. J. Life Cycle Assessment*, Vol. 3, No. 1, pp. 43–46.

Owens, J.W. (1999) Why Life Cycle? Impact Assessment is now described as an indicator system. *Int. J. Life Cycle Assessment*, Vol. 4, No. 2, pp. 81–86.

Panjwani, P.U. (1998) *Municipal Solid Waste – A Case Study Mumbai*. Municipal Corporation of Greater Mumbai, India.

Panter K., De Garmo R. and Border D. (1996) *A Review of Features, Benefits and Costs of Tunnel Composting Systems in Europe and in the USA. The Science of Composting*, Part 2. Blackie Academic & Professional, Glasgow, pp. 983–986.

Papworth, R. (1993) *An Assessment of the Adur District Council Kerbside Collection Scheme Utilising a Simplistic Analytical Approach Based on the ERRA Codification Programme – April 1993*. Report CR 3815, Warren Spring Laboratories, Stevenage, UK.

Patel, N.M. and Edgcumbe, D. (1993) *Observations on MSW Management in Japan*. *Waste Management (Journal of the Institute of Wastes Management)*, April, pp. 27–31.

Patyk, A. (1996) *Balance of Energy Consumption and Emissions of Fertilizer Production and Supply*. In: Proc. International Conference on Application of Life Cycle Assessment in Agriculture, Food and Non-Food Agro-Industry and Forestry: Achievements And Prospects, Ed. Ceuterick, D. Vlaamse Instelling voor Technologisch Onderzoek, Energy Division, Brussels.

Pearce, D.W. and Turner, R. K. (1993) *Externalities from Landfill and Incineration*. HMSO, London.

Pearce, D.W. and Turner, R.K. (1992) *Packaging Waste and the Polluter Pays Principle – A Taxation Solution*. A working paper, Centre for Social and Economic Research on the Global Environment. University of East Anglia and University College London.

Perry, R.H. and Green, D.W. (1997) *The Chemical Engineer's Handbook*, 7th edition. McGraw Hill, London.

Petts, J. (1995) Waste Management Strategy Development: A case study of community involvement and consensus-building in Hampshire. *Journal of Environmental Planning and Management*, Vol. 38, No. 4, pp. 519–536.

PFGB (1999) Fact Sheets. Paper Federation of Great Britain, www.paper.org.uk Swindon, UK.

Platt, B. (1998) Cutting the waste stream in half: record setting communities show how. *Resource Recycling*, February, pp. 13–15.

Poll, A.J. (1991) *Sampling and Analysis of Domestic Refuse – A Review of Procedures at Warren Spring Laboratory*. Report LR 667 (MR), Warren Spring Laboratories, Stevenage, UK.

Porteous, A. (1991) *Municipal Waste Incineration in the UK – Time for a Reappraisal?* Proceedings of 1991 Harwell Waste Management Symposium, Harwell, Oxfordshire, UK, May.

Porteous, A. (1992) *LCA Study of Municipal Solid Waste Components*. Report prepared for Energy Technology Support Unit (ETSU), Harwell, Oxfordshire, UK, July.

Porteous, A. (1994) *Dioxins and MSW Combustion*, Briefing Note 3; Energy from Waste Association, UK.

Porter, R. and Roberts, T. (1985) *Energy Savings by Wastes Recycling*. Elsevier Applied Science, Oxford.

Potting, J., Schopp, J., Blok, K. and Hauschild, M. (1998) Site-dependent life cycle impact assessment of acidification. *Industrial Ecology*, Vol. 2, No. 2, pp. 63–87.

Powell, J. (2000) The potential for using Life Cycle Inventory Analysis in Local Authority waste management decision making. *Journal of Environmental Planning and Management*, Vol. 43, No. 3, pp. 351–368.

Powell, J., Steele, A., Sherwood, N. and T. Robson. (1998) Using Life Cycle Inventory Analysis in the development of a waste management strategy for Gloucestershire, UK. *Environmental and Waste Management*, Vol. 1, No. 4, pp. 221–234.

PPIC (1999) Pulp and Paper Information Centre, Swindon, UK.

Pritchard, L. (1995) *Technical and Economic Study of Incineration Processes* Department of the Environment, Warren Spring Laboratories, Stevenage, UK.

PricewaterhouseCoopers (1998) *The Facts: A European Cost–Benefit Perspective*. Management Systems for Packaging Waste, November.

Procter & Gamble *Hard Surface Cleaners*. Internal report.

Procter & Gamble (1992) *Life Cycle Inventory for Consumer Goods Packages*. A copy of this spreadsheet can be obtained from Procter & Gamble European Technical Center, Temselaan 100, B-1853 Strombeek-Bever, Belgium.

Pulp and Paper (1976) Secondary vs Virgin Fibre Newsprint. *Pulp and Paper*, Vol. 50, No. 5.

PWMI (1993) *Eco-profiles of the European Plastics Industry, Report 3: Polyethylene and Polypropylene*. Report by Dr I. Boustead for The European Centre for plastics in the Environment (PWMI), Brussels, May.

Quinte (1993) *The First Year*. Report of Quinte Blue Box 200 project, Quinte, Ontario, Canada.

Ramke, H.G. (1991) *Die Bedeutung der Rotte vor der Deponierung*. I: Ministerium für Umwelt und Gesundheit, Rheinland-Pfalz (Hrsg.): Aktuelle Deponiekonzepte: Fachtagung Mainz, 21.1.1991. Mainz: MfU, pp. 46–71.

Rand, R. (1997) Source separation of organic waste and the paper sack. *Waste Management*, May, pp. 39–40.

Rattray, T. (1993) Fixing plastic recycling. *Resource Recycling*, May, Recovered plastics supplement, pp. 65–71.

RCEP (1993) *Incineration of Waste*. Royal Commission on Environmental Pollution 17th Report, HMSO, London.

RDC Coopers & Lybrand (1997) *Eco-balances for Policy-making in the Domain of Packaging and Packaging Waste*. Reference No. B4-3040/95001058/MAR/E3.

Reidy, R. (1992) *Municipal Solid Waste Recycling in Western Europe to 1996*. Elsevier Science, Oxford.

Reimann, D.O. (1992) *Basiskonzept einer ökologischen, integrierten Abfallwirtschaft*. Pres. at Integrierte Abfallwirtschaft, Forum Zukunft e.v. conference, Kloster Banz, Germany.

Reimann, D.O. (1998) Verbrennungskosten und deren Beeinflußbarkeit. *Müll und Abfall*, Vol. 7, pp. 452–457.

Rheinland-Pfalz Ministry of Environment (1989) *Leitfaden zur Kompostierung organischer Abfäll*. Cited in ORCA (1992).

Richards, K.M. and Aitchison, E.M. (1991) Landfilling in a greenhouse world. In: *Biogas Disposal and Utilisation, Choice of Material and Quality Control, Landfill Completion and Aftercare, Environmental Monitoring*. Third International Landfill Symposium, Sardinia, 14–18 October, pp. 33–44.

RIS (1996) *Ontario Blue Box Recovery Project – Final Report*. Resource Integration Systems Ltd and McConnell–Weaver Communication Management.

Rohe, T. (1998) Fraunhofer Institute, Web Page http://www.ict.fhg.de/english/ident.html

Rousseaux, P. (1988) *Les Métaux Lourds Dans Les Ordures Ménagères Origines, Formes Chimiques, Teneurs*. R&D programme on recycling and utilization of waste, E.E.C.D.G. XII.

Ruggeri, R., Chiampo, F and Conti, R. (1991) Economic analysis of biogas production from MSW landfill. In: *Biogas Disposal and Utilisation, Choice of Material and Quality Control, Landfill Completion and Aftercare, Environmental Monitoring*. Third International Landfill Symposium, Sardinia, 14–18 October, pp. 263–276.

Russell, D., O'Neill, J. and Boustead, I. (1994) *Is HDPE Recycling the Best Deal for the Environment?* Dow Europe S.A., Horgen, Switzerland.

Rutten, J. (1991) *Gescheiden inzameling en verwerking van GFT–afval te Diepenbeek*, syllabus studiedag 7 maart, 1991. Verwerkingsmogelijkheden en scheidingsregelsvan groente- fruit- en tuinafval, Koninklijke Vlaamse Ingeneurs-vereniging (K.V.I.V).

Scarlett, L. (1999) *Product Take-Back Systems: Mandates Reconsidered*. Policy Study No. 153, Centre for the Study of American Business, Washington University, St Louis, Missouri 63130-4899, USA.

Schauner, P. (1995) *Industrial Composting of Home-sorted Biowaste – High Yield Refining Systems District of Bapaume– France*. Procter & Gamble France S.N.C., Neuilly, France.

Schauner, P. (1996) How to get "best and most" from composting: an original experience from Bapaume, France. *Biocycle*, June 1996.

Schauner, P. (1997) *Sustainable Biological Treatment for the Next Millennium*. In: Proc. Organic Recovery & Biological Treatment into the Next Millenium, Ed. Stentiford. Zebra Publishing, Manchester, pp. 279–281.

Schleiss, K. (1990) Übersicht über die Kompostanlagen in der Schweiz. *Abfall-spektrum*, April, pp. 12–14.

Schmitt-Tegge, J. (1998) *Economic Instruments in German Waste Management*. Presentation at EWC/WRF Seminar on Economic Instruments & Waste Management, London, 15 July.

Schneider, D. (1992) Das Wabio-Verfahren der Deutschen Babcock Anlagen. In: *Biologische Abfallbehandlung; Kompostierung – Anaerobtechnik – Kalte Vorbehandlung*, Eds: K. Wiemer and M. Kern. Reihe (book series) Abfall-Wirtschaft – Neues aus Forschung und Praxis, University of Kassel, pp. 863–875.

Schön, M. (1992) Das Kompogas-Verfahren der Firma Bühler. In: *Biologische Abfallbehandlung; Kompostierung – Anaerobtechnik – Kalte Vorbehandlung*, Eds K. Wiemer and M. Kern, Reihe (book series) Abfall-Wirtschaft – Neues aus Forschung und Praxis, University of Kassel, pp. 821–829.

Schroll, M. (1998) *The Challenge of the German Industry in Solid Waste Management: The Example Of Packaging*. Institute for Applied Innovation and Research (IAI), Bochum, Germany.

Schweiger, J-W. (1992) *Planung, Genehmigung und Betrieb von Recyclinghöfen*. Presented at Integrierte Abfallwirtschaft, Forum Zukunft e.v. conference, Kloster Banz, Germany.

Schwing, E. (1999) *Bewertung der emissionen der kombination mechanisch-biologischer und thermischer abfallbehandlungsverfahren in Südhessen*. Schriftenreihe, WAR 111. Institut WAR, Wasserversogung Abwassertechnik Abfalltechnik, Umwelt und Raaumplanung der Technischen Universität Darmstadt, Germany.

Selle, M., Kron, D. and Hangen, H.O. (1988) *Die Biomüllsammlung und Kompostierung in der Bundesreplublik Deutschland.Situationsanalyse 1988*. Schriftenreihe des Arbeitskreises für die Nutzbarmachung von Siedlungsabfällen (ANS) e.v., Heft 13.

SETAC (1992) *Life-cycle assessment*. Report of the SETAC Workshop, Leiden, December. Society of Environmental Toxicology and Chemistry, Brussels, pp. 57–70.

SETAC (1997) *Simplifying LCA : Just a Cut? Final report from the SETAC – Europe LCA Screening and Streamlining Working Group*. The Society of Environmental Toxicology and Chemistry – Europe, Brussels.

Shell Petrochemicals (1992) Cited in *EEWC* (1993).

Shields, S. (1999) *A Feasibility Study of the Introduction of Process and Quality Standards for Compost in the UK*. The Composting Association, pp. 107–110.

Shotton Paper Company, (1992) Cited in Porteous (1992).

Simon, J.L. (1996) *The Ultimate Resource 2*. Princeton University Press. Princeton, NJ.

Six, W and De Baere, L. (1988) *Dry Anaerobic Composting of Various Organic Wastes*. Proc. Fifth International Symposium on Anaerobic Digestion, Italy, 22–26 May, pp. 793–797.

Skumatz, L. Truitt, E. and Green, J. (1997) The state of variable rates: Economic signals move into the mainstream. *Resource Recycling*, August.

Smith, C.W. and White, P.R. (1999) Life Cycle Assessment of packaging. In: *Packaging, Policy and the Environment*, Ed. Levy, G.M. Aspen, Chapter 8, pp. 178-204.

Spinosa, L., Brunetti, A., Lorè, F., and Antonacci, R. (1991) Combined treatment of leachate and sludge: lab experiments. In: *Biogas Disposal and Utilisation, Choice of Material and Quality Control, Landfill Completion and Aftercare, Environmental Monitoring*. Third International Landfill Symposium, Sardinia, 14–18 October, pp. 961–968.

SPI (1988) The Society of the Plastics Industry, Inc. (SPI) About the industry. Washington DC, 20005-4006, USA. www.socplas.org

SPMP (1991) *Thermal Recycling, Cornerstone of Rational Waste Management*. Paper prepared by the Syndicat de Producteurs de Matières Plastiques, October.

SPOLD (1995) *Directory of Life Cycle Inventory Data Sources*. Society for the Promotion of Life Cycle Development (SPOLD), Brussels.

Stalmans, M. (1992) *LCA Studies for Chemical Substances: Major Detergent Surfactants and their Raw Materials*. Proc. Third CESIO Congress, London, 1–5 June, pp. 237–250.

Stanford, J. (1999) *The Role of IWM Planning and Evaluation in London, Ontario*. Presentation at Society of Environmental Toxicology, Philadelphia, PA, November 1999.

Staudt, E. (1997) Die Verpackungsverordnung – Auswirkungen eines umweltpolitischen Großexperimentes. *Innovation: Forschung und Management*, Vol. 11, Institut für Angewandte Innovationsforschung, Ruhr-Universität Bochum, Germany.

Staudt, E. and Schroll, M. (1999) The German packaging ordinance: the questionable effects of a fragmentary solid waste management approach. *J. Material Cycles Waste Management*, Vol. 1, No. 1, pp. 17–24.

Stegmann, R. (1990) Sanitary landfilling in the Federal Republic of Germany. In: *International Perspectives on Municipal Solid Wastes and Sanitary Landfilling*, Eds J.S. Carra and R. Cossu, Academic Press, London, pp. 51–66.

Steinsvaag, D. (1996) Overview of electrostatic precipitators. *Plant Engineering*, Vol. 50, No. 10, pp. 16–18.

Stessel, R.I. (1996) *Recycling and Resource Recovery Engineering: Principles of Waste Processing*. Springer, New York.

Stieglitz, L., Zwick, G., Beck, J. Roth, W. and Vogg, H. (1989) On the de novo synthesis of PCDD/PCDF on fly ash of municipal waste incinerators. *Chemosphere*, Vol. 18, pp. 1219–1226.

Studley, B. and Moyer, R. (1997) *Start-up Experience at the Robbins Resource Recovery Facility While Focusing on the Environment*. Proc. of Fifth Annual North American Waste-to-Energy Conference, 22–25 April, 1997, Research Triangle Park, North Carolina, USA.

Sushil (1990) Waste Management: a systems perspective. *Industrial Management and Data Systems*, Vol. 90, No. 5, pp. 7–66.

Svedberg, G. (1992) *Waste Incineration for Energy Recovery*. Final report.

TA Siedlungsabfall (1993) 4.c. Dritte Allgemeine Verwaltungsvorschrift zum Abfallgesetz (TA Siedlungsabfall). Vom 14. Mai 1993, BAnz. Nr.99a.

Tabasaran, O. (1976) Überlegungen zum Problem De Poniegas. *Müll und Abfall*, Vol. 7, p. 204.

Tchobanoglous, G. (1993) *Integrated Solid Waste Management: Engineering Principles and Management Issues*. McGraw-Hill, New York, pp. 291–300.

Textile Recycling Association, (1999) *Why You Should Recycle Your Unwanted Household Textiles, Clothing and Shoes*. Boxworth, Cambridge, UK.

Thorneloe, S.A. (1991) US EPA's global climate change program – landfill emissions and mitigation research. In: *Biogas Disposal and Utilisation, Choice of Material and Quality Control, Landfill Completion and Aftercare, Environmental Monitoring*. Third International Landfill Symposium, Sardinia, 14–18 October, pp. 51–68.

Thorneloe, S.A., Cosulich, J., Pacey, J., Roqueta, A. and Bottero, C. (1997) *Database of Landfill Gas to Energy Projects in the United States*. Sixth International Landfill Symposium, Sardinia, 14–18 October, Vol. 2, pp. 593–600.

Thurgood, M. (1998) Modeling Waste Management. An environmental Life Cycle Inventory and economic cost analysis model for municipal solid waste management. *Warmer Bulletin*, Vol. 58, pp. 4–7.

Tidden, F. and Oetjen-Dehne, R. (1992) Modellversuch Bioabfallvergärung Ismaning-Sammlung und Vergärung von Bioabfällen aus dem Geschoßwohnungsbau. *Abfallwirtschafts Journal*, Vol. 4, No. 10, p. 787.

Torres, E.B., Subida, R.D. and Rabuco, L.B. (1991) *The Profile Of Child Scavengers in Smokey Mountain, Blaut, Tondo, Manila*. University of Philippines, pp. 1–38.

Tosine, H. (1983) *Chlorinated Dioxins and Dibenzofurans in the Total Environment*, Eds G. Choudhary, L.M. Keith and C. Rappe. Butterworth, London.

Toussaint, A., (1989) *Proceedings of ENVITEC 89*, Kongreßband, p. 61.

UBA, (1993) Umweltbundesamt. (Federal German Environmental Agency). Annual statistical report. 1993.

UCPTE (1994) *Jahresbericht 1993*; UCPTE-Sekretariat, Vienna, Austria.

Udo De Haes, H.A., Jollier, O., Finnveden, G., Hauschild, M., Krewitt, W. and Mueller-Wenk, R. (1999) Best available practice regarding impact categories and category indicators in Life Cycle Impact Assessment. Background document for the second working group on Life Cycle Impact Assessment of SETAC Europe (WIA-2) Part 1 and 2. *Int. J. LCA*, Vol. 4, No. 2, pp. 66–74; Vol. 4, No. 3, pp. 167–174.

Uehling, M. (1993. Keeping rubbish rotten to the core. *New Scientist*, Vol. 1888, 23 August, pp. 12–13.

UNDESA (1999) *Sustainability through the Market – A Business-based Approach to Sustainable Consumption and Production*. Background paper #11. Commission on Sustainable Development, Seventh Session, 19–30 April 1999, New York. A paper prepared by the World Business Council for Sustainable Development for The United Nations Department of Economic and Social Affairs.

UNDP (1998) *Human Development Report 1998*. Oxford University Press, for the United Nations Development Programme, p. 4 .

UNEP (1996) *International Source Book on Environmentally Sound Technologies for Municipal Solid Waste Management*. International Environmental Technology Centre Technical Publication Series (6). United Nations Environmental Programme.

US EPA (1995) *New Municipal Waste Incinerator, Standards of Performance*. New Standards, Fact Sheet.

US EPA (1997) *Application of Life Cycle Management To Evaluate Integrated Municipal Solid Waste Strategies*. Appendix F, Combustion process model. Research Triangle Institute, Research Triangle Park, NC, USA.

US EPA (1998) *Guidance for Landfilling Waste in Economically Developing Countries*. EPA-600/R-98-040.

Van der Vlugt, A. J. and Rulkens, W.H. (1984) Biogas production from a domestic waste fraction. In: *Anaerobic Digestion and Carbohydrate Hydrolysis of Waste*, Eds G.L. Ferrero, M.P. Ferranti and H. Naveau, Elsevier Applied Science, London, pp. 245–250.

van Mark, M. and Nellessen, K. (1993) Neuere Entwicklungen bei den Preisen von Abfalldeponierung und -verbrennung. *Müll and Abfall*, January, pp. 20–24.

van Santen, A. (1993) *Incineration: Its Role in a Waste Management Strategy for the UK*. Paper presented at the Institute of Wastes Management annual meeting, June 1993, Torquay, UK.

Van Zyl, M., Legg, P.A. and Du Preez, J.J. (1996) *Community Involvement in the Rehabilitation of the Boipatong Landfill site*. Proc. Wastecon 96, Durban, Institute of Waste Management of Southern Africa, pp. 473–484.

Verstraete, W., De Baere, L. and Seeboth, R-G. (1993) *Getrenntsammlung und Kompostierung von Bioabfall in Europa*. Entsorgungs Praxis, March.

Vesilind, P.A. and Rimer, A.E. (1989) *Unit Operations in Resource Recovery Engineering*, Prentice Hall, NJ.

Vogg, H. (1992) *Arguments in Favor of Waste Incineration*. Annual European Toxicology Forum 1–5 June 1992, Copenhagen, Denmark.

Vogg, H. (1995) *PCDD/PCDF und abfallverbrennung*. Organohalogen Compd. No. 22, pp. 31–48.

Vogg, H., Merz, A., Stieglitz, L., Vehlow, J.(1989) VGB-Kraftwerkstechnik 69 S.795. Vol. 43, No. 3, p. 2000.

von Schoenberg, A. (1990) Waste disposal in East Germany – an overview. *Warmer Bulletin*, No. 27, November, pp. 4–5.

WCED (1987) *Our Common Future*. World Commission on Environment and Development. Oxford University Press, Oxford.

Warmer Bulletin (1990) Waste incineration. Warmer Bulletin Factsheet. *Warmer Bulletin*, January.

Warmer Bulletin (1991) Fuel from waste. Warmer Bulletin Factsheet. *Warmer Bulletin*, January.

Warmer Bulletin (1993a) Batteries. *Warmer Bulletin*, Vol. 39, November, p. 7.

Warmer Bulletin (1993b) Refuse-derived fuel. Warmer Information Sheet, *Warmer Bulletin*, No. 39, November.

Warmer Bulletin (1993c) Retrofitting waste incineration plant – below detectability limits. *Warmer Bulletin*, Vol. 36, February.

Warmer Bulletin (1997a) Pay as you throw. *Warmer Bulletin*, Vol. 55, July, p. 14. July.

Warmer Bulletin (1997b) Landfill information sheet. *Warmer Bulletin*, No. 57.

Warmer Bulletin (1997c) Paper making and recycling. Warmer Information Sheet. *Warmer Bulletin*, Vol. 55.

Warmer Bulletin (1997d) Plastics. Warmer Information Sheet. *Warmer Bulletin*, Vol. 54.

Warmer Bulletin (1998a) Pyrolysis and gasification set to grow. *Warmer Bulletin*, Vol. 59, March.

Warmer Bulletin (1998b) Signs of change? *Warmer Bulletin*, Vol. 58, January.

Warmer Bulletin (1999a) Waste paper composting in the US. *Warmer Bulletin*, Vol. 65, March, p. 19.

Warmer Bulletin (1999b) *Advanced Thermal Processing*. *Warmer Bulletin*, Vol. 68, September.

Warmer Bulletin (1999c) Incinerator ash recycling in Denmark. *Warmer Bulletin*, Vol. 68, September.

WCED. (1987) *Our Common Future*. World Commission on Environment and Development, Oxford University Press, Oxford, 1987.

Weaver, E. and Azzinnzri, C. (1997) *Design, Operation and Performance of a Modern Air Pollution Control System for a Refuse Derived Fuel Combustion Facility*. Proc. Fifth Annual North American Waste-to-Energy Conference. April 22–25, Research Triangle Park, NC, USA.

Weber, B. and Holz, F. (1991) Disposal of leachate treatment residues. In: *Biogas Disposal and Utilisation, Choice of Material and Quality Control, Landfill Completion and Aftercare, Environmental Monitoring*. Third International Landfill Symposium, Sardinia, 14–18 October, pp. 951–960.

Weitz K.A., Nishtala, S. and Thorneloe, S. (1999) *Using A Life Cycle Approach to Achieve Sustainable Municipal Solid Waste Management Strategies at Local, State, and National Levels in the United States*. Available from US Environmental Protection Agency, Office of Research and Development, Air Pollution Prevention and Control Division (MD-63), Research Triangle Park, NC, USA.

Wells, C. (1995) Managing solid waste in Brazil. *Biocycle*, June, p. 52.

White, P.R. (1993) Waste-to-energy technology within Integrated Waste Management. In: *Proceedings of Cost Effective Power and Steam Generation from the Incineration of Waste*. Institute of Mechanical Engineers Seminar, London.

White, P.R, Hindle P. and Dräger, K. (1993) Lifecycle assessment of packaging. In: *Packaging in the Environment*, Ed. G. Levy. Blackie Academic & Professional, Glasgow, pp. 118–146.

White, P., De Smet, B., Owens, J. and Hindle, P. (1995b) Environmental management in an international consumer goods company. *Resources, Conservation and Recycling*, Vol. 14, pp. 171–184.

White, P.R., Franke, M. and Hindle, P. (1995) *Integrated Solid Waste Management: A Lifecycle Inventory*. Blackie Academic & Professional, Glasgow.

WHO (1993) *Urban Solid Waste Management*, 1991–1993 edition. World Health Organization, p. 130.

Willumsen, H.C. (1991) The problematics of landfill gas technology. In: *Biogas Disposal and Utilisation, Choice of Material and Quality Control, Landfill Completion and Aftercare, Environmental Monitoring*. Third International Landfill Symposium, Sardinia 14–18 October, pp. 77–86.

Wilson E. (1998–1999) Personal communication regarding LCI and waste management.

Wilson, E. (1997) *Incorporating Environmental Variables into Municipal Solid Waste planning: A Case Study of the Pamplona Region, Spain*. Masters thesis, Vrije Universiteit, Brussel.

Wilson, E. (1998) Life Cycle Inventory tools in Pamplona. *Warmer Bulletin*, No. 58, January, pp. 13–15.

Waste Management International (1994) *Waste Management International, Landfill Gas Data*. Waste Management International, London.

WMIC (1995) *A Clean Land ... The Danish Solution*, No. 3, *Incineration*. Waste Management Information Center, Rendan A/S, 376 Gladsaxevej, DK-2860 Soborg, Denmark.

World Bank (1999) *What A Waste: Solid Waste Management in Asia*, Eds D. Hoornweg and L. Thomas. Urban Waste Management, East Asia & Pacific Region. Urban & Local Government Working Paper Series, Washington DC 20433, USA.

Wright, M. (1998) Home composting: real waste minimisation or just feel good factor? *Wastes Management*, September, pp. 27–28.

Würz W. (1999) Vergessene Forderungen zur Abfuhr Und Reinigung Der Biotonne. *Müll und Abfall*, Vol. 4, pp. 218–222.

Yhdego, M. (1991) Scavenging solid wastes in Dar Es Salaam, Tanzania. *Waste Man. Res.*, pp. 259–265.

Young, C.P. and Blakey, N.C. (1991) Emissions from power generation plants fuelled by landfill gas. In: *Biogas Disposal and Utilisation, Choice of Material and Quality Control, Landfill Completion and Aftercare, Environmental Monitoring*. Third International Landfill Symposium, Sardinia, 14–18 October pp. 359–368.

Zabalza, J. (1995) International News, Guatemala. *Warmer Bulletin*, August, No. 46, p. 2.

Index

References to figures are in **bold type**; tables in *italics* and boxes in **_bold italics_**

REGISTRATION

To register your copy of IWM-2 please send a blank e-mail to
the following address:

waste.management@blacksci.co.uk

This will allow us to provide you with further information
regarding revisions in inventory data.

Your e-mail address will not be sold, traded or rented to others.